Galileo:

*HIS SCIENCE AND HIS SIGNIFICANCE
FOR THE FUTURE OF MAN*

Albert Di Canzio

ADASI Publishing Company
1465 Woodbury Avenue, Suite 261
Portsmouth, New Hampshire 03801

This book does not bear an *imprimatur* from any hierarchy, nor has it been approved by an editorial committee of any university press. The reader will be exposed to the viewpoint of an individual.

Publisher's Cataloging Data:

Di Canzio, Albert
Galileo: His Science and His Significance for the Future of Man

Topics discussed in this book include:

1. Science 2. History of Science
3. Astronomy 4. Physics
5. Mathematics 6. Literature 7. Poetry
8. Philosophy 9. Theology 10. Cosmology

Includes bibliographical references.

ISBN 0-9641295-6-6

PRINTED IN THE UNITED STATES OF AMERICA
Printed on acid-free paper

XX XX 9 8 7 6 5 4 3 2 1

(Portrait: courtesy,
The Observatories of the Carnegie Institution of Washington;
Galileo's signature on a deposition: from Favaro's National Edition)

"Philosophy is written in this grand book, the universe, which stands continually open to our gaze. But the book cannot be understood unless one first learns to comprehend the language and read the letters in which it is composed. It is written in the language of mathematics, and its characters are triangles, circles, and other geometric figures without which it is humanly impossible to understand a single word of it; without these, one wanders about in a dark labyrinth."

... Galileo Galilei, from *The Assayer*, 1623, translation by Stillman Drake

Illustration Credits

Cover Montage

Cover Design	by author
Leaning Tower	Photo by author (*)
Moon	Photo by author
Earth	Photo by NASA
Star Background	Photo by author(*)
Solar Corona	Photo by author
Sun and spots	Courtesy, Mr. John Sanford;(*)
	Photos by Dr. Jean Dragesco (*)
Galileo Portrait	Courtesy, The Observatories of the Carnegie Institution of Washington
Graphics Imaging	by Monique Berry

* (altered in graphics imaging)

End Sheet (Inside Back Cover)

Photo of 1991 solar eclipse by author
Photo of author by Dr. Antonio Naples

Title Page

The title page illustration is a sketch (which the author has modified in graphics imaging based on a prototype by Krissy Maria Di Canzio) of the Galilei coat of arms as mounted on the wall of the Great Hall of the University of Padova.

Frontispiece

Portrait of Galileo appears to be derived from the engraving by Ottavio Leoni in 1624.

Line Diagrams in the Text

Where indicated in the list of line diagrams (p. v) by the initials *dr*, the corresponding named line diagrams in the text were drawn by Don Rado based on a prototype by the author. Those tagged with initials *agd* were drawn by the author. That tagged with initials *ved1* is the author's drawing based on a prototype sketch by Vincent Di Canzio and inspired by Stillman Drake's earlier prototype. That tagged with initials *ved2* is Vincent Di Canzio's sketch based on the prototype by Vincenzio Viviani.

Index of Line Diagrams

Index of Halftone Illustrations

Dedication

This book and the following verse are dedicated to the memory of

Berenice Finarelli Di Canzio

alias "Grandmom"
who first educated the author about Galileo's "recantation".

To Grandmom
At The Parkview Nursing Home

Trapped in the confines of this mortal frame
By a ladder of code yet undeciphered by men,
Held hostage of the fate in its billion rungs,
Still from your wheelchair you may inspire
The harmony of thoughts that flow from a pen
Or the conquest and harnessing of distant suns.

Sparking the tinder of a young imagination
Your account of Galileo makes its mark though succinct;
To hold wrinkled hands that worked toils of generations
Is to feel the love to which my being is linked.

Your life, like a delicate flower poised
Between evening's twilight and the chill of night,
Bending 'gainst the wind to bring its petals in sight
In its lonely defiance of nature's harshness
Bestows on my dark journey a caressing light.

Powerless are the princes who can enslave
A life but not a mind,
The race is to the weak whose thoughts and love
To their object do bind
The indomitable will to seek and to find.
You are stronger by far than all the rulers of the earth,
For as long as we live, until we must part,
And beyond this frail life through science and art
You still have the power to bring joy to my heart.

... Written on an airplane
(a machine that defies nature's harshness)
6 November 1983

Prologue ... xv

The Author's Bias • The Author's Disqualification • The Author's Sinister Motivation • The "Man in the Street" • How Galileo Has Been Misunderstood • Sources and Acknowledgments • Why This Book? • Prologue to the First Edition and Requiem for the Zeroth

Introduction ... xxv

Chapter 1. Shadow of A Pyramid ... 1

Ancient Natural Philosophy ... 1

Thales of Miletos, the Founder • Pythagoras of Samos • Empedocles • Zeno • Democritos • Hippocrates of Chios • Plato • Eudoxos • Aristotle • Euclid of Alexandria • Aristarchos • Menaechmos and Apollonios • Archimedes • Eratosthenes • Hipparchos of Nicaea • Ptolemy

The End of the Grand Era of Greek Science ... 17

Pappos and Diophantos • A Monstrous Madness • A New Beginning

Chapter 2. From Bedrock Heavenbound ... 21

The Young Galileo (1564 - 1585) ... 21

Toscana before Galileo • Galilei Family • An Unremarkable Home • Music, Machines, Michelangelo, Medicine and Mathematics • The Turning Point • Through an Open Door • Confrontation with Vincenzio • The Crossing of the Bridge

Chapter 3. The Rebirth of Science ... 29

Assayer of Gold, Surveyor of Hell, Tutor of Math (1585-1588) ... 29

The Royal Crown Takes a Bath • La Bilancetta, The Proportional Hydrostatic Balance • Archimedean Mathematics and Aristotelian Induction • Centers of Gravity of Solids • Moment of Force • Measuring the Inferno

Flashback: The Dark Ages ... 36

Sundials, Numerals, Twilight and Cubics • Positional Notation

Pre-Galilean Renaissance ... 38

Oresme, da Vinci • Nicholas of Cusa • Tartaglia • Copernicus • Bruno • Vieta • Benedetti • Tycho • Kepler

Chipping Away at the Rust (1587-1589) ... 43

Chapter 4. The Spiral Staircase   45

Professor Galilei at Pisa (1589-1592)   45

The Pisan De Motu • Aristotle's Law of Fall • The Concept of Instantaneous Velocity • The Continuous Abscissa • Falling and Floating

The Mechanics   50

How Machines Work • The Perpetual Screw • Conservation of Motion: an Inertial Concept • One Idealization, Two Principles of Nature

The Spiral Staircase   53

The Leaning Tower of Pisa • The Cathedral Lamp • Galileo's Law of the Pendulum • Leaning Tower Demonstration: a Legend? • The Scattering of Loose Dust • Against the Wearing of the Toga • Conflicts with Pisan Philosophers and Royalty • Promotion for the "Troublemaker"

Chapter 5. The Royal Road to Geometry   61

Professor Galilei at Padova (1592 - 1610)   61

Galileo's Padovan Science   62

Levers, Inclined Planes, and Falling Bodies • Quod Movetur, ab Alio Movetur • Inclined Planes and Pendula • Rolling Balls Gather No Moss But Leak Acceleration • Greasing the Skids • Galilean Impetus • Impetus and Inertial Mass • Impetus and Gravitational Mass • Leaning Tower Revisited • Experimental Derivation of the Law of Falling Bodies

Galileo's Inventions   72

Galileo's Telescopes • Opera Glasses and the Telescopic Helmet • Galileo's Handmade Lenses • The Jupiter Satellite Ephemeris and Analog Computer • Galileo's Micrometer • A Revolutionary General-purpose Computer • Archimedean/Galilean Stamps on Mechanical Devices • Compound Microscope • Thermometer • Astrolabe • Magnetic Lift • Medical Innovations • Other Pendulum Devices: Clock and Vibration Counter

Attacks on Invention   85

Galileo's Defense Against a Plagiarizer • Rewards and Risks of the Renaissance

Chapter 6. The Celestial Beacon   93

Galileo Ventures into Astronomy (1604 - 1610)   93

The Supernova of 1604 • New Columbus of the Heavens • Johannes Kepler, Cosmic Theologian • Trials, Tribulations, and Telescopes • The First Astronomical Telescope • Demonstration of the Telescope • While the Slow Moon Climbs • Jupiter's Companions • Invention of the Telescope • Galileo, Kepler, and the Telescope

Chapter 13. Path of a Pyramid — 277

Prologue

Probably you will not be surprised to hear me speak of Galileo as a scientist, a philosopher, an astronomer, a mathematician, a science writer or an educator. Did you know that, besides being all of these, he was a lover of poetry? A literary interpreter? A businessman? A prolific inventor? Central of all the facets in his brilliant and sharp-edged personality was his passion for epistemology, the science of validating one's claims to knowledge. It fueled his drive to arm seekers of truth with methods of dismantling mystery and corralling uncertainty, and to disarm dogmatists who would dominate the mind and spirit of otherwise free individuals. More than his scientific breakthroughs, his innovations in methods of discovery and validation through measurement and calculation led him into conflict with the authoritarian "philosophers" and Churchmen of his day.

In spite of the necessity, incident to writing of that conflict, of describing the actions of Church officials in what was certainly not their finest hour, I deny any personal bias against the Church, though I do express a reasoned reaction to an experience of the Jesuit educational system. Over a period of nine years, I received instruction from numerous Jesuits. Several of them I remember with respect and with warmest personal regards; especially, the late Fr. Joseph Cosenza, S.J., who taught trigonometry and was affectionately known as "Father Cosine"; Fr. Francis Heyden, S.J., Astronomer and former Director of the Georgetown University Observatory, who encouraged my fascination with celestial mechanics; Fr. Vincent M. O'Brien, S.J., a great teacher of Latin and English, a man of wisdom, courage and incomparable sensitivity whose instruction left indelible imprints, and who encouraged me to set forth my thoughts on what has been called the historical conflict of science and theology. My first such thought is that it is nothing of the kind. It was (in Galileo's time) a conflict of science and theocracy; it is now a conflict of science and indoctrination.

The Author's Bias

I will admit to one bias. Though I have made a conscious effort not to exaggerate his achievements, Galileo is one of my personal heroes. If he were not, I would hardly be inclined to offer you an interpretation of his work that, by setting forth facts in their proper context, distinguishes itself from the mountain of mythology published by his detractors. Who can deeply appreciate the impact of his science and remain immune to a bias favorable to this hero of the Renaissance? In his foreword to Galileo's *Dialogo*, Einstein follows his praise of Galileo's will, intelligence and courage, with the following remark: "In speaking this way, I notice that I, too, am falling in with the general weakness of those who, intoxicated with devotion, exaggerate the stature of their heroes." If a man of no less stature than Professor Einstein can admit to this tendency, why should I deny it?

The Author's Disqualification

A dean at a university that had conferred on its former student an advanced degree in science considered him an unlikely candidate to teach the history of science because his formal background was in science and not in humanities. By this same rationale, I am unqualified to write this book. Apparently, I am weird to think that the history of science is mainly about science. Professionally I have worked on the production of various industrial processes, having been neither a historian nor an academic. There may be nothing kinder to say about my qualifications to produce this work than that I am not vulnerable to their professional biases.

The Author's Sinister Motivation

Some readers may misinterpret observations about Galileo's mistreatment by the Inquisition, and by the Church in our own time, as a sinister attempt to heap opprobrium on the Church. Nothing could be farther from the truth. The suppression of Galileo was a disaster for all parties; it retarded scientific progress and tended to discredit the Church as well. My choice of Galileo as a topic is grounded in the conviction that his contemporary importance has been generally understated by historians. The conflict between science and institutional indoctrination, of which his case is the most outstanding example in history, is one that affords an opportunity to examine, reflect upon, and act upon the causes of all conflicts of this nature and the means to remedy them. In this work I infer compatible roles for individual scientists and theologians, given certain role modifications. I am a signpost, pointing toward that compatibility. If this is a sinister motivation, then let "sinister" be my epitaph.

The "Man in the Street"

Surviving the shock waves of a massive collision, centered around Galileo, between the Renaissance and the Inquisition, the effect of lingering indoctrination has spread over 360 years to elements of modern society. It's about time that the "man in the street"[1] speak up on this matter. Neither a historian, nor an academic, nor a theologian, nor recognizable by the mainstream of scientists, and unconcerned about consequences to his reputation (because he has none) of exposing the inept response of these professional groups to the Galileo affair, he lacks the "ax to grind" that many of them wield. Having met these qualifications, I am the "man in the street". Only a curious onlooker, I marvel at the complacency toward this crucial collision — among those who ought to know better. The attitude that I detect goes something like this: "It's all over. It's history. Galileo was right. Anyone can plainly see in NASA video footage that the earth turns on its axis daily.[2] The Church learned its lesson. It withdrew from science, declaring

---

1. By this phrase I refer to a curious, literate, non-academic individual operating independently of any organization that dominates the thought processes of its members.

itself infallible only in matters of faith and morals. People aren't burnt at the stake for heresy anymore. There is no more Inquisition. Science and theology are now divorced, and have nothing to offer each other. Scientists can thrive on their victorious epistemology and need no longer be concerned with this affair."

How Galileo Has Been Misunderstood

What's wrong with this picture? It is this. Primarily in the form of institutional indoctrination, the Galileo conflict remains alive and, on account of being less obvious, is more dangerous now than in 1632. The principle[3] then used to suppress Galileo now subjugates educated people, scientists among them. Those who preach its use, and those who apply it, have with devastating effect sharpened and re-employed the tools of 17th century churchmen. The conflict of Galileo and the Inquisition, which is that between right thinkers and those in authority, has degenerated to the relationship of a well-fed vassal to his patronizing master. Unjoined parts of a solution to a dysfunctional relationship between critical societal elements remained unrecognized until a far-seeing visionary, Pope John Paul II, made a stunning announcement following the report of a Papal Commission on the Galileo case.

That interlocking parts of a solution have gone unrecognized is mainly due to faulty historical interpretation. Despite apparently competent researchers who tackled this subject, an important component of Galileo's message remains clouded by generations of academic detractors focussed on small issues and engaged in petty, perverted criticism or in misrepresentation of his work. As a result, his popularity among academics arises from their perception of him merely as an interesting subject on which to write scholarly papers, and not as a key to understanding the future of man.

Sources and Acknowledgments

If by the foregoing polemic I seem to be introducing a perspective on Galileo that marks a discontinuity from past accounts of his life and work, my intent is equally to extend the previous interpretation on this subject that has taken a right and constructive direction. Except as otherwise noted, I derived information about Galileo from his writings and from sources listed in my bibliography and footnotes. Though relying on historical research, I made a practice of checking sources against original writings of the subject philosophers (e.g., Galileo himself), or of historians who have done seminal research. Despite considerable authoritative writing in secondary sources, I was disappointed at the missing coverage of Galileo's analysis of Dante's *Inferno*, and of Favaro's

---

2. On the other hand, observing the earth from a geosynchronous satellite can be expected to give the superficial impression that the Ptolemaic stationary-earth model is the correct one.
3. identified in Chapter 8. See "Galileo's Dictum of Bird Droppings".

Preface, both scholarly and passionate, to *Galileo e l'Inquisizione*, and have therefore supplied translation or paraphrased from my translation.[4]

Having heard that I was already obsessed with astronomy, my Italian immigrant grandmother took me aside, when I was nine years old, and told me the story of what has falsely been called Galileo's recantation.[5] The shock of hearing how this great man was "rewarded" by society for his accomplishments, some of which were already known to me, never left me. At the age of ten, I was befriended by a kind and scholarly astronomer, Dr. Edgar W. Woolard, then on the staff of the U.S. Naval Observatory, who invited me to his house, gave me a lesson on the number π, and supplied me with several astrophotographs, one of which appears in this book. Later, after having studied celestial mechanics under a Jesuit astronomer (Fr. Heyden) and its foundations under the mathematicians Malcolm Oliphant and Richard McCoart, I formed the intention of investigating the scientific inquiries of Galileo, focussing on the underpinnings of his conflict with various "theologians" and philosophers. Soon, I thought, I would visit the sites of his work and see firsthand the source of my subject. Many years elapsed until I made these visits.

Among his many courses, one given by Professor Andrew J. Galambos jointly with Mrs. Suzanne J. Galambos was based upon their European travels. That course helped inspire me to end a long period of procrastination about the thoughts, notes and plans that had too often lain dormant, and that have evolved into this book. In particular, I wish to thank Mrs. Galambos for her review of an early draft of this book, a review that implies neither her endorsement nor her disendorsement of the finished product. Unfortunately, I am unable to say anything about a lecture series that Professor Galambos gave concerning Galileo. (My attendance was precluded by a business commitment in England, where I used free time to gather background materials for a planned sequel to the present writing.)

I am grateful to Professor Ayer for his book *Language, Truth, and Logic*, which I read in my university student days. His statement of the Principle of Verification as a criterion of meaningfulness caused me to draw for the first time a distinction between theology as the study of God through nature, and "theology" as a body of untestable, metaphysical claims. I am grateful to my unofficial mentor Professor Diamandopoulos, who has been Chairman of the Departments of Philosophy and History of Ideas at Brandeis University, for pointing me toward Ayer's work.

Mr. Robert Eklund of the Mount Wilson Observatory Association graciously invited me to address the Association, thereby providing the occasion to organize materials for an original lecture from which this work derives. Mr. John Sanford,

---

4. See Appendix A for the complete translated Preface to *Galileo e l'Inquisizione* and the acknowledgment "In Grateful Memory" to Professor Jardini.

5. Cf. Epilogue, "Myth Number Six: The Dishonor Myth".

President of the Orange County Astronomers, took it upon himself to reprocess my photographs. I wish to thank Mr. Sanford for the appreciation of my research implied in his voluntary and laborious attention to this task.

Also not to be overlooked is the assistance of research librarians and others who pointed me toward source materials, and of persons whose insights, comments, references, or encouragement proved helpful. The following individuals (listed alphabetically by surname) are remembered: Dr. Leo S. Abrams; Mr. Besim Amado; Dr. Sallie Baliunas; Mr. Robert Bauer; Sig. Stefano Casati; Mr. and Mrs. Royal Crowell; Mr. Victor Anthony Crowell; Mrs. Florence Drake; Mrs. Laura Woodard Eklund; Professor and Mrs. Victor Elconin; Mrs. Ann Hammer; Frau Esther Hammer; Professor Doris Helfer; Mr. Seyhan Hizal; Professor Robert E. Houston, Jr.; Mr. Michael J. Huxtable; Professor Trevor Levere; Miss Amy Luedecke; Professor James MacLachlan; Mr. Geoffrey Masaki; Frau Senta Matschak; Mr. Michael Mendola; Ms. Dona Morgan; Mr. Joseph B. Moriarty; Dr. Antonio Naples; Rev. Fr. Vincent O'Brien, S.J.; Mr. Don Rado; Dr. Eric Sit; Mrs. Adelia Spann; Sig. Danilo Tani; Mr. Lawrence Tyson; Mr. Byron Wall; Ms. Deanna Wood; Dr. Edgar Woolard, Miss Barbara Zajac. And, though I lost track of you before conceiving of this project, my friend and former classmate Dr. Ed Hook, wherever you are, thanks for laying those mathematical challenges on the table.

My father and mother, Albert and Cecelia Di Canzio, encouraged my childhood interest in science and assisted my journey through the wonderland of Jesuit education. In these ways they helped to produce this book, as well as having collaborated in the production of its author.

The topic of my sources would be incomplete without brief observations on two resources: research time and information. Given so prolific a biographee as Galileo, the research is never completed. If I had to finish it, I would have to abandon this project altogether. Though I will continue to study and marvel at his work after this book is printed, I hope to spread my enthusiasm about this man and this episode in the history of man, so that you and I can walk together the trail of one of the grandest adventures that the history of science has to offer.

Uncomfortable, too, is the feeling that writing a historical account of developments in physics can resemble speculation ungrounded in physics. I have not witnessed the events nor known the persons about whom I write. Most are long gone. What remains is a paper trail, with gaps, overlaps, and paradoxes. Silence about probable but unprovable matters is, however, as sterile as it is safe. I choose instead to examine the data, interpreting and amplifying them for the reader. When new data are consistent with the rest, I tend to accept them, and to look at original writings when necessary to validate a claim. My duty to give the reader a complete picture also requires walking on the insecure ground of my information sources.

Why This Book?

Hundreds of books have been written about Galileo. Why, then, am I writing yet another? I intend this book to have these qualities, which are not jointly and sufficiently characteristic of writings known to me:

1. It should be precise without being pedantic, and interpretive without being unbalanced. It is my personal observation that authors of interpretive works on the history of science tend to gravitate toward one of two extremes. Professor Publish-or-Perish is more concerned about getting his book into print than about attracting readers. He writes with historical validity but habitually lapses into leaden and lifeless prose. Novelist-in-Disguise is an entertainer who, in exaggerated zeal to attract readers, casts off the yoke of fidelity to probable truth. Where facts are available, he may select only those that conform to his biases. Or, where facts are lacking, he may supply imaginative conjecture without alerting the reader, add a superficial veneer of validity by citing copious references, and produce a historical novel under the guise of a biography. As the delivery vehicle for a fresh perspective on the still unfolding drama of Galileo, this book is intended to avoid both extremes.

2. It should offer insight into Galileo's motivations — those which you can share whatever your field of endeavor — by examining their object: his science. Many commentators misrepresent Galileo as motivated by his Catholic religion, by Copernican zeal, or by this or that philosophical alignment. Why stereotype him as a "club" member: a Catholic, a Copernican, a Platonist, etc.? These labels are substitutes for thinking and create the false impression that science advances when one collective body of thought wins over another by the relative weight of accepted opinion. It is a central theme of this book that Galileo acted as an individual motivated not to force-fit nature to a pre-conceived philosophy but to fit philosophy to nature. This aim induced him to revolutionize scientific method.

3. Properly recounted, Galileo's science and his personal example suggest future applications by cosmologists, physicists, educators, historians and entrepreneurs. Among them are the following:

· Questions of cosmology include those of the origin and remote future of our species and do not yield to analysis of fables. To build a tower of knowledge stretching to the stars requires data-gathering systems that operate across the full spectrum of detectible wavelengths. Aided by the vision that Galileo supplied to astronomy and the method he supplied to science, the way is open for cosmologists to do just that.

· Unified theories require future insights into the properties of force fields. Galilean Units, because they are adapted to the study of gravitational phenomena, offer the potential for such insights acquired through reduction and analysis of data expressed in dimensionally commensurable units.

· Not geocentricity but a related anachronism, the substitution of false doctrine for truth, massively thwarted Galileo's progress through forced attention to nonscientific issues, for example, that of defending his science from charges of deviating from pathways of Catholic expression. The divergence between what was demanded of the faithful and what could be observed and deduced through application of Galilean science contributed to the later demise of the theocratic system of his persecutors. The story of this conflict flashes to the designers of future educational systems an alert about freedom of inquiry and of scientific communication. Though the process of attaining such freedoms is not yet complete in our own time, the lessons of the Galileo affair bring into sharp focus the dangers of censorship and indoctrination to all parties, including the censors and indoctrinators.

· Historians are by no means done with his story. There are unsolved questions as to the content of his lost papers: on the mathematical theory of continuity and indivisibles, and on light and color. Among unresolved issues relating to his inventions, little is known about the structure of the early compound microscopes that Galileo made.

· Restorer of Thalesian science, Galileo developed a scientific method applicable to business. The analysis of the human body as a machine, begun by his disciple Borelli, holds the potential for radical advancement in health engineering through two outgrowths of Galilean science: general-purpose computing technology and the use of statistics to establish correlations. Systems engineering and the data-processing industry have emerged following application of probability theory and the calculus to military problems in WWII. Operations research is in its infancy and, with the trend away from military and toward business applications of Galilean scientific method, industries of the future are being born.

4. In this book, there should be no attempt at "political correctness". Aside from being inherently nauseating, "political correctness" is distinctly un-Galilean. This should be a book that, if Galileo were transported near the speed of light so as to return to earth in our time, he might choose to read in order to inform himself how his work and influence have unfolded four centuries later.

5. Our story should have a self-extensive quality, as does Galilean science. A sufficient number of pieces of a puzzle must be in place before a recognizable picture emerges, and sometimes concentration is needed to avoid missing an important clue. This account of Galileo's work and significance ought to offer the highest reward to a reader who stays with it to its conclusion — and perhaps, in self-study, beyond it. Though there is no "royal road" to Galilean science, I have tried to smooth the paving.

6. This book should present those heroes about whom I write as examples for emulation to men and women of the millenium of which the 21st century is the start. The history of science is as much about science as it is about history; I aim to present science through its history, and to make our heroes real by connecting their unfinished work to the continuity of civilizations.

Prologue to the First Edition and Requiem for the Zeroth

The evening wind is whipping through hollow spaces of trees that have not yet enjoyed the first bloom of Spring. Across the street from my hotel room stands an old A-frame house painted yellow, with red shutters, as seemingly immobile against the dusky sky as was the earth itself to medieval churchmen. After long days of travel, it now seems strange, almost derelict of duty, to be sitting here motionless (relative to the earth's surface — a qualifier necessitated by Copernican/Galilean astronomy) as if to move on were life. But moved on I have, to the first edition, and I now bury the zeroth.

The zeroth edition is an unpublished manuscript from which this first edition has emerged. That revision is largely the product of my considering the comments of four editors: Mr. and Mrs. Royal Crowell, Professor James MacLachlan and Mr. Don Rado. By calling this first published edition a new one, I emphasize the effect that each has made by painstakingly detailed examination of the original text, and by insightful comments. To say that without these individual contributions, I would not have not published this book is no exaggeration.

Although the thought content of this published edition is close to that of the zeroth, with the addition of certain intellectual gleanings from my conversations with these editors,[6] there have been major enhancements to the book due to each of them. Mr. Rado, in particular, who in my opinion displays an outstanding faculty of spatial visualization, was justifiably critical of some of the original drawings, and has been instrumental in redrawing these pictures. Mrs. Crowell's observations and encouragement were such that I might not have published at all had it not been for her pointing out the relationship of certain personal circumstances known to me but unrelated by me. Her, and Mr. Crowell's, extensive discussions with me of relevant Renaissance history were most helpful in suggesting many sources,[7] topics, and additional research. Beyond the usual editing, their cross-checking of my writing versus source documents, and of my calculations, proved invaluable. Of Mr. Crowell, whose knowledge of geometry and its pre-Galilean history was most helpful, it can fairly be said that he has led me through an on-line workshop in the proper arrangement and support of my points. From him I discovered that I had sometimes suppressed portions of my thinking as if parenthetical, whereas expressing them would prove vital to the flow of the text. In my zeal to provide the reader a quick understanding, I had written only half a book. He also pointed out my disturbing tendency to take huge mental leaps in time and space, without

---

6. Throughout the footnotes, the symbols (rdc), (dc), (jm) or (dr) will tag observations or ideas based on private conversations with Royal Davis Crowell, Diane Crowell, James MacLachlan or Don Rado, respectively, on the content of this book, and which I attribute to one of them.
7. including the books by A. D. White , G. F. Young, Durant vol. vii, and Dantzig.

providing a proper cue to the reader. Apparently, in the original manuscript, I had kept my promise much too well after telling the reader in the Introduction that we would move freely between past, present, and future.

To Professor MacLachlan I am grateful for his careful reading of my manuscript and his numerous suggestions that substantially improved this book. A colleague of Galileo biographer Stillman Drake, MacLachlan was a reviewer of the manuscript of Drake's *Galileo: Pioneer Scientist* and is himself an expert who has written articles about Galileo as well as textbooks in physics and in the history of science.

Nothing said about these pre-publication reviews is meant to imply any disclaimer of responsibility for this book. Having conscientiously attempted to avoid error, I am still 100% responsible for any mistakes in its content. Unlike the holy Fathers against whose opinions Galileo was accused of philosophizing, I do not claim divine inspiration or guidance. Therefore, if I have erred or offended anyone with anything I've said or not said in this book, I offer my apology in advance. Although not infallible, I don't do things intentionally that are harmful or wrong. The price of being sure that you will offend no one is to waste time doing nothing. The price is too high. I hope the reader will enjoy and profit from this book.

Albert Di Canzio

Earth In Space (NASA photo)

Introduction

Orbiting its parent dwarf star, a majestic blue and white biosphere pirouettes on an invisible axis in the vast, near emptiness of its surroundings. Questions of cosmology, of our origin and destiny, loom more visibly before us now that humans have looked back at the Earth and seen it as a planet. Immensely larger now than in 1609, when the first astronomical telescope had already expanded the horizons of our vision by five orders of magnitude, the apparent universe is devoid of physical boundaries. The travel time of light from the farthest reaches of our telescopes has begun to approach the calculated age of the physical universe. From this perspective, what can a single human, seemingly insignificant in time and in space, do that matters in the context of our civilization, our species, and our universe? To approach this question, that links you and me individually to the rest of the universe, one may examine the work of one special individual, who stands at the center of the historical conflict between science and what has, at times, been thought to be theology. Most of us know him simply by his first name: Galileo.

The seeds of the conflict were planted some twenty years before Galileo's birth by Copernicus, who revived, refined and mathematically elaborated the ancient idea of Aristarchos[1] to the effect that the sun, not the earth, is the center of planetary motions. That was difficult to believe in those times. Even now, it appears that I am standing on a very solid, steady earth. I see the sun rising, moving across the sky, and setting. I see planets wandering among the stars. And it is easy to accept that the earth is motionless and the center of all these perceived motions. Certainly, Copernicus declared an important shift in thinking, away from the notion of man's occupying some favored place in the universe. But not only was he not the first to declare it, he had no proof based on direct physical evidence. It was Galileo, along with his contemporary Kepler, who revealed the principal evidence — and the link to corroborating evidence — that led to the overturn of the general belief in the geocentric world system.[2]

And so this modern vision that you and I share of the double motion of Earth in space, around its axis of rotation and around the balance point shared with the star to which our planet is bound, survived the scrutiny of crucial observational tests to which these two men, working independently, submitted it. Newton then created a general physical theory to complete the picture. Of these three men, it was conspicuously

---

1.   Where feasible, I choose to transliterate the Greek names directly to English rather than to imply, using Romanized Greek names, that the Roman culture should be some sort of filter for the Greek; hence, "Aristarchos", not "Aristarchus", for example. As exceptions, I use the familiar transliterations "Aristotle", "Euclid" and "Plato", rather than the more literal "Aristoteles", "Euclaides" and "Platon".

2.   Certainly, Tycho produced data useful and necessary to Kepler's investigation, but it was Kepler who converted the data into evidence contradicting the geocentric world view.

Galileo who ventured far beyond his new science to create the intellectual climate through which modern man has begun to invade the sanctuary in which nature hides its deepest secrets.

Apparently, the notion that humans must bind themselves to authoritative traditions of the past and to institutions in which they are embodied persists among seemingly educated people nearly 400 years after Galileo's bold forays into independent and fruitful thinking. Although he often tempered his iconoclasm with tolerance for church tradition, Galileo did not generally act as if bound by the same traditions that he tolerated. Instead he treated the institutions of his day as harbors from which he could depart on a voyage to more rational traditions. His example puts a non-eschatological twist on these centuries-later words of "The Chambered Nautilus" in contrast to what its author, Oliver Wendell Holmes, seems to have intended in an implicit reference to afterlife:

> *"... Build thee more stately mansions, O my soul,*
> *As the swift seasons roll! Leave thy low-vaulted past!*
> *Let each new temple, nobler than the last,*
> *Shut thee from heaven with a dome more vast,*
> *Till thou at length art free,*
> *Leaving thine outgrown shell by life's unresting sea!"*

Stretching in our newly found home and peering back through the keyhole of an earlier idle door, we are about to see one man lighting, in a dark chamber dominated by the Inquisition, our way into the chamber of the present. Looking to the chamber of the future, we can see, under a dome more vast, the unfinished business that he left for others, breaking through barriers to progress in the span and focus of life.

The story of the Renaissance is a story of individual heroes who restored the progress of civilization after a thousand years of its wandering about in what Galileo, one of its principal heroes, might have called a "dark labyrinth". To see the high drama, the heroism and the achievement of that restoration is to see what it was that had to be restored: the advanced scientific culture of ancient Greeks, obliterated in the Dark Ages. In quest of this vision, let's set out on a journey back to a time before Galileo, when natural philosophers are laying the foundation on which he will build a more stately mansion. Our journey through time takes us first to the eastern Mediterranean seacoast around 2600 years ago. In our travels, we shall frequently connect past, present, and future by moving freely among them.

Chapter 1. Shadow of A Pyramid

Ancient Natural Philosophy

"He determined the height of the Great Pyramid by measuring the shadow it cast at an hour when a man's shadow was equal to his height." [1]

Alternately nourisher and destroyer of two cities, a mighty river [2] meandering through fertile valleys toward the shores of Asia Minor meets the calm waters of the Aegean in an outpouring of silt it has carried from the high desert of Anatolia. At this delta in the sixth century B. C. stood proudly the city of Miletos, the ancient Venice, commercial gateway to Asia and the Orient. Today, after 30 centuries of ongoing silt deposits from its inception in the 11th century B.C., the remains of Miletos can be found nine kilometers inland. Having flourished during the Ionian, Hellenistic and Byzantine civilizations, and having then lost its port, it likewise lost its importance: to commerce, that is, and since commerce was its lifeblood, it lost its life. A similar catastrophe occurred at nearby Ephesos, home of the philosopher Heraclitos and site of one of the Seven Wonders of the Ancient World.[3] But the name of Miletos, first mentioned in Homer's *Iliad*,[4] will not pass from the writings of historians. For however devoid of human habitation, Miletos can never lose its historical importance: it is here that science [5] was born.

That event is a consequence of Thales (c624-c546),[6] the first scientist of ancient Greece, one of the Seven Sages of the Ancient World, and probable author of the

Thales of Miletos, the Founder

Delphic inscription: ΓΝΩΘΙΣΕΑΥΤΟΝ (*Know Thyself)*.[7] Unlike Newton and other scientists of recorded history, Thales performed an act of creation in a void of science.[8] Despite being called the ancient Venice, Miletos is better linked conceptually to Pisa. For at Pisa occurred the rebirth of science born at Miletos, a

---

1. Written by Aristotle's pupil Hieronymos in reference to Thales and the Pyramid of Cheops, quoted by Tobias Dantzig, *The Bequest of the Greeks*, Charles Scribner's Sons, NY, 1955, p. 50.
2. The river was named "Meandros" by ancient Greeks.
3. The Temple of Artemis, renowned as the first temple of Ionian architecture.
4. Hakks Gultekin, *Miletus*, Archaeological Museum of Izmir, 1961.
5. In this context, I use the term "science", as distinct from the technology that preceded science, in the sense of Clagett: "the orderly and systematic comprehension, description and/or explanation of natural phenomena". M. Clagett, *Greek Science in Antiquity*, Abelard-Schuman, London, 1957, p. 4.
6. All biographical date ranges of the Greek philosophers' lives are negative or positive (B. C. or A. D.) relative to the base year of the Gregorian calendar widely used in the English speaking world at the time of this writing, depending on whether the absolute magnitude of the birth date exceeds that of the death date or not. Source of these dates is *Encyclopedia Brittanica*; most dates are approximate (hence, 'c' for 'circa'). Hence (c624-c546) is an approximate B.C. date range.
7. In Thales' time, words were written in upper case without punctuation or separation by spaces; cf. William Dugan, *The Golden Book of How Our Alphabet Grew*, Golden Press, NY, 1972, p. 5. (dc)

rebirth due in large measure to Galileo. Because of Thales, Miletos is more durably the ancient Pisa, a status that no river can wash away.

A visitor to the site of ancient Miletos today will find no remaining structure that dates from the time of Thales, with the possible exception of the wall on the southern slope of the hill Kalabaktepe.[9] A short walking distance from the still standing theatre and the baths built by the Romans more than seven centuries after Thales, a grove of olive trees complements the bucolic setting, reminding us that science flourishes in the setting of an entrepreneurial challenge. For legend holds that, challenged by a skeptic who said of science that it is useless to a businessman, Thales applied his knowledge of astronomy and of weather patterns to predict a bumper crop for olives the next year, and having cornered the market on the available olive presses, accrued a fortune by renting them out at high prices.[10] But the path from the olive grove to modern business mathematics encountered a nearly fatal discontinuity that required a restoration, and that is where our story will lead. Galileo was the restorer of Thales' creation, lost in the Dark Ages.

From his travels in Egypt, Thales brought back primitive geometry in a "maze of rules, recipes and rites amassed without rhyme or reason"[11] by Egyptian priests. On this maze he imposed a logical structure to create the first system of knowledge built on definitions and postulates. That imposition of structure renders his science of geometry[12] deductive and mathematical in nature and yet, when informed by observation of the real world, fundamental to physics.

Physics began with mechanics, the study of motion, and more specifically with kinematics, the study of motion without reference to forces. How do we express

---

8. "That it took the Greeks more than a century to absorb the work of the Founder (Thales) becomes less surprising when we contemplate the grandeur and revolutionary character of his achievement and remember that there were no geometers when Thales began: Thales the *teacher* produced the first geometers, even as Thales the *thinker* founded the first geometry worthy of the name." Tobias Dantzig, *The Bequest of the Greeks*, op. cit., p. 43.

9. Cf. Suzan Bayhan, Archaeologist, *Priene.Miletus.Didyma*, trans. Anita Gillett, Keskin Color Kartpostalcilik Ltd., Ankara, 1993. In 494 a Persian naval fleet attacked Miletos and, not content to destroy the Milesian fleet, seized and demolished the entire town, removing the survivors; cf. Ord. Prof. Dr. Ekrem Akurgal, *Ancient Civilization and Ruins of Turkey*, Turistic Yayinlar, 1993, p. 207. Thanks to this gang of rowdies, visitors today can see nothing from the era of Thales but a miserable hunk of wall and, while from nearby Priene we have a well preserved bust of Alexander "the Great", we have no surviving image of the truly great Thales. Miletos was rebuilt to the plan of Hippodamos using a right-angled geometric pattern, of which he was a pioneer in systematizing (ibid., p. 210). Aristotle, who called Hippodamos a strange man and noted that he wore expensive jewelry with cheap clothing, yet admitted that he invented the art of planning cities; Sarton, *A History of Science*, Harvard U. Press, 1952, p. 295.

10. Cf. J. M. Cook, *The Greeks in Ionia and The East*, Thames and Hudson, London, 1962. This story was originally related by Aristotle; cf., Carl Boyer, *A History of Mathematics*, Wiley, 1968.

11. This quotation is from Tobias Dantzig, *The Bequest of the Greeks*, op. cit., p. 34.

12. "Geometry is the science that treats of relations among forms." (rdc)

motion, or change of position of an object relative to another? The separation of objects in space is measured as a distance (or length). Distance is one of three axiomatic concepts of mechanics; the other two are time and mass. In a rectangular coordinate system, the distance between two points is treated as a hypotenuse and calculated by the so-called "Pythagorean theorem" that we learned in school:

> *The square of the hypotenuse of a right triangle equals the sum of squares of the sides containing the right angle.*[13]

Can it be that Thales proved this principle before Pythagoras existed? The answer lies in the shadow of a pyramid.

In the quotation heading this Chapter, Hieronymos tells us how Thales may have surveyed the Pyramid of Cheops. Plutarch, another ancient historian, gave a different answer. In Plutarch's version, Thales measured the shadow of a stick and inferred the height of the pyramid from the similitude of triangles. In either case, the measurement cannot be made directly because the pyramid itself intervenes between the tip of its shadow and the center of its base. That Thales' could calculate the inaccessible portion of that path implies a knowledge of the "Pythagorean theorem" or of the principle of similar triangles on which it is based. While the rule expressed by this theorem was known through experiment before Thales,[14] he inferred it deductively and gave original proofs of other theorems: [15]

- An angle inscribed in a semicircle is a right angle.
  (This theorem is known as the "Theorem of Thales".)

- A circle is bisected by any diameter.

- The base angles of an isosceles triangle are equal.

- Opposite angles made by intersecting lines are equal.

- The sides of equiangular triangles are proportionals.

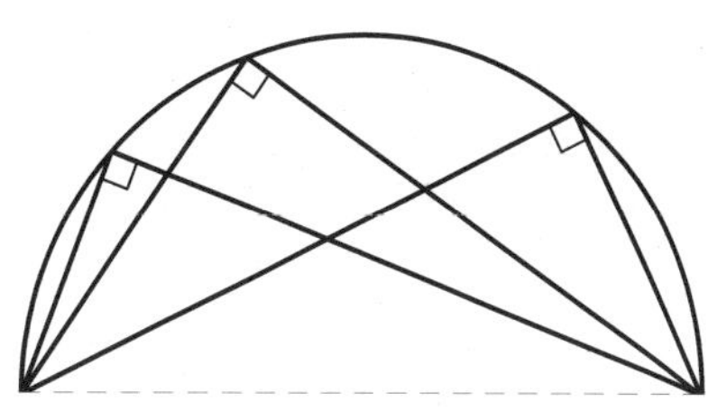

THEOREM OF THALES

Thales was also concerned with the structure of the physical world. He attempted to explain the composition and origin of matter by pointing to a single fundamental

---

13. ibid., p. 30. Dantzig has made an informed and well supported conjecture that the theorem named for Pythagoras was known earlier to Thales and probably demonstrated by him on the similitude of triangles, with which Thales was "fully conversant". Even today, generalizations of this relationship continue to be studied; e.g., Fermat's Last Theorem (that $x^n + y^n = z^n$ if n is an integer > 2 has no solution in positive integers) has recently been the object of intense effort to create and validate a proof.

14. "... twentieth-century examination of cuneiform baked clay tablets excavated in Mesopotamia has revealed that the ancient Babylonians of well over a thousand years prior to Pythagoras' time were aware of the theorem." cf. Howard Eves, *Great Moments in Mathematics Before 1650*, Mathematical Association of America, 1983.

15. See Carl Boyer, *A History of Mathematics*, p. 46 (see Bibliography) for a discussion of evidence that Thales proved these theorems.

element: water. Reducing the multiplicity of substances that humans commonly encounter at the level of their perception to a single underlying reality, he revealed a cosmology that, unlike its predecessors, is not mythological but observational. With an observational cosmology, it becomes possible to use deductive reasoning, aided by the abstract forms of geometry, as the basis of describing the real world in order to predict events. Though his answer was mistaken, his question was fundamental and provocative of later rational cosmologies.

Under the old mythological cosmology, for example, an eclipse of the sun may have been conceived as a random and whimsical act of an angry god, impossible to predict. But Herodotus credited Thales with successfully predicting a solar eclipse, that of May 28 in the year 585 B. C., noting that the eclipse "caused suspension of a battle between the Medes and the Lydians, followed by a treaty of peace". There is no evidence that he developed the mathematics required for general prediction of eclipses, nor is that likely. What is likely is that, assuming some rational pattern, Thales looked at records of past eclipses kept by the Chaldeans and learned about the 18 year 11 day cycle of solar eclipses.[16] The saros, as this cycle is called, can be derived from the modern theory of eclipses. In the time of Thales, it had emerged not from mathematical deductions but from observation and accurate historical record keeping. It was the same sort of analysis of past observations, applied to weather patterns, that permitted Thales to make his fortune in a commercial venture. Renting olive presses was not the only such venture.

While the river system of Asia Minor could nourish or destroy commercial cities, Thales demonstrated human power over nature by diverting the flow of the Halys River.[17] To achieve the diversion successfully, he had to predict the outcome of altering the river's course. Twenty-two centuries later, Galileo used Thalesian geometry to predict the consequence of redirecting the flow of the Bisenzio River. Over the ages that separate Thales from the Renaissance, we can see predictive power unleashed by rational cosmology and applied to commercial ventures that fended off famine and forged flourishing civilizations.

Thales applied science to business and not principally to warfare or other forms of cultural folly to which science has been perverted even in our own time. The emergence after World War II of operations research, the application of advanced mathematics to business management, is a post-Galilean restoration of a Thalesian development. This recent development gives new and deeper modern meaning to the olive branch as a symbol of peaceful human interaction.

---

16. Sir John F. W. Herschel, *Outlines of Astronomy*, Part Two, American Home Library Company, NY, 1902, p. 838. W.W. Rouse Ball suggests that Thales had learned about the saros from the Chaldeans; cf., *A Primer of the History of Mathematics*, MacMillan, London, 1927.
17. Cf. Dr. Cemil Toksoz, *Ancient Cities of Western Anatolia*, Alas Ofset, Istanbul, 1979, p. 115.

The geometry of Thales is the model for all science.[18] It is the foundation of radical new concepts of space and time, of natural philosophy, and of the potency of man over nature. It permitted the substitution of orderly systems of reasoning accessible to the individual for convoluted mythological cosmologies that gave priests and mystics a powerful and unmerited monopoly over thinking. Conveyed over two millenia by Euclid to the young student Galileo, it influenced Galileo's choice of a career, and through that career, all science after Galileo.

How was it conveyed to Galileo? Thales left no parchment that survived to the present day or even to the time of Galileo. Not a direct transfer of knowledge to Galileo from Thales or even from his immediate successors, the basis of restoring Thalesian developments came through much later Greek successors. Eudoxos, Aristotle, Euclid, Archimedes, Apollonios and Ptolemy were among the students of Thales and, through their surviving works, teachers of Galileo.[19] These great thinkers flourished in the intellectual climate of Ionian civilization.

Ionian civilization spread itself over more than a dozen settlements on the mainland and at least two islands (Samos and Chios) in an area a few hundred kilometers wide and centered at the ancient city of Smyrna, renamed Izmir by the Turks, an early Ionian Greek colony, and probably the birthplace of Homer. Today, with its magnificent harbor, it has become the chief port of the Aegean region, exporting raisins, figs, and cotton grown there, as well as handmade rugs. In the northern part of this area is Nicaea, renamed Iznik by the Turks, the home of the great astronomer Hipparchos; near the southwestern corner is Knidos (Cnidus), home of Eudoxos the mathematician; to the south and east is Perga, home of Apollonios the geometer. Away from urban areas, past and present merge in the eye of the mind and body, respectively, and while watching a goat herder drive his flock, one easily visualizes Xenophon driving his hoplites through winter snowdrifts across the barren desert in retreat from the advancing Persians.

Just offshore of Miletos lies the Greek island of Samos, from the time of Thales the habitat of a series of great thinkers.

Pythagoras of Samos (c580-c500) discovered the first known law of mathematical physics: the relation of the pitch made by a string of given diameter and tension to its length,[20] a relation first extended over two millenia later by Galileo's father. Pythagoras sought to

Pythagoras of Samos

---

18. rdc argues and holds: "Thales is the originator of Western civilization."

19. "... the more one studies the period of Thales — the more one compares the knowledge he bequeathed to posterity with the one he had found when he began his work — the more does his mathematical stature grow, until one is impelled to range Thales with such figures as Archimedes, Fermat, Newton, Gauss, and Poincaré." Dantzig, *The Bequest of the Greeks*, p. 32.

interpret all phenomena through number. Thales was about sixty years of age and of widespread fame when Pythagoras was a young student, and Samos is a two hour boat ride from Miletos.[21] It is circumstantially plausible and even likely that Pythagoras became a student of Thales.[22] In general, it is nearly impossible to assign any discovery personally to Pythagoras because he operated his school "communally" and in secrecy. Much of Pythagorean research was preoccupied with mysticism and numerology. Pythagoreans discovered the five regular polyhedra, and in particular, Hippasos discovered the "mystical" dodecahedron,[23] a closely guarded secret of the Pythagoreans. Demonstrations that the sum of angles of a triangle is equal to two right angles, and that $\sqrt{2}$ is irrational,[24] are products of the Pythagorean school. The discovery of irrationals is a profound one because it exposes the incompleteness of the rational number system.

Empedocles

Beginning with Thales and his water cosmology, philosophers tried to identify elements of which everything on Earth consists. Various guesses culminated in the opinion of Empedocles (c490-c430) that all things on Earth are combinations of earth, air, fire, and water. The striving to look beneath superficial differences in objects of ordinary experience, to find common elements that keep their identity through change, led to a paradoxical duality in human perception of nature: permanence versus change. "In the same rivers, we step and we do not step ..."[25] said Heraclitos of Ephesos (c540-c480), articulating the paradox. In this paradox the science of chemistry has its origin.

Zeno

Between those first conjectures of Thales and Empedocles about the elements of nature came Zeno of Elea (c490-c425) who connected change with the continuity of space and time. He denied motion through space in four famous paradoxes as puzzling to ancient philosophers as irrational numbers. Although modern mathematical demonstrations of infinite series convergent to finite sums have answered and thus obsoleted it, who can forget the paradox of

---

20. Cf. G. D. Birkhoff, *Aesthetic Measure*, Cambridge, 1933, p. 90 as quoted in James R. Newman, *The World of Mathematics*, Vol. 4, Simon and Schuster, NY, 1956, p. 2183 footnote 7.

21. Modern boats with engine-driven propellers may ply the Aegean faster, but the time gained is now lost by having to clear passport control and baggage inspections at both ends (after 2,600 years the progress of technology has been eradicated by the increase of political borders and their collectivist gatekeepers). At Samos, an elegant monument to Pythagoras adorns the Pythagorion marina. According to museum officials there, no monument to Aristarchos exists on the island.

22. "Pythagoras, from his early age, went to Lesbos where he became a student of the philosopher Ferekides; later he became a student of Euridamas, at Samos, and at a later stage, of Thales and of Anaxandrides at Miletus." D. G. Davaris, *Samos*, trans. by Michael Heath, Davaris, Athens, 1981.

23. See W. W. Rouse Ball, *A Primer of the History of Mathematics*, MacMillan, London, 1927.

24. See G. H. Hardy, *A Mathematician's Apology*, Cambridge University Press, 1967. Republished in Turkish translation as *Bir Matematikçinin Savunması* by Tübitak, Ankara, 1994.

25. from fragment 81, per M. C. Nahm, *Selections from Early Greek Philosophy*, 1962.

Achilles and the tortoise that we all (didn't we?) learned as children? If Zeno's tortoise has a 5000 meter head start, then when the fleet-footed Achilles, let's say 50 times as fast as the tortoise, reaches the initial position of the tortoise, the latter has crawled ahead 100 meters, and as Achilles dashes that 100 meters, the tortoise advances 2 meters, etc. Zeno's denial of motion assumes that the resulting infinity of such intervals, during which Achilles approaches but never overtakes the tortoise, cannot elapse in a finite time.[26] His implied question, whether space and time are infinitely subdivisible, launched the quest for continuity: for a compact, infinitely dense number domain that maps into physical space and time. Leaping toward it, Galileo introduced time as a coordinate and summed immeasurably small time intervals to a measurable interval of fall. Zeno showed us that it is not necessary to be right in order to have a positive effect on the progress of knowledge; it suffices to pose well chosen questions that are difficult to answer given the state of knowledge in one's own time.

A Greek visionary named Democritos of Abdera (c460-c370) collapsed onto the visible world the whole scope and compass of **Democritos** imaginative perspective, from the astronomical to the submicroscopic, creating a synopsis of physical reality at all levels, seen and unseen. With an ideal telescope made only of his insight, Democritos envisioned a Milky Way of myriad individual stars, and a mountainous, cratered moon. Peering conceptually into unseen depths of the material world, he postulated immutable, invisible and indivisible ("atomic") particles common to all matter and explaining the properties of material substances.[27] Two millenia later, Galileo extended man's optical vision with the telescope and with the microscope into the corresponding reality and found Democritos' conceptual vision uncannily accurate still.

How many acres of real estate are in the lighted area of a crescent moon, assuming the moon were a flat disk? Such **Hippocrates of Chios** seemingly whimsical considerations by ancient Greeks created conditions for the scientific base of the industrial revolution. If a crescent is quadrable, meaning that a square of equal area can be constructed with straightedge and compass, then the problem of calculating its area reduces to the trivial one of squaring a side of its associated square.

---

26. If the tortoise moves $50 + 25 + 25/2 + 25/4 + ... = 100$ meters in $5 + 5/2 + 5/4 + ... = 10$ minutes, while Achilles moves $2500 + 1250 + 625 + 625/2 + ... = 5000$ meters, the race finishes at the $5000/(1-1/50)$ meter mark in less than 10 minutes, 13 seconds. As far as I have been able to determine, it was not known before Eudoxos, whom we will shortly visit, that an infinite series can sum to a finite number in this way.

27. Leucippos before him had proposed an atomic hypothesis. Currently, one of the aircraft in the Olympic Airways fleet has the name of Democritos painted on its hull (but spelled "Democritus" — surprisingly even modern Greeks tolerate the unwarranted Romanization of Greek names). The ten drachma coin in current circulation bears a portrait of Democritos and the symbol of an atom.

A few hours by boat north of the island of Samos, we come to the island of Chios. Here in the fifth century B. C. Hippocrates of Chios [28] examined the conditions for quadrability of crescents.[29] It took over two millenia for his study of crescents to realize its potential. By extending the concept of area from rectilinear to curvilinear figures, that study led to the calculation of areas and volumes of shapes and surfaces, pioneered by his successors Eudoxos and Archimedes, and fundamental not only to geometry but to industrial engineering, aerospace engineering and other applied mathematics that would not exist for 2400 years.[30]

What makes the calculation of an area fundamental to science? The question is hard to answer if we think of an area as so many acres of real estate. It turns out that many an important relationship in physics can be represented as an area under a curve denoting another, more directly measurable, relationship; but that is getting ahead of our story. Let's resume that story in Athens, where the trend of extending geometry to the curvilinear domain gained momentum after Plato founded a school from which Eudoxos emerged to bring new tools to this advance.

Plato Unlike Galileo, who emphasized measurement and calculation, Platon or Plato (c428-c347), the broad-shouldered philosopher of Athens, viewed ideal forms as reality while distrusting sense experience. He fostered in his students such admiration for the circle that his student Eudoxos, whom we will shortly visit, modelled the motions of heavenly bodies as circular. With a set of nested spheres, Eudoxos explained irregularities in observed planetary motions so ingeniously that his astronomy, refined by his successors, endured in a theory of Ptolemy for nearly 1500 years until the combined work of Copernicus, Galileo and Kepler overturned it. Plato prepared the way for Kepler's conclusive role in a dialectic overturn of Plato's perfect planetary circles — he described, in the *Timaeus*, all five possible regular solids with which Kepler struggled to fit the planetary orbits into the scheme of Copernicus. The experience of this trial run served Kepler well in his later successful assault on the problem, as did his generalization of faulty circular orbital models (promoted by Plato's students) into the correct notion that a real celestial orbit is conformable to an ideal geometric shape.

---

28. not to be confused with Hippocrates of Cos, his contemporary, who founded a rational school of medicine and is known today as the father of medicine.

29. Cf. Dantzig, *Bequest of the Greeks*, Ch. 10.

30. In addition, Hippocrates' work on the quadrability of crescents led to extensions in number theory that begin around 1840; thus it was finally shown by Dorodnov in 1947 following the work of Claussen, Landau and Tchebotarev that there are five and only five quadrable Hippocratean crescents, a discovery reminiscent of the five (and only five) regular solids in nature (see Dantzig, ibid.). This example suggests the danger of equating unpopularity with obsolescence, and of failing to care for whatever medium encapsulates an unpopular work. Librarians who preserve works viewed as useless or out of fashion may be enabling the technology of a remote future civilization.

To reconcile with observation his teacher's perfect circles in which heavenly bodies were supposed to move, Eudoxos of Knidos (c408-c355), a student of Plato, placed the planets in a system of nested rotating spheres. Although Galileo and Kepler eventually converted the whole system of rotating spheres into a scientific relic, Eudoxos bequeathed to posterity certain enduring legacies. Among these were the theory of proportions invoked by Galileo to deal with continuous variables *without the real number system*, and the method of exhaustion, a means of calculating the limits of infinite processes *without the calculus*. Remember Zeno's tortoise race? The distances run by tortoise and Achilles over an infinity of increasingly small intervals sum to a finite number. Another way of saying this is that the distance traversed by either participant is a *convergent variable*.

With his method of exhaustion, Eudoxos revealed the secret of calculating limits of convergent variables, trapping them like hunted quarry between increasing lower bounds and decreasing upper bounds. The tricky business of calculating areas with peripheral curvature illustrates the value of this technique. School children are taught that the area A of a circle of radius r is expressed by $A = \pi r^2$.

How many of them learn that this is ancient knowledge? The method of Eudoxos was applied by Archimedes,[31] whom we will visit a bit later in this story. Jump ahead to the third century B.C., take a front row seat and watch Archimedes as he uses Eudoxian technique to prove an equivalent relationship between the area A of a circle of radius r and circumference C; that is, $A = rC/2$.

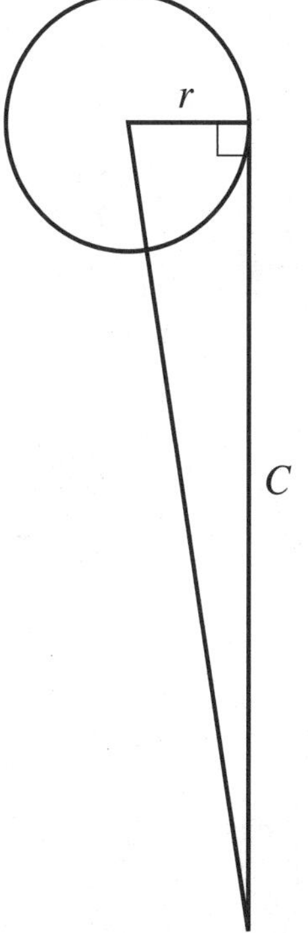

Archimedes notes that A is the area of a right triangle with sides of length r and C including the right angle. By Aristotle's Law of the Excluded Middle, the area of the circle is either equal to that of the triangle, or is larger or smaller. Now Archimedes asks us to assume that the circle is larger in area than the triangle. From this starting premise it would follow that if he subtracts the triangle's area from that of the circle, some positive difference, D, would result. As we look on, he inscribes a sequence of regular polygons in the circle, beginning with a square and successively doubling the number of sides. This process may be repeated until the portion of the circle's interior not enclosed by the polygon is

---

31. (rdc)

smaller than any given positive number, in particular, the difference D. In this way, he fits the polygon close enough to the circle to be *larger in area than the triangle*.

Then he proceeds to prove the opposite. Consider that any regular polygon of n sides is composed of n equal triangles formed by segments connecting the center with the vertices. Each triangular segment of the polygon that Archimedes constructed has an altitude smaller than the circle's radius r, and a base shorter than the arc C/n it spans; hence, the polygon is *smaller in area than the triangle*. He has arrived at a contradiction: the polygon is both larger and smaller than the triangle. Validly derived from a premise, a contradiction implies the falsity of the premise. Therefore, the premise that the circle is greater in area than the triangle is overturned. By a like argument trapping the circle inside a series of circumscribed polygons, then tightening the noose, so to speak, he overturns the opposite assumption that the circle is smaller than the triangle. Neither larger nor smaller, the circle can only be equal in area to the triangle.

To reduce circular area measurements to child's play, one can define a quantity $\pi \equiv C/2r$ so that $A = rC/2 = \pi\, r^2$. This trick lets us calculate area by measuring the straight line distance r without knowing the curved circumference C, *provided that we have an accurate value for the constant π*. Again employing Eudoxian exhaustion, Archimedes will squeeze out a rational approximation to π, trapping the circumference between inscribed and circumscribed polygons as the number of their sides approaches infinity.[32] Wouldn't Hippocrates of Chios, the great crescent squarer, have cheered to see him do that?

Returning to the story of Eudoxos, we find that he contributed another tool directly useful to Galileo. Among quantities inexpressible as whole numbers are the diagonal of a square in terms of its side and the circumference of a circle in terms of its radius. Both turned up in Galileo's pendulum timings of falling bodies. The time required for a pendulum to swing from turnaround to the vertical is a multiple of the time taken to freely fall through a distance equal to the pendulum length.[33] And that multiple is:

$$\frac{\pi}{2 \cdot \sqrt{2}}$$

Irrational quantities are not limited to the ideal constructs of man. They are diversely abundant in descriptions of natural processes, from the arrangement of leaves about plant stems to the energy levels of atoms in hydrogen gas. Not all

---

32. For an exposition of Archimedes' method and Huygens' improvement on it, see Dörrie, *100 Great Problems of Elementary Mathematics*, trans. by David Antin, Dover, NY, 1965, p. 184 ff.

33. Cf. Ch. 5, "Inclined Planes and Pendula". This quantity is called "Galileo's constant" by Stillman Drake; see his *Galileo: Pioneer Scientist*, University of Toronto Press, 1990, p. 237.

occurrences of them in nature were known to the Greeks, but the Pythagoreans exposed the incompleteness of the rational numbers. Only Eudoxos was able to bridge the gaps. Through the theory of proportions he extended the triumph of a finite mind over the problem of infinity, defining ratios with sufficient generality to include an infinity of magnitudes inexpressible as fractional multiples of one another.[34] His method of expressing relations among rational and irrational quantities, and his method of exhaustion anticipating the modern limit concept,[35] were essential tools for Archimedes' in developing mathematical physics. Two millenia after Eudoxos, Galileo's reasoning relied on Eudoxian ratios to describe continuously changing processes, like the fall from the Tower of Pisa.

Another of Plato's students, Aristoteles or Aristotle (c384-c322), **Aristotle** left Athens to tutor Alexander "the Great", returning later to form his own "peripatetic" school, the Lyceum. His catalogued lectures eventually filled over a hundred volumes on science and philosophy. Although mathematics is not included, in these volumes he laid its foundation in logic, the orderly procedures for reasoning from premise to conclusion and for validating conclusions through observation. He promoted the duality of Earth and heavens, holding the heavens to be perfect and unchangeable, while the Empedoclean elements (earth, air, fire, and water) are changeable and corrupt. For the heavens, he added a fifth element, an "aether" permeating space, in which natural motion is circular. This system requires two sets of natural laws, one for Earth, another for the celestial realm. Although his physics was accepted as "gospel" for more than a millenium, it is seriously defective and much of it had to be overturned by Galileo.

Still, Aristotle's contribution to the Renaissance through his logic is not to be overlooked. In rebutting "Aristotelians" who used their master's own logic very badly, Galileo made splendid use of Aristotle's system of logic. Aristotle's insistence on scrupulous observation led Galileo to remark that Aristotle put experience at a higher priority than any reasoning based on authority.[36] Aristotle's organization and archiving of knowledge served as a model for Euclid's *Elements* and the Library of Alexandria. About the time of his death, the Alexandrian school took over the progress of mathematics from the Athenian school.

---

34. For Heath's translation of the Eudoxian definition, see Ch. 11, "Galileo, Mathematician to the End". It is generally acknowledged that Eudoxos discovered most of what will be contained in Euclid's *Elements* V.

35. The modern concept of continuity derives from the calculus of Augustin-Louis Cauchy, which can be applied to the continuity of falling speed. For any arbitrarily small interval we may choose around the speed at any instant of fall, there exists a time interval such that the difference between the speed of fall at every instant within it, and the speed of fall at the given instant, is bounded by the chosen small interval.

36. See Galileo's 15 Sept. 1640 letter to Liceti, quoted by Drake in *Galileo At Work*, Ch. 21, p. 409. Aristotle is commemorated by modern Greeks on the five drachma coin.

Euclid of Alexandria

Fusty and forgotten, like an old dusty locket or string of pearls, our most precious thoughts can remain locked up in the attics of our minds until they are organized, recorded and transmitted. The main vehicle that carried Greek science across the Dark Ages to the Renaissance was built by a scholar at the Library of Alexandria. Euclid of Alexandria [37] (c325-c270) catalogued knowledge about geometry, from Thales through Democritos, much of which is the work of Eudoxos, and wrote the definitive textbook on the subject, the model for those that remain in use in schools in the 20th century. In Euclid's technical language we find a one-to-one correspondence between words and ideas, an important step toward algebraic symbolism. [38] In the 16th century, Ricci's lectures on this book helped convince young Galileo to switch his studies in medicine to mathematics.

Aristarchos

Before the Earth is measured, Aristarchos of Samos (c310-c230) measured the distance and size of the moon and sun, inaccurately but correctly in principle. [39] His outstanding achievement is the sun-centered, or heliocentric, postulate. [40] Before him, Heraclides of Pontus declared the rotation of Earth on its axis; Aristarchos added that Earth too orbits the sun. [41] By adopting a heliocentric postulate, Aristarchos conceived the core of the "Copernican" hypothesis, some 18 centuries before Copernicus.

Thus was born the concept of Earth as a planet.

Menaechmos and Apollonios

When Greek geometers sliced through a cone, they freed a secret of dynamic motion wrapped up inside it. Originally, geometers studied relations among simple figures: line segments and circles. By intersecting a cone with a plane at various angles to its base, Menaechmos (c350) extended geometry to include the ellipse, parabola, and hyperbola. On this foundation, Apollonios of Perga (c262-c190) further extended

---

37. Not to be confused with Euclid of Megara, a logician and student of Socrates, whose name appeared incorrectly on some editions of the *Elements*. Boyer, pp. 100-101.
38. Cf. Giuseppe Peano, "Importanza dei simboli in matematica", *Scientia*, 18 (1915); English translation by H. C. Kennedy, *Selected Works of Giuseppe Peano*, Allen & Unwin Ltd., London, 1973.
39. In *A Short History of Astronomy*, Berry tells us that Aristarchos estimated an angle of $3°$ subtended at the sun between Earth and the moon, by taking the complement of a measured angle at Earth between the sun and the half moon. The inaccuracy is mainly due to the difficulty of determining the instant of half moon; if this were done correctly, then the tangent of the angle subtended at the sun would give the Earth-to-sun distance as a multiple of the Earth-to-moon distance. To infer the sizes of the sun and moon, he pointed out that their ratio of size is equal to their ratio of distance, since they have the same apparent angular diameter (as is easily seen at the time of a solar eclipse).
40. For the statements of both Archimedes and Copernicus acknowledging this, cf. Sir Thomas Heath, *Aristarchus of Samos: The Ancient Copernicus*, Dover, NY, 1981, p. 301ff.
41. See Angus Armitage, *The World of Copernicus*, Henry Schuman, Inc., NY, 1947, p. 40.

geometry to conic sections. The extension proved crucial to the discovery of Kepler and Newton that all orbits are conic sections, and to Galileo's law of parabolic fall. Apollonios replaced the nested spheres of Eudoxos with a system of epicycles and deferents as the preferred method of "saving the appearances" of Plato's perfect circles. An epicycle is a small circular motion within a large circular orbit, called the "deferent". If we observe Mars against background stars over a long enough time, we see it regressing now and then, looping back and forth in the night sky and tracing out a curve that Eudoxos called a hippopede. ("Hippopede" is a transliteration of the Greek word for "horse-fetter", not to be confused with the result of attempting to cross-breed a hippopotamus with a centipede.) [42] But what if Mars were executing little circles as it moves on its assumed path around Earth? Apollonios invoked this complex motion to explain the "back and forth" motion of a planet as seen from Earth. Hipparchos and Ptolemy later imported his epicycles into their models of planetary motion.

Let's leave Ionia for an island known later as Sicily. Colonized by Greeks and under attack by Romans who would conquer it, it was home to the city Syracuse.

In our story of his application of Eudoxian exhaustion, we have already met Archimedes of Syracuse (c287-c212). The **Archimedes** mathematical foundation of physics sprang from his mind as he extended and buillt ingenious applications of the work of his predecessors, emulating nature itself in his inventions. The most noted scientist of antiquity, he designed and built machines that held off the Roman army, gave an accurate estimate of the number π, and expounded the theory of basic machines: the lever, screw, and pulley.

Similarly applying the exhaustion method of Eudoxos, Archimedes found a way to calculate the number π by encapsulating a circle between a circumscribed and an inscribed polygon. Beginning with a square and progressively doubling the number of sides, he found upper and lower bounds for the circumference of a circle with unit diameter; viz., the perimeters of the two polygons. Not until sometime around 1670 did James Gregory discover [43] that the area under the curve $1/(x^2 + 1)$ in the interval $(0, x)$ is just $\arctan(x)$.[44] Since $\arctan(1) = \pi/4$, π can be enumerated by the converging series:[45]

---

42. The more fantastic the image underlying a concept, the easier it is to remember. Intended as a mnemonic, this remark is an application of Galileo's observation, stated thus in the preface to his first book: " For the nature of the human mind is such that unless it is stimulated by images of things acting upon it from without, all remembrance of them passes easily away."

43. See Petr Beckmann, *A History of* π, Golem Press, NY, 1971.

44. Arctan(x) can be thought of as an angular radius. For example, if when you observe the moon from the Earth-moon distance B, the moon's diameter 2A subtends an angle 2α at your eye, then $\arctan(A/B)$ is the angle α.

$$4 \cdot (1 - \tfrac{1}{3} + \tfrac{1}{5} - \tfrac{1}{7} + \dots)$$

as Leibniz realized upon seeing Gregory's result. Computing machines of the 20th century calculate π to hundreds of thousands of decimal places using faster converging arctangent functions designed by later mathematicians, but still following the method of Archimedes and Eudoxos — rational approximation.[46]

From Archimedes' calculations of areas and volumes it can be inferred that he developed an early form of the integral calculus. Some of his discovery appears in a letter to Eratosthenes, in which he illustrated his application of *the mechanical method of finding theorems*. In *The Method* he treated geometric figures as attached to a lever so that the lever remains in equilibrium; he viewed an area as the totality of line segments traversing a bounded two dimensional space in a uniform direction, and the volume of a solid as a totality of such areas traced out by a combination of translation and rotation of a bounded planar surface.

This happy marriage of mathematics, physics, and practical applications ended suddenly when a Roman soldier, against the order of his commander, rushed upon and killed Archimedes, while the latter was drawing geometrical figures. Would-be military commanders can find here exemplified the futility of good intentions while operating a war machine. Sadly, many of Archimedes' writings are lost. Topics covered in the remaining writings include the circle, the sphere, the cylinder, the cone, spirals, centers of gravity of plane figures, the quadrature of the parabola (finding a square equal to its area), polyhedra, and the theory of floating bodies.[47] This last became the starting point of the scientific career of Galileo, who adopted him as a personal ideal model of a mathematical scientist.

In Chapter 4 we will see Galileo using Archimedean machines to create the archetypal concept of work and to anticipate the conservation laws of physics. Starting with a hunch that free fall can be described by extending Archimedean hydrostatics, he created dynamics from that extension and from Aristotle's science of motion. In Chapter 12, we will see how the derivative science of rocketry was launched in the catapults of Archimedes, and propelled by Tartaglia to Galileo, whose analysis of projectile paths put it into a modern framework.

Combining Archimedean mathematical physics with the logic of Aristotle, Galileo overcame a medieval mindset that would have blocked the progress of these

---

45. Gregory derived the series expansion $\arctan(x) = x \cdot \sum_{k=0}^{\infty} (-1)^k \cdot \dfrac{x^{2k}}{2k+1}$, using a formula of Galileo's disciple, Cavalieri. (ibid. p. 132).
46. Due to computer storage limitations, the symbolic representation of any irrational number is a rational approximation.
47. See Ivor Thomas, *Greek Mathematical Works* (see Bibliography).

sciences. For it is Archimedes who gave Galileo the leverage to raise natural philosophy above the metaphysical speculation of his day by restricting the language of natural philosophy to well defined, measurable, and testable propositions. Aristotle emphasized qualities; Archimedes, quantities. Galileo put them together in his philosophy of science. While calculating the number of grains of sand that can fit inside a sphere the size of the known universe, Archimedes prepared for a successor who expanded that universe by orders of magnitude.

Geometry, the measurement of Earth, finally achieved the literal justification of its name with the work of Eratosthenes of Cyrene (c276-c194). In a letter to him, Archimedes says:

Eratosthenes

"... I know that you are diligent, an excellent teacher of philosophy, and greatly interested in any mathematical investigations that may come your way..." [48]

What did Eratosthenes do to justify this praise from the great Archimedes? Suppose you are in Alexandria at this time. You have knowledge of geometry, a straight stick, a means of measuring angles, and you know that at Syene, a measurable distance to the south, the sun's reflection can be seen at the bottom of a vertical well at noon on the summer solstice. Would it occur to you to use these resources to measure the circumference of the Earth? It does to Eratosthenes, and he succeeds to within 1% of the accepted modern value. Additionally, he works out an algorithm to generate prime numbers (the Sieve of Eratosthenes), invents the leap year, determines the *obliquity of the ecliptic* [49] as 23° 51′ 20″, and prepares a star map with over 600 stars.

If there were no trigonometry, what would motivate you to see a need for it? For a great astronomer of antiquity, Hipparchos of Nicaea (c180-c125), the motivation arose from a need to measure

Hipparchos of Nicaea

the sizes and orbital parameters of the sun and moon. And so he developed functions of right-angled triangles and their relationships. With these tools he formulated the coordinate systems of astronomy in close to the modern form and estimated the length of the year to an accuracy of 6½ minutes. First noticing a 2° change in zodiacal longitude of the star Spica, he checked other fixed stars; this project led him to discover and to measure the precession of the equinoxes, the rate at which Earth's axis rotates, like that of a spinning top. In the technology of measurement and calculation, he advanced beyond Aristarchos, while regressing to geocentricity. To preserve the circles cherished by Plato and Aristotle against new observational data, Hipparchos refined the coefficients of Apollonios' epicycle model.

---

48. *Archimedes* by E. J. Dijksterhuis (see Bibliography).
49. This is the angle that the plane of Earth's orbit makes with Earth's equatorial plane.

Over 2100 years later, on a secluded hillside a few kilometers north and east of the western shore of Lake Iznik, the rain begins to fall and the dirt road turns to mud as I am arriving. Here, among sprays of white chamomile and red poppies, two parallel right-triangular rock walls brace the hill. Amid a rectangular earth retainer spanning them is a small door. One must crouch to enter. Inside, an ancient mural adorns the tomb where Hipparchos and his wife were buried from antiquity until 17 years ago, when archaeologists responded to the discovery of a local boy digging on the hill. The remains of the bodies and the jewelry by which they were identified have been removed to the local museum.[50] There are no signs, no markings, no inscriptions. Only the right-triangular retainers built into the hillside suggest that here lay the remains of the inventor of trigonometry, the science of right triangles.

Coincidence would suffice to explain being in the right place and time to take the photograph, shown at the end of this chapter, of the 1991 solar eclipse. But the probability of my getting this shot would have been near zero without Hipparchos among others. In Chapter 7, we will see how Galileo used a Hipparchan metric to survey lunar mountains. Even more remarkably, we will see how he used Hipparchan parallax theory to counter-demonstrate the epicycles of Hipparchos. He did it by designing an experiment[51] carried out after Kepler and Newton, building on the theories of Copernicus and Galileo, brought new data and new theory against epicycles. The abandonment of epicycles after Newton illustrates that an explanation fitting all currently known facts is not equivalent to a certainty.

Ptolemy

Functionally analogous, though less grand in scope and originality, to the role Euclid had filled in geometry, the role of Claudius Ptolemy (c100-c170) was to preserve Apollonian/Hipparchan astronomy in a book called the *Almagest* containing stellar coordinates that were largely observed by Hipparchos or a contemporary (to Hipparchos) and extending the Hipparchan register of about 850 stars to 1,027.[52] Ptolemy preserved Apollonian/Hipparchan astronomy so well that the system is called "Ptolemaic" and lasted around 1,500 years, until overturned by Copernicus, Galileo, and Kepler. But that overturn does not limit the utility of Ptolemy's work. In the 18th century, Edmund Halley discovered the proper motion of the fixed stars by comparing his own observations with those recorded in the *Almagest*.[53]

---

50. Source: Museum of Iznik. In modern times, Iznik has been a principal center for production of mosaic tiles. In 1609, the year of Galileo's first great telescopic discoveries, the architect Mehmet drew upon Iznik as the source of the magnificent interior decoration of the imperial mosque of Sultan Ahmet I (also known as the "Blue Mosque") in Constantinople (now Istanbul).
51. See "Bessel's Demonstration" near the end of Chapter 12.
52. See Gerd Grasshoff, *The History of Ptolemy's Star Catalogue*.
53. ibid.

The End of the Grand Era of Greek Science

Because circular motion viewed in the plane of the circle is seen as a back-and-forth cycle, Hipparchan epicycles suffice to explain the apparent reversals in planetary wanderings. It is largely this convincing mental illusion, along with its conversion to dogma, that accounts for the long life of Ptolemy's *Almagest*. The *Almagest* did not mark the end of an era; it was followed by at least two other great works whose longevity did not depend even in part on illusion or dogma.

Pappos and Diophantos

The *Collection* of Pappos (fl. c300) can be considered the culmination of Greek geometry. This work summarized and extended the classical mathematics of the Greek period. It helped inspire Vieta's algebra through the procedural assumption that the unknown is given and the consequent deduction of a conclusion from which it can be determined. Ancient Greek mathematics, having started with geometry, also ended with it, but not without introducing a new mathematical discipline: algebra. The *Arithmetica* of Diophantos (fl. c250) gave rules for solving linear and quadratic equations in which the unknown is treated symbolically, a work that "inspired the number-theoretical discoveries of Fermat".[54] Hence it can be considered a treatise on algebra, a word that had not yet been coined.

A Monstrous Madness

And then it happened. A fear-crazed mob took over. The mob incinerated that symbol, model and extension of Aristotle's engine of civilization, the Library at Alexandria. As if the heat of the fire had melted the innovative spirit of man for nearly a thousand years, the holocaust of books marked a regression in civilization, a setback for later generations, including our own. Searing flames destroyed not just a library, but channels and archives through which the wisdom of this age had been transmitted, preserved, and extended.[55] The dark smoke rising from its ashes flashed a signal of danger to thinkers and achievers everywhere. The Library was rebuilt, only to be destroyed by yet another mob. Through the ensuing Dark Ages, a monstrous madness dissipated the momentum of Greek science. Incalculable are the effects of that loss on you and me.

A New Beginning

In our journey, we advance now to 1564, the year of Michelangelo's death and Shakespeare's birth, skipping over the intervening centuries of darkness, and over the pre-Galilean

---

54. cf. Dantzig, *Bequest of the Greeks* (see Bibliography).

55. Besides the world's greatest book collection, the Library housed a museum and a combined graduate school and institute of advanced research in science and mathematics.

Renaissance, to which we will return after picking up the main thread of our story. On February 25th, in a section of Pisa visible from the Leaning Tower, a child is born who will grow up to use the lean of the tower as a tool for raising the civilization of man, phoenix-like, from the ashes of Alexandria.[56] He will accomplish this task by beginning where the ancient Greeks whom we have just visited concluded their work. His intellect, blazing hotter than the fires that destroyed a scientific culture, will fuse Greek science and the Renaissance into a continuum of knowledge almost as if the intervening millenium had not happened.

His name: Galileo Galilei.

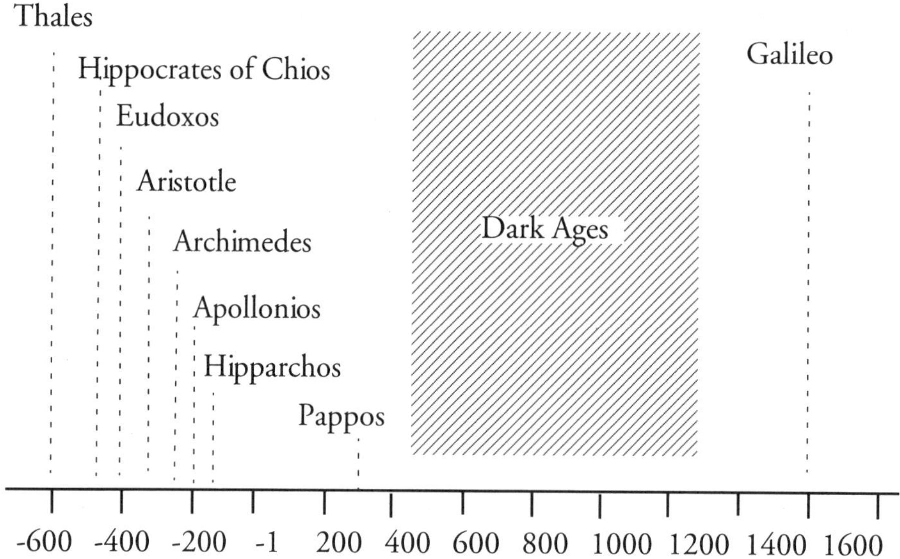

---

56. February 15 is the date of Galileo's birth according to the old calendar system in use at the time and place of his birth. In the year 1582, when Galileo was 18 years old, ten days were dropped to establish the current Gregorian calendar. Hence, Galileo's birthday is February 25 according to the Gregorian calendar, which even the English accept today.

Three Millenia Old Wall Fragment
at Kalabaktepe Hill (Miletos),
Home of Thales,
Founder of Science
(photo by author)

Monument to Pythagoras,
Founder of Mathematical Physics,
on the Island of Samos, Birthplace of
Aristarchos and Pythagoras
(photo by author)

Solar Eclipse at Cabo San Lucas, 91.07.11 18:53:56 UT.
This photograph relies on eclipse-prediction technology
pioneered by Hipparchos of Nicaea.
(photo by author)

Birthplace of Galileo: Like the Earth,
Looking Neither Central Nor Special
(photo by author)

Chapter 2. From Bedrock Heavenbound

The Young Galileo (1564 - 1585)

*"High mountains loom ahead, threatening the heavens.
Whoever sits upon them sees beneath his feet all of Italian Gaul,
And in front of the Alps, the boundaries of Italy.
A thousand species of birds and wild animals live here and roam the sacred forest;
A cool stream flows in the shadows, and along its irrigating wanderings
The flourishing grass grows."*

... Petrarch, from *"Epistole Metriche"* [1]

Like Petrarch's flourishing grass, great ideas of great men have grown in the once fertile intellectual climate of Pisa, a small city nestled in the plain between the mountain ranges of Northern Italy. Visitors wandering today through its timeworn streets and back alleys encounter a blend of the medieval, the renaissance, and the modern woven together by threads of history that are visible to the eye. Here, and in the plain that stretches to Florence, Medieval cathedrals, renaissance villas, and modern centers of science and learning stand juxtaposed. Carried aloft by fresh ocean breezes at fiesta time, vibrant melodies played by street musicians recall the musical revolution of Monteverdi, in which Galileo's father played a part, while the music of the spheres, a conceptual music composed by ancient Greeks, was being reorchestrated by Galileo's contemporary, Kepler. There is music, magic, and mystery in the air.[2] The source of these is not in the occult, but in the mathematics of Pisa's native sons.

Let's visit the Renaissance before Galileo, and the duchy of Toscana, to which Pisa belongs at that time, and where the Galilei family will dwell for many generations. It is the **Toscana before Galileo** year 1174 and construction is under way on the bell tower of the great Cathedral at Pisa. The tower will not be completed in the lifetime of the then four-year-old child who will one day revolutionize commercial mathematics: Leonardo of Pisa. By the time it has reached the third story, the tower is leaning and construction is suspended as its architects debate how to correct the lean. Meanwhile, Leonardo

---

1. My translation from Enrico Bianchi's version on the leaf of Cardini's Italian translation of Galileo's *Sidereus Nuncius*, published in Venice by Marsilio, 1993. Galileo wrote *Sidereus Nuncius* in Latin, as Gino Loria implies, to render it internationally readable.

2. '"Magic" here refers to logical analysis that reduces events having an aura of the miraculous to a natural chain of consequences. For examples, see Aaron Bakst *Mathematics: Its Magic and Mastery*, Ch. 37. "Mystery" here refers to an unknown (not an unknowable) that gives rise to scientific curiosity. In this sense, mystery lies outside but in the path of expanding knowledge, and is subject in principle to an empirical test; it is not to be confused with mysticism.

(better known to us as "Fibonacci") will be popularizing Arabic [3] numerals and the use of the number zero throughout the Western World. Moving forward to the 13th century, we find the bell tower completed, but still leaning. In the 15th century, it is still standing despite its lean, and another Leonardo is studying mathematics under the Franciscan Fra Luca Pacioli. Leonardo da Vinci, as we know him, returns the favor by illustrating Pacioli's book *Divina Proportione*. Known also as the "Father of Bookkeeping", Pacioli introduces the double-entry system [4] that is to propel commercial advances in Europe and remain permanently in use by accountants worldwide. Early in the 16th century, we find the physician Girolamo Fracastoro classifying diseases and leading medicine away from the superstitious "cures" of the Middle Ages. [5] Among famous and learned men of Toscana, Fracastoro is an elder compatriot of Signor Galilei.

Galilei Family Who is this Renaissance man, Signor Galilei? He has formulated a law of mathematical physics. He is argumentative, and an instigator of disputes about mathematical matters. He has an experimental attitude toward the validation of theory. He publishes his theory in dialogue form. His thinking is fiercely independent, and in one of his books he writes: "*It appears to me, that they who in proof of any assertion rely simply on the weight of authority, without adducing any argument in support of it, act very absurdly. I, on the contrary, wish to be allowed freely to question and freely to answer you without any sort of adulation, as well becomes those who are truly in search of truth.*" [6] If you have some knowledge about Galileo from sources other than this book, you might guess that he is the author. And an astute guess it would be, though a wrong one. These are the words of his father Vincenzio Galilei (1520-1591),[7] whose approach to validating one's claims to knowledge will be made more durable by his most famous son.

Vincenzio is an accomplished musician whose sporadic musical engagements prove inadequate to the demands of a growing family. Turning to opportunities for trade that flourish under the new mercantile arithmetic [8] and the infusion of capital still provided by the Medici bankers, he earns a living as a cloth merchant. But he never abandons his musical practice, and in quest of his goal "to revive through monody the ancient Greek ideal of the union of music and poetry",[9] he authors

---

3. "Arabic" here refers to the Hindu system of numeration which Leonardo learned from Muslim teachers schooled in the *Al-jabr* of al-Khwarizmi, or from one of four translations of that text to which he had access. (cf. Gillispie, *Dictionary of Scientific Biography*)
4. described in his *Summa de arithmetica*, pub. 1494.
5. Cf. Irwin Shapiro, *The Golden Book of The Renaissance*, Golden Press, NY, 1961.
6. trans. by Mrs. Allan-Olney, *The Private Life of Galileo*, MacMillan, London, 1870, p. 2.
7. His name is spelled variously in the literature as "Vincenzo" or "Vincenzio".
8. i.e., Treviso arithmetic; cf. Frank J. Swetz, *Capitalism and Arithmetic*, Open Court, LaSalle Illinois, 1989.

books on counterpoint and other musical topics. He begins his *Dialogue on Ancient and Modern Music* with a definition of the liberal arts:

> "Music was numbered by the ancients among the arts that are called liberal, that is, worthy of a free man, and among the Greeks its masters and discoverers, like those of almost all the other sciences, were always in great esteem." [10]

While his teacher Zarlino defends and extends the Pythagorean doctrine that the only true musical consonances would be produced by string lengths in the ratio $(n + 1)/n$, Vincenzio substitutes an empirically established acoustical law: "a musical interval between similar strings is produced either by different lengths, or by tensions inversely as the squares of those lengths." [11] Vincenzio's may be the first law of mathematical physics to result from systematic experimentation and to replace a doctrine universally accepted on authority. His readiness to question doctrine will soon be recognizable in his first son, Galileo (1564-1642).

An Unremarkable Home

In an otherwise unremarkable section of 20th century Pisa, an inconspicuous plaque by the drainpipe of a building declares this to be the birthplace of Galileo. Tucked away in a phalanx of row houses, looking neither central nor special, it was a fitting home for a young man who would grow up to demonstrate that our planetary home, Earth, occupies just such a place in the cosmos: neither central nor special. The upper stories are residential units; the ground floor a commercial area in which, 420 years after Galileo's birth, an Olivetti computer driving a line printer suggested connections between his work and the development of computing machines. If we return to the 16th century, we find no such machines but an intensely curious boy tinkering with old toys and fashioning new ones for his siblings, using creative arrangements of pulleys, wheels, and levers. It's a good bet that he is a visitor to the local blacksmith, for his experiments with the pendulum will require hooks and fasteners.

Music, Machines, Michelangelo, Medicine and Mathematics

Young Galileo excels in Latin and Greek, is an artist and musician, accomplished on the organ and lute, and is a scholar noted for his questioning spirit. So much is this true that he earns the nickname of "The Wrangler" among fellow students and teachers, a derogatory name implying

---

9. *The New Grove Dictionary of Music and Musicians*, Stanley Sadie ed., MacMillan Publishers, London, 1980, p. 96.

10. from the "Dialogo della musica antica e della moderna" by Vincenzio Galilei, as translated by Oliver Strunk ed., *Source Readings in Music History*, W. W. Norton & Co., NY, 1950, p. 302.

11. IVincenzio Galilei, *Discorso*, pub. 1589; see *Dictionary of Scientific Biography*, ed. by Gillespie, pub. by Charles Scribners & Sons, 1973.

"troublemaker". Taunting does not affect his behavior, however. Undaunted by disapproval, he prefers working with his father (whom he assists with acoustical experiments) to pursuing popularity with peers.

For the perspective in his early drawings, he earns praise from the Renaissance masters at Florence.[12] Vincenzio can point out that his son was born in the same month and year that Michelangelo died, hinting that perhaps God could not leave the Earth devoid of such an artist. In Galileo's later illustration for the cover of his manual on the geometric compass, we will see evidence that he could have had a career like that of da Vinci. But he isn't inclined to pursue art as a career. Nor will he remain on the path of an ancestor, Galileo Bonaiuti, a medical doctor whose reputation was such that the family had changed their name to Galilei. Instead, this 18-year-old student is about to discover a passion that will revolutionize his life and the progress of science, and draw him into massive conflict with the philosophers and theologians of his day. From what history tells us about the turning point that drove Galileo into a scientific career, and filling gaps in the record with plausible detail, we can construct a scenario of what happened next.[13] Let's set the clock of our imagination to the winter of 1582-1583, just after Pope Gregory XIII dropped ten days from the so-called "Julian" calendar.

The Turning Point

A cathedral spire is an ambiguity that can symbolize reaching for heaven or reaching for the stars. Strolling pensively down the Borgo Stretto, a university student who one day will alter the direction of theology pauses at the steps of a tenth century cathedral, built on the remains of a heathen temple, near which his father Vincenzio Galilei was born. To please Vincenzio, he enrolled last year at the University of Pisa as a medical student.[14] Now during tedious lectures on Galen and Hippocrates, an inner voice whispers to him: "Do your own experiments!" How had the ancient authorities, he asked his professors, drawn their conclusions — what were their methods of proof? Though these questions would probably have been expected, and even welcomed, by Galen and Hippocrates themselves, they bring Galileo ridicule, not answers.

---

12. e.g., Ludovico Cigoli, Bronzino, Passignano and Jacopo da Empoli. "Cigoli, in particular, was wont to say that Galileo alone had been his teacher in the art of perspective, and that whatever credit he enjoyed as a painter was owing to his advice and encouragement." Fahie, p. 7 (see Bibliography).

13. For the next two sections only, I ask the reader's indulgence to lapse into the style of a historical novel, supplying some details which are contrived yet consistent with the historical record.

14. After Galileo's death, his son wrote: "Galileo, occupied in the study of medicine, for some time showed himself disinclined to mathematics, although his father, who was highly proficient in it, exhorted him to it; finally to satisfy his father he applied his mind to it; but no sooner did he begin to taste the manner of demonstrating and the way of reaching the knowledge of truth, than, abandoning any other studies, he gave himself entirely to mathematics."; trans. by Fermi and Bernardini, *Galileo and the Scientific Revolution*, Basic Books, NY, 1961, p. 18.

If all this were not trouble enough, his tuition fees have become a burden to his father. A few years prior at the monastery of Villambrosa, Galileo learned logic from a monk. Now, a longing to return to the life of a scholar, sheltered by the peaceful woods from city life, wells up in his soul. Or, might it be in his musical talent, with the organ and the lute, that he could at last please his father? Had not Vincenzio suggested that Galileo might play his music at the Court of the Grand Duke? As if drawn by this thought, the young man bolts the cathedral steps, turns the corner, and heads for the summer Court of the Grand Duke. Hoping perhaps to find a friend who would invite him to sit in on a rehearsal of court musicians, he picks up his pace, unaware of being about to discover a new direction for his life.

Through an Open Door

Echoing through the corridors of the Court, with its high arched doorways and vaulted ceilings, the footsteps of young Galileo pause at each corner as he follows musical notes wafting through the corridors and flowing in a harmony created by Cristofano Malvezzi, a 35-year-old composer of flute music. Outside a lecture hall, a familiar voice rises above the ensemble: the voice of Ostilio Ricci, Court Geometer and friend of the elder Galilei. Recalling that his father had recommended Ricci's lectures to him,[15] Galileo peers through a partly open doorway just at the instant when Ricci, lecturing on the geometry of Euclid, slices through the air an imaginary arc of a circle. Transfixed by the confident style of the lecturer, Galileo stands motionless in the doorway as Ricci moves gracefully, like a prince with flowing robes, through a reasoned argument from definitions and postulates to *Quod Erat Demonstrandum*. Here there is no appeal to authority, no memorization of results to be accepted on faith, but an elegant and imposing logical structure, a Thalesian proof that stands nobly and unassailably on its own merits, rising like a cathedral spire from bedrock heavenbound.

Galileo has taken the bait and, without knowing it, Ricci has captured him for mathematics forever. Vincenzio wants him only to appreciate mathematics but not to make it his career. After hearing Ricci's lecture, while pretending in the presence of his father to be studying Galen, what Galileo really studies is Euclid. When he can no longer conceal his curiousity about questions arising from his private study, he consults Ricci about them. Ricci detects in Galileo's questions a mathematically precocious student and asks a unexpected question: Who is your

---

15. "It was at one of the court's temporary stays in Pisa that Galileo, on his father's insistent advice, went to hear Ostilio Ricci explain the geometry of Euclid", ibid. Nevertheless, Mrs. Olney (based on analysis of correspondence) goes so far as to say "Unknown to his father, Galileo applied to him [Ricci] for instruction" (Mrs. Allan-Olney, *The Private Life of Galileo*, MacMillan, London, 1870, p.6). This ought to be a challenge worthy of historians, to set the record straight concerning so important a figure (in his own right, as well as through his influence on his son) as Vincenzio Galilei.

teacher? Though anxious about the consequences of his intrusion, Galileo confesses having attended only Ricci's own lectures. But Ricci is impressed that he could have learned so much geometry mostly through self-study. And so he encourages Galileo to pursue his new interest.

Confrontation with Vincenzio

Upon hearing of Vincenzio's desire that Galileo undertake a medical career, Ricci intercedes with Vincenzio. And though insisting that his son should first finish his medical studies, Vincenzio "looks the other way" as Ricci tutors Galileo in mathematics while Galileo pretends to study Galen. In the winter of 1584, Vincenzio hears that his son is in academic trouble, but upon checking with his business partner with whom Galileo is boarding, learns that Galileo is deeply immersed in study of something. At last, there ensues an honest, direct confrontation between father and son. Galileo confesses his preference for mathematics. Fearing that his son would be facing a life of unproductive theorizing and meagre employment, Vincenzio threatens to withdraw financial support. Galileo asks for another year, after which he would be on his own. Vincenzio acquiesces and Galileo seeks a scholarship. But the faculty are not supportive of his application. Has he not repeatedly demanded reasons for claims that they could support only by quoting an authority? Galileo quits the university without graduating and, preferring to experience a path he has not traversed, leaves student life forever.

The Crossing of the Bridge

The central bridge in Pisa spans the Arno near Galileo's birthplace, its supporting members forming a cycloidal arch. A cycloid is a shape you can generate with a fixed point on the rim of a wheel by rolling the wheel. So, every cycloid has a generating circle. Like Hippocrates of Chios who puzzled out the quadrature of the crescent, young Galileo asks: How does the area under the cycloid relate to that of the associated circle? Lacking a mathematical tool, he chooses a physical solution. He makes a model of a cycloid, cutting out its underlying area and the disk bounded by its circle. As he finds the former about three times heavier than the latter, Galileo estimates the ratio of areas as 3:1, and thus discovers the relationship.[16] Today, using the integral calculus, it can be shown that the ratio is exactly 3, the difference in Galileo's experiment being a small measurement error.[17] Here we see the young Galileo trying to observe a mathematical characteristic in nature, as fellow Pisan Leonardo has done.[18] But he does not stop with observations. By 1590, we will find him applying his studies of the cycloid to designing the arch of a bridge.

---

16. "Soon after its discovery, and on Galileo's recommendation, it [the cycloid] was applied in the formation of the arches of the bridge (Ponte di Mezzo) over the Arno in Pisa." Fahie J.J. (see Bibliography), p. 22.

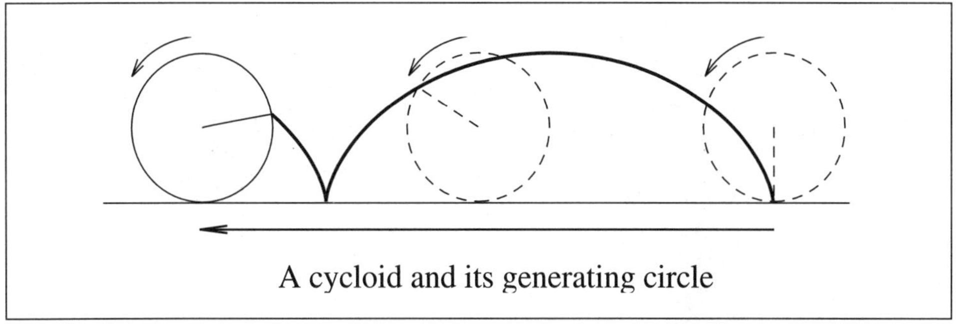

A cycloid and its generating circle

Like the Leaning Tower that remained incomplete and unused for nearly a century while its construction was suspended, the magnificent edifice of Archimedean physics has remained incomplete through the Dark Ages of our civilization, awaiting a Thalesian successor. Though perhaps unaware of it, the young Galileo has been prepared by the geometry of Thales and by a truth-seeking father, who would not rely on authority, to build upon an edifice leaning away from the tradition of the Aristotelians. He crosses a bridge near his home, bound for Florence and Siena where he will seek students in need of a mathematics tutor. Three years later, he will return to Pisa to assume the chair of mathematics. How will Galileo, who has no degree, wind up three years later, at the age of 25, as a professor of mathematics at the same university which refused him a scholarship?

---

17. If as this wheel of radius r rolls, the angle of rotation is α, the cycloid may be represented parametrically by $x = r(\alpha - \sin\alpha), y = r(1 - \cos\alpha)$. Using post-Galilean methods, a mathematically inclined reader may wish to evaluate the area expressed by

$$\int_{\alpha=0}^{2\pi} y \cdot dx \quad \text{and see if it evaluates to } 3\pi r^2.$$

Hint: You may want to begin by differentiating x with respect to α .

Based on early integration methods, Torricelli first published a solution of the area under a cycloid.

18. The Fibonacci sequence 0,1,1,2,3,5,8 ... $(x_{n-1}+x_{n-2})$ expresses relationships found not only in mathematics, but in architecture and in nature as well; e.g., the tendency to arrange in this sequence as shown by leaves along a plant stem, or scales along a fir cone, or the genealogy of bees, or the ideally simplified atoms of a quantity of hydrogen gas. The "golden number" ϕ, which by definition differs from its reciprocal by unity, is known as the limit of ratio of the (n+1)st to the nth term of the Fibonacci sequence as n approaches infinity; cf. H. E. Huntley, *The Divine Proportion*, Dover, NY, 1970. This last fact was noted by Kepler (rdc). Leonardo's fame was such that Emperor Frederick II came to Pisa in 1224 to test his skill, but did not succeed in stumping him. "The first question was to find a number of which the square when either increased or decreased by 5 would remain a square: Leonardo gave a correct answer: 41/12. Another question was to find by the methods used in the Tenth book of Euclid a line whose length x should satisfy the equation $x^3 + 2x^2 + 10x = 20$: Leonardo showed by geometry that the problem was impossible but he gave an approximate value of the root of this equation, namely, ... 1.3688081075..., and is correct to nine places of decimals." (cf. W. W. Rouse Ball, *Primer of the History of Mathematics,* pp. 46-47.)

Cycloid Bridge Spanning the Arno River Near Galileo's Birthplace.
Design of the bridge was influenced by his study of the cycloid.
(photo by author)

City of Pisa Seen from Summit of Leaning Tower
View Toward Galileo's Birthplace
(photo by author)

Chapter 3. The Rebirth of Science

Assayer of Gold, Surveyor of Hell, Tutor of Math (1585-1588)

*"... as the sun was rising from the fair sea
into the firmament of heaven
to shed light on mortals and immortals ..."*

... Homer, The Odyssey, Book III

He will do it by reviving the intellectual legacy of Archimedes. Unlike a biological resurrection, this is neither a miracle nor a temporary reversal of state, as in the parable of Lazarus, in which a mortal regains his physical life only to lose it again later. Galileo's initial move to breathe life into the static body of knowledge that Archimedes had left for a successor is to invent an experimental method of revealing hydrostatic principles of Archimedes. You have noticed that bodies less dense than water float, like a duck in a pond. But your observation would not be equivalent to the knowledge of floating gained by combining the observation with theory to create a proof. Carving out his arguments with the scalpel of geometry, Galileo will later prove theorems fundamental to hydrostatics and buoyed by Archimedes' principle.[1]

Probably you've heard some version of the legend about King Hieron's gold crown, actually a wreath. The Roman architect Vitruvius tells us that the King of Syracuse

The Royal Crown Takes a Bath

suspected a goldsmith of having alloyed it with a base metal. Since the wreath was consecrated, his priests would have frowned upon subjecting it to chemical analysis. So Hieron called on Archimedes to determine its purity in some other way. While bathing, the philosopher realized that if he were to immerse the wreath in a full basin of water, and divert the amount spilled over into a container of regular shape, he could measure the volume of the wreath. Its volume and weight would yield a density comparable to that of gold. The legend holds that, excited by this thought, he ran unclothed through the streets shouting "Eureka!" Concluding the story with the observation that Archimedes found the wreath to have caused a greater overflow of water than an equal weight of gold,[2] Vitruvius leaves to his reader's imagination the

---

1. In modern terms, this principle states that a solid object floating or submerged in a fluid is buoyed upward with a force equal in magnitude to the weight of the displaced fluid, and directed vertically through the original center of gravity of the now-displaced fluid. In his *Discourse on Bodies Floating On or In Water*, published later (1612), Galileo extends Archimedes' work to objects in finite containers. This extension leads to the surprising result that a body of specific gravity greater than that of water may still be supported by watery "ridges" while floating somewhat below the surface. In this work he refers to a concept of resistance at the surface of a fluid (that later will be called surface tension), and cites experimental evidence against the notion of Aristotelians that bodies float or sink because of differences in the resistance of the medium to being divided; cf. Drake's *Galileo at Work* (Bibliography). Drake's translation of Galileo's 1612 Discourse is included in *Cause, Experiment and Science*, U. of Chicago Press, 1981.

consequences to the goldsmith.

La Bilancetta, The Proportional Hydrostatic Balance

What surprises Galileo in this legend is the measurement technique, not the streaking incident. After all, public baths and nudity were common in ancient Greece, and Galileo would soon be writing his own defense of nudity.[3] Would Archimedes have tested the wreath this way? In 1586, Galileo suggests an answer by improving a device called *the hydrostatic balance.* Having positioned equal weights, one being the object to be tested, at opposite ends of a balance and equidistant from the fulcrum, the assayer can level the balance arm. When he immerses a test object in water, its weight will be reduced by the weight of an equivalent volume of water. In order to bring the mechanism back into balance, he moves the counterpoise. Galileo realizes that after repeating this procedure with test weights of pure gold and silver, he can mark calibration points on the balance arm. Using a test metal of gold alloyed with silver, he moves the counterpoise to some intermediate point. Then the ratio of distances between the intermediate point and the calibration points for gold and silver, respectively, is the ratio of silver to gold in the test metal. Having wound a spiral of fine wire between the calibration points, one can determine the ratio by counting the turns of the spiral with a knife. In this way, Galileo turns an ordinary balance into a tool for accurately determining the gold/silver ratio of an alloy.

Archimedean Mathematics and Aristotelian Induction

During this period he tutors students privately at Florence and at Siena. Continuing a self-directed education, he studies Aristotle as well as Archimedes. Although it might be supposed that he would prefer the more mathematical approach of Archimedes, he finds that many rigorous mathematical deductions have no counterpart in the real world of experience. On the other hand, Aristotle approved induction from observations and insisted that to be valid, conclusions must be consistent with observations. This feature of his philosophy seems to get scant attention from his followers. Seeking to combine experiment with theory, Galileo treats the logical and physical questions of Aristotle as susceptible of experimentation that would remove them from the realm of abstract speculation and make them testable, an approach foreign to the mindset of his contemporaries. Within a year after writing his little treatise on *la bilancetta,* he discovers a theorem of mathematical physics; i.e., one that can be tested by experiment.

---

2. "*... invenit plus aquae defluxisse in coronam quam in auream eodem pondere massam*". The Oxford English Dictionary confirmed my suspicion that this story is the origin of the English word "eureka", a crude transliteration from the Greek for "I have found it."
3. Cf. Chapter 4, "Against the Wearing of the Toga".

His discovery is an original theorem on centers of gravity of solids. He shows it to Guidobaldo del Monte and to Father Clavius,[4] a Jesuit mathematician to whom he may have introduced himself on his 1587 excursion to Rome:[5] "If [equal] weights in arithmetical progression are equally spaced along the arm of a balance, their center of gravity divides the balance arm in the ratio 2:1".[6]

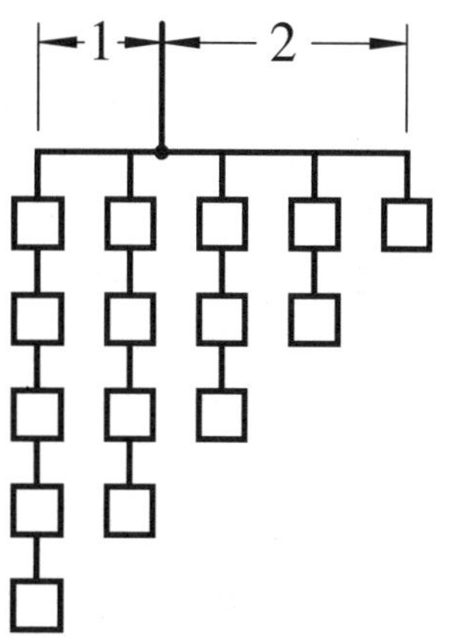

Although Clavius and del Monte initially quarrel with his theorem, Galileo defends it successfully and del Monte honestly admits that Galileo has been right all along. (Nothing is known of Clavius' response.)

Wherever he goes, Galileo is an observer of nature. During a hailstorm he questions the "Aristotelian" notion that heavy objects fall faster than lighter ones, as he recognizes that smaller and larger hailstones reach the ground simultaneously. Blended with his theoretical creativity, his incisiveness as an observer empowers the young Galileo to extend the conceptual base of natural philosophy and even to coin new words in the process. Concepts named by Galileo's new words apply universally, and in particular, at a playground.

A child at play learns by induction that bigger children are heavier and often more dangerous. Galileo's proofs from this period include a well-known relationship that may seem obvious in the 20th century — after Galileo proved it in the 16th:

Moment of Force

> *If two bodies are equally dense (i.e., have the same specific gravity), their weights are directly proportional to their volumes.*

The same child may notice that he or she can lift a heavier child at a seesaw by moving farther from the point around which this extended object, the board, is fixed and allowed to turn. This, of course, is a simple case of an Archimedean lever. To be able to measure the turning effect of a force relative to some point, Galileo spoke of the product of a weight and a length, and coined the word known in English as

---

4. Father Cristoforo Clavio, leading mathematician of the Collegio Romano, wrote various mathematical treatises principally in Latin under the Romanized form of his name, Clavius.

5. Archimedes had bequeathed Galileo a basic lever law contained in Propositions 6 and 7 of *On the Equilibrium of Planes*. Taken together, they imply that arbitrarily chosen weights suspended from a lever are in equilibrium at reciprocally proportional distances from the fulcrum.

6. from Drake's trans. of Galileo's *Two New Sciences*, p. 261 ff. Thanks are due to Drake for having the good sense to include the theorems on centers of gravity of solids, whereas, incredibly, the translators of the Crew and de Salvio edition of Galileo's *Discorsi* omitted them "as being of minor interest" !!!!!!

"moment", as in "moment of force". By turning the force of his mind to Archimedes, Galileo thus created a concept of statics, the science of forces on an object in equilibrium, distinct from kinematics as we defined it on page 2.

Later, if our hypothetical boy or girl takes up surfing, he or she learns that the speed and size (mass) of a wave affect the distance traversed by a surfer. As Galileo turned his thoughts from statics to kinematics, he began to speak of a product of weight and speed attached to the modern concept of "momentum". Though this concept did not appear until his later writings, it emerged then from his study of Archimedes. Years later, his new words were added to an Italian dictionary, and his name to its list of authorities on the Italian language. In the meantime, Galileo proved his originality at literary interpretation by analyzing the *Inferno*.

Measuring the Inferno (1588)

Where in hell are Dante's compartments? In Galileo's time, this is a hot issue that has puzzled literary interpreters. Unable to solve related problems of interpretation and calculation, the Florentine Academy consults Galileo on the configuration, positions, and sizes of the compartments of hell in Dante's *Inferno*.

His solution of the problem may help us to grasp a medieval universe — that Galileo would later help overturn — in which every entity, like Earth, has its divinely appointed special place. Appearing before the Academy and its guests, this red-haired, stocky, 24-year-old youth must have cut an unusual figure while addressing a host of elderly intellectuals. Let's hear his opening as he warms them to his topic:[7]

> "In an undertaking both arduous and astonishing, demanding long and continuous observations and risky reckoning, men have managed to measure and determine spaces of the heavens, varying motions of its bodies and their proportions, and separations of the stars, not to mention spatial arrangements and positions of distant lands and of the seas. To a large extent, these things are within the realm of sense experience. How much more then must we confront in awe the investigation and description of the position and shape of hell? Buried in the bowels of Earth, it lies concealed from all the senses and familiar to no one through ordinary experience. Even if the descent into it is made well and easily, most perilous is the exit back, as is taught by our Poet in this quote:

> *"Abandon all hope, you who enter."*

and by his guide in another quote:

> *"It's easy descending into hell;*

---

7. What follows is my rather literal translation of an introductory excerpt from Galileo's lesson on Dante's *Inferno* as contained in Favaro's National Edition, vol. IX (31-57). At the end of this quotation and for the remainder of this section, I paraphrase excerpts from my rather free translation of portions of this lesson; likewise, the quotations from Dante are my translations of Galileo's rendering of them.

> *But the return again from the realm of the dead*
> *that path of returning to the pure breeze,*
> *This, this is an uphill and formidable undertaking!"*

Its isolation from other parts of the world poses a challenge for anyone to describe hell. Against this challenge it was left to explain an infernal scene, mapped out and architected by an exalted guide, who in reality was our Dante. His effort permits me to reveal the amazing construction of the afterworld and to describe meticulously its site on Earth. ...

Dante writes about his Inferno, but leaves it in its eclipsed darkness, which to those who come after him has given cause to toil for a detailed explanation of its architecture. Among them, two have done extensive analysis: Antonio Manetti and Alessandro Vellutello."

Galileo then announces his objective: to specify the shape, size and position of the Inferno, the number and distribution of its levels, the number of compartments within each level, the intervals separating one level from another, and the distance across each level and compartment.

Hell is located in an inverted conical cavity, he says, whose vertex is at the center of Earth and whose base is toward its surface, centered under Jerusalem. Here he agrees with Manetti's interpretation and supports this conclusion with a reference to the second canto of Purgatory:

> *"Already the sun had reached the horizon whose*
> *meridian circle covers Jerusalem with its summit. ...."*

Beginning at Jerusalem he draws an arc along the surface of Earth, extending through a 12th part of its circumference. From the end of this arc to the center of Earth he projects a straight line. In his imagination, he fixes the extremity of the line segment at Earth's center and moves the extremity at the surface so as to cut it in a circular arc until it returns from where it leaves. We can thus imagine carved out of Earth a conical cavity in which the Inferno is situated.[8] Its ceiling is a disc beneath Jerusalem whose diameter is a chord equal in length to Earth's radius and intercepting the arc of a sixth part of Earth's circumference. Using this geometric figure, Galileo proceeds to correct Manetti's estimate that hell occupies a sixth part of the whole Earth; rather, "considering the things demonstrated by Archimedes in his book *On the Sphere and Cylinder*"[9] Galileo finds that the Inferno occupies something less than 1/14 part of the

---

8. If we thus set a compass at Jerusalem to draw a circle of π inches radius on a globe of 12 inch diameter, we find that it crosses Spain, France, Germany, the Baltic Sea, Russia, Afghanistan and Pakistan, and after crossing the Indian Ocean, it enters the African continent at Somalia and continues across Africa to the Mediterranean north of Algeria. Galileo tells us that the Inferno of Dante's imagination lies in a cone stretching to the center of Earth beneath the portion of Earth's surface bounded by this circle.

whole earth.[10] He then denies that its space extends to Earth's surface. On the contrary, the opening of hell remains covered by the vault of Earth whose summit is Jerusalem, and whose height is 1/8 of Earth's radius (nearly 800 km).[11]

Dante's Inferno resembles an amphitheater that contracts as it descends, except that an amphitheater has in its base a piazza, while the Inferno ends in a point. The internal surface contains eight levels which go round the cavity like the rows of an amphitheater. In order of increasing depth, they are assigned ordinal numbers and occupants:

First:	Limbo [12]
Second:	The Lustful
Third:	The Gluttons
Fourth:	The Wasteful and Avaricious

The fifth level is divided into two rings: the Stygian quagmire, the assigned place for punishment of the Wrathful and the Lazy, and the city of Dis, where heretics are punished, surrounded by the quagmire. Because Dante puts two rings at the fifth level, Galileo counts 9 rings within the 8 levels. At the sixth level (ring 7), the violent are tormented. This ring is divided into 3 compartments with rather uninviting names: [13]

- The River of Boiling Blood
- The Forest of Harpies
- The Field of Burning Sand

The fraudulent are punished in the Evil Pits of the 7th level (ring 8). The eighth and last level (ring 9) encloses four spheres of ice in which traitors are suspended.

---

9. Galileo does not identify his premise or his method of calculation, but I note, for example, Prop. *2* in Book II of the cited work of Archimedes: *"Any segment of a sphere is equal to a cone which has the same base as the segment and for height a straight line which has to the height of the segment the same ratio as the sum of the radius of the sphere and the height of the remaining segment has to the height of the remaining segment."*; trans. by E. J. Dijksterhuis, *Archimedes*, p. 189. The domed vault of Earth covering hell is just such a spherical "segment".

10. We can calculate an equivalent to Galileo's unpublished calculation. Taking a planar slice through Earth intersecting its center and Jerusalem, one can inscribe a regular hexagon inside the resulting circle in such a way that Jerusalem is centered on the arc that intercepts the vertices of one triangular slice of the hexagon. The right conical cavity occupied by the Inferno is produced by a rotation of that equilateral triangle about its bisector passing through Jerusalem and the center of Earth. Splitting the triangle with this bisector produces two equal right triangles. Their sides comprising the chord that intercepts the arc of Earth are each half Earth's radius, r, at Jerusalem. Therefore the conical cavity has a base of radius 0.5r and height kr where $k = \sqrt{3/4}$ or, if you prefer trigonometric notation, k = cos(π/6). Approximating Earth as a sphere of radius r, one finds the ratio of the volume of the right conical cavity [π(0.5r)^2(kr)]/3 to Earth's volume (4πr^3)/3 to be (0.25k)/4 = 1/18.475 to five significant places. Galileo's "something less than 1/14 part of the whole" differs from this result by something less than 2% of Earth's volume. His later denial that the Inferno should include the spherical-sector vault over the cone implies, as rdc has pointed out, that his result includes the volume of this vault, about 1.3% of Earth's volume.

11. Galileo says "405 15/22 miglia". From this comment it is evident that he puts Earth's radius = 3245.45 miglia = 6378.14 km; ergo, one miglia = 1.9653 kilometers.

Having thus described the levels, Galileo now specifies the distances that separate them. In depth the first six including Limbo, domed as it is by the arc of Earth, are equal to each other and to 1/8 the radius of Earth.[14] There remain the two last intervals, comprising 1/4 of Earth's radius; namely, the distance between the ring of the violent and the Evil Pits (the depth of the ravine of Gerione) and that from the Evil Pits to the ice (the height of the wall of giants). After further calculation, he corrects Manetti (who holds that all the levels are 1/8 Earth radius in depth) by putting the depth of the ravine at 1435 km (0.225r, where r = Earth's radius), and the height of the ice wall of giants at 159 km (0.025r).

The analysis continues, but we leave it now, having seen how Galileo applies the Archimedean theory of conic structures to Dante's medieval conception of an afterworld in which every kind of creature has his special assigned place, like the special place at the center of the universe that philosophers had assigned to Earth. This exercise in creative lecturing is one of Galileo's final steps on the footpath that leads him to a professorship and to a spiral staircase — identified in the next Chapter — that he will ascend to look out and see his position in the universe and to abolish the medieval method of partitioning the universe by flight of fancy.

A current of knowledge, flowing beyond but growing out of the science of the early Greeks, now washes into the receptive harbor of Galileo's ponderings on physical propositions as he assimilates the work of Renaissance precursors. Before continuing with his career, we can survey medieval science as it will present itself to him by picking up the story where our first chronicle of pre-Galilean science left off: at the destruction of the Library at Alexandria.[15] Certainly that event was not the only one marking the global decline of human achievement before the Dark Ages. The oriental train of applied mathematics had already (in the time of Apollonios) been derailed by the founder of the Ch'in dynasty, who considered it his imperial prerogative to burn all existing books and to bury alive 460 scholars who protested.[16] But it was the disaster at

---

12. Dante describes Limbo as a place where unbelievers who lived otherwise meritorious lives dwell in a dim fog separated from the joys of paradise but otherwise untormented. Here are masses of heathens who achieved nothing of great significance in their lives; hence, the most spacious level is reserved for them. But within limbo there is a green meadow within an illuminated castle where non-Christian philosophers, scientists and men of great achievement can hobnob or resume their solitary contemplations as they wish. Thales, Empedocles, Heraclitos, Zeno, Democritos, Socrates, Plato, Aristotle, Euclid and other great thinkers are specifically mentioned as occupants of this castle. (Cf. Dante's *Inferno*, Canto IV, trans. by I. C. Wright, pub. by Henry G. Bohn, London, 1859.)

13. Here in order to give these names in their uninviting fullness, I depart from Galileo's abbreviations of Dante's names in favor of the translation by Dorothy Sayers, pub. Penguin Books, 1949 (dc, rdc).

14. Apparently, Galileo uses the fraction 1/8 as a rough approximation to $(1 - \sqrt{3/4}) \cong 0.134$

15. This is a journey with unannounced retrograde motion; the retrograde tendency has been infused on the mind of the author by Ptolemaic clingers (rdc).

16. David Eugene Smith, *History of Mathematics*, Ginn & Co., Boston, 1923, pp. 24, 32-33, and 139.

Alexandria which, by attacking the Thalesian creation of science, nearly pulled up western civilization by its roots.

Flashback: The Dark Ages

As the ashes of the burnt library cooled, the dynamism of the once flowering Greek science froze into near stasis. Knowledge expansion slowed to a crawl during the Dark Ages. Remnants of Greek discovery lived with scholarly Catholic monks in the Benedictine monasteries of western Europe. These remnants sparked the cultural awakening of the Muslim empire beginning in the 7th century. While Bhagdad strove to replace the Alexandrian center of learning, culture migrated to the Middle East. Arab, Indian and Persian mathematicians drew upon remnants of Greek mathematics to introduce a few advancements of their own, flexing the sinew of Greek mathematics that Galileo would later restore to prominence. Who were these achievers?

Sundials, Numerals, Twilight and Cubics

The first of them is Mohammed al-Khwarizmi (Latin: Algoritmi, c780-c850), from whose name the word "algorithm" derives, and author of the *Al-jabr*,[17] from which the word "algebra" derives. Besides expressing rules of algebra, al-Khwarizmi published astronomical treatises on the sundial and the astrolabe. He popularized Hindu numerals in a text to which Leonardo of Pisa would later have access, so that today they are misleadingly known as Arabic numerals. Another astronomer of the ninth century, al-Battani (Albatenius), will become — after improving the accuracy of tables showing apparent diameters of the sun and moon — one of those posthumously criticized by Galileo for needlessly overestimating the apparent diameter of fixed stars. Ptolemy had asserted the impossibility of annular solar eclipses (in which our moon, when interposed between sun and Earth and smaller in angular diameter than the sun, is crowned by a ring of sunlight); al-Battani corrected him, showing that annular eclipses are possible[18] and improving on Ptolemy's measurement of the tropical year.

Mentioned in the same context by Galileo,[19] Thabit ibn-Qurra (Thabit ben Korah) preserved Greek mathematics in the 9th century by translating works of Archimedes, Euclid, Apollonios and Ptolemy into Arabic, including the *Conics* that would have been lost to us. In the tenth century, Avicenna and Alhazen undertook novel applications of

---

17. abbreviated form of expressions variously transliterated "Al-jabr wa'l muqabalah", or "Al-jebr w' almuqabala". "The word *Al-jabr* presumably meant something like "restoration" or "completion" and seems to refer to the transposition of subtracted terms to the other side of an equation; the word *muqabalah* is said to refer to "reduction" or "balancing" — that is, the cancellation of like terms on opposite sides of the equation" (cf. Boyer, *History of Mathematics*, op. cit.)
18. Observing a mid-day annular eclipse in 1994, I saw (through aluminized mylar) a ring of sunlight whose mean thickness I estimated at 1/15 the sun's angular diameter, or about 2 minutes of arc.
19. See *Dialogo*, Drake trans., p. 360.

mathematics to physics and to astronomy; Alhazen calculated the height of Earth's atmosphere from the sun's position at the end of twilight. Meanwhile, the intellectual climate of Persia had prospered under the potentate Khosrú I (Anôschirvân) who, in the sixth century, encouraged an influx of Greek scholars. In the remote aftermath of this influx, the Persian poet and mathematician Omar Khayyam (c1050-c1123), author of the *Rubaiyat*, advanced algebra to cubic equations, decomposing them into quadratics and solving for the intersections of their conic sections. Omar originated graphical methods and anticipated Pascal's development of the binomial formula,[20] later generalized by Newton to negative and fractional exponents. Following his lead, Persian mathematics flourished in the 12th century with the geometry of al-Rázi and of al-Tûsi; these developments culminated in astronomical tables, prized by European astronomers and worked out by the Persian Prince Ulûg Beg, founder of an observatory at Samarkand.[21] Ulûg Beg's assistant, al-Kashî, enumerated π to better than ten places and, as Galileo would later do, developed a special-purpose calculator for astronomy.

Have you ever tried to do arithmetic with Roman numerals? Fortunately for us, the system of consistent positional notation in use today was invented in India in the Middle Ages. To express the number of days in a year, rounded down to the nearest integer, in a way that facilitates calculation, we (consciously or not) collapse the polynomial:

Positional Notation

$$3 \cdot r^2 + 6 \cdot r^1 + 5 \cdot r^0$$

into the positional form "365", where r is the radix of our numeration system (which for general uses is ten, probably because most humans have ten fingers). Today, this notation is taken for granted by everyone except bankers who prefer to write this number as "360" for the purpose of calculating interest due them.

Positional notation reached its culmination with Bhaskara (b. 1114), who brought out the cipher zero as an additive identity and introduced negative numbers. He was first to state that division by zero cannot be performed in a domain of finite numbers. Bhaskara is noted for his elegant geometric proof of the Thalesian "theorem of Pythagoras":[22] taking each side (of length c) of a square as the hypotenuse of a right triangle with sides a, b and c built into the

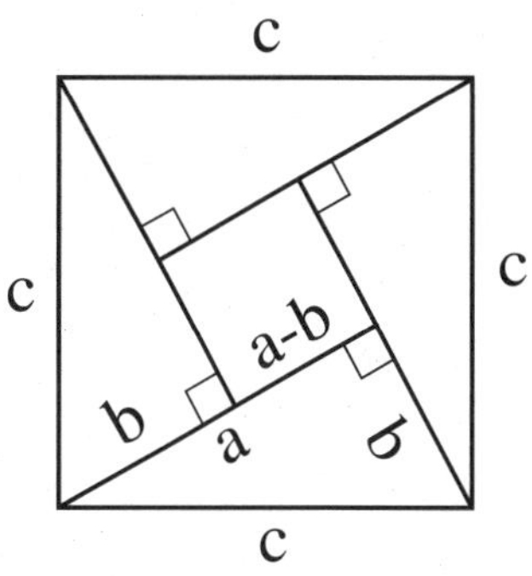

---

20. Omar was justifiably proud of discovering the binomial expansion of $(a+b)^n$ "for all (integral positive) exponents n, which no one had been able to accomplish before him." Quoted in Heinrich Dörrie's *100 Great Problems of Elementary Mathematics*, op. cit., p. 37.
21. Cf. David Eugene Smith, op. cit., pp. 286-289.
22. See H. E. Huntley, *The Divine Proportion*, Dover, NY, 1970.

square, the four triangles cover all but a small square of side (*a-b*) inside the original square; hence, the area of the original square is $4ab/2 + (a-b)^2 = a^2 + b^2 = c^2$. Although his positional notation facilitates arithmetic, Roman numerals hung on long past Bhaskara's invention largely because of the fear of fraudulent alteration of characters in business transactions, an alteration difficult to perform with Roman numerals. Fortunately for us, the Roman system gradually gave way to the enormous computational advantages of positional notation.

Caught in a crater dug by the impact of systematic knowledge-eradicators, those few brave Arab, Indian and Persian thinkers merely sustained a tradition of uniting algebra and geometry that had begun with the work of Archimedes and Pappos. Centuries later, Vieta (whom we will visit in this chapter) filled that crater and sparked the development of analytic technique used by Newton to build his system of the world — in part on the principles of Galileo.

The same Medieval Church that helped preserve that once dynamic Greek science froze it into stasis through its conservative and authoritative operation. It required the genius of Galileo to separate Greek science from this static framework and to show that a new synthesis could be built upon it. To see how he did this, let's leap over the intervening centuries to the early Renaissance.

Pre-Galilean Renaissance

Major Renaissance thinkers paved the way for a scientific question once thought the exclusive concern of the priesthood: What natural mechanism has supplied the energy to form the physical world and to drive the cycle of life? Today, we know that the energy required to evolve and sustain life, including human life, is the energy emitted by a gravitational machine called a star. Modern cosmologists present a chain reaction scenario of energy propagation. Beginning with the infalling of matter from interstellar clouds of gas and dust, continuing as that same matter collides, fuses and ignites a star, and propagating itself through the radiation of some of the resultant energy, gravitational energy is seen as the source and sustaining power for life on a planet formed under suitable initial conditions. But before man could come to an understanding of this key link between the progress of life and the way that things fall, it was necessary to know how things move in our local vicinity, particularly things that fall. This understanding is included in Galileo's laws of motion, laws that he will develop against the backdrop of men and events of the early Renaissance.

Oresme, da Vinci   From Bradwardine and the Mertonian school at Oxford to the Parisian school, independent efforts were made in the 14th and 15th centuries to understand how objects freely fall. In particular, Nicole Oresme (1325-1382) and Leonardo da Vinci (1452-1519) logically preceded Galileo in that each formulated a description of accelerated motion. Oresme published a theorem on general accelerated motion like one that Galileo used to describe bodies in free fall: the space traversed by a body in uniformly accelerated motion is as if the body had moved uniformly at the speed it had at the middle instant of the elapsed time of fall. In this formulation, it resembled the "mean speed theorem" that has been attributed to the Mertonian school: "A uniformly accelerating body will cover a distance equal to what it would have covered in the time, if it had been moving uniformly at its mean or average velocity".[23] Leonardo took a step further and related uniform acceleration to a falling body, implying that its speed of descent increases somehow arithmetically with elapsed time intervals, a step in the direction of Galileo's later precise quantitative formulation.[24] Certainly, the concept of constant acceleration was a key ingredient in Galileo's correct formulation of his law of falling bodies.

Nicholas of Cusa

While da Vinci and the Oxford and Paris physicists grappled with the problem of objects falling near the Earth, the question of the motion of Earth itself began to gain attention. Nicholas of Cusa (1401-1464) was perhaps the first Renaissance man to publish the idea of Heraclides and others[25] that Earth spins on its axis and of Aristarchos that it revolves around the sun. Well aware of his predecessors, Nicholas offered no proof from observation or from established theory but built an intuitive conception of the universe on the conjecture, derived from his application of the *coincidentia oppositorum*[26] concept, that Heraclides and Aristarchos had been right. His insight, similar to and perhaps a source of that for which Giordano Bruno was later murdered by the Inquisition, included the infinity of space and the sun-like nature of the stars. Yet Nicholas escaped persecution and was appointed a Cardinal of the Church. One wonders if anyone read his book [27] at that time.

---

23. Rom Harré, *Great Scientific Experiments,* Oxford U. Press, p. 70. Galileo later applied this important theorem to falling bodies.

24. Cf. Gillespie ed., *Dictionary of Scientific Biography*, pp. 231-232. For Galileo's precise formulation, see Ch. 5 "Levers, Inclined Planes and Falling Bodies". Da Vinci also generalized Archimedes' mechanical laws in his investigations of statical moments of rigid bodies. See Moulton, Forest Ray, *An Introduction to Celestial Mechanics*, 2nd ed., Ch. 1.

25. From Berry (cf. Bibliography) we learn that Copernicus had found a reference of Cicero to the opinion of Hicetas that the Earth rotates on its axis daily. And from Boyer's *History of Mathematics* we learn that another Pythagorean, Ecphantos, also explained day and night by a rotating Earth.

26. An example of the coincidence of opposites is identification of a circle of infinite radius with a straight line which is, for example, what Earth's horizon approaches as the observer shrinks. In the geometric domain of continuous forms, the point is the unit that generates those forms. As infinity is generated through unity in the number domain, so in geometry an infinite sphere is generated by a point. (One may conceptually generate a infinite number of lines in all directions from an arbitrary point to infinity, thus generating a unique sphere from that point.) Such considerations led Nicholas to conclude that "there could be no cosmic mechanism or center point for the motions of the heavens, since such a point of necessity included the whole universe"; cf. Gillespie ed., *Dictionary of Scientific Biography*, op. cit.

27. Nicholas Cusano, *De docta ignorantia*, pub. 1440.

Tartaglia

Known as the "Stammerer" and less (but better) known as the principal deliverer of Greek geometry to Galileo, Tartaglia (1499-1557) was an early Renaissance Archimedean. A soldier from an invading army had slashed Tartaglia's face when he was a boy, cutting open his skull, jaw and palate and leaving him for dead. His mother, having no resources, licked his wounds and saved his life. The wounds created a speech defect, but the boy was a genius who would later communicate with Galileo through his textbook. Professor Drake holds it probable that the text of Euclid used by Galileo as a student was Tartaglia's translation published in 1543, reprinted in 1565. Because Ricci[28] taught in the vernacular Italian, this text would have been preferred by him to the Latin texts of the universities, which were lacking in their treatment of Eudoxian proportion theory. In this period prior to analytic geometry, later invented independently by Fermat and Descartes, Galileo uses the theory of proportions to describe continuous change, as in the speed of a falling object. Eudoxos becomes his link between mathematics and physical events.

Copernicus

To convince learned men of Earth's motion, against the geostatic universe mathematically grounded by Ptolemy and supported by the Church, required more than merely re-asserting the views of Aristarchos. The argument had to begin with a trigonometric demonstration that a heliocentric universe would explain observed planetary motions more elegantly than had the system of Ptolemy and had to be consistent with those annual cycles of nature observed on Earth that we call seasons. Nicolaus Copernicus (1473- 1543), a central figure in the Renaissance, produced just such a reinforcement of the Aristarchan[29] heliocentric hypothesis. While retaining the circular orbits of the Ptolemaic system, he moved their center close to that of the sun. In so doing, he moved his Church and his species away from the center of the world. Moving our species off center is a fundamental act of the Renaissance: it affects the way man views his relationship to the world around him. When Galileo furnished massive evidence[30] for Copernicus and contrary to Ptolemy, Copernicus could neither be ignored nor easily disputed. Galileo was punished instead of rewarded for his great discovery. Copernicus, though no longer ignored, continued to be seriously disputed until Newton combined his concept of a force field with Kepler's and Galileo's independent results. With this combination, Newton proved that the solar system has a gravitational barycenter located in the sun. Perhaps wisely, Copernicus had waited until his last days to publish his radical theory.

---

28. Galileo's teacher Ostilio Ricci (not to be confused with Matteo Ricci, S. J., a famous Jesuit who propagated Western mathematics to China beginning in 1577.)
29. In view of this connection it is going a bit too far to say of his achievement "that he overturned the astronomical world view handed down from the ancients." (cf. Dieter B. Hermann, *The history of astronomy from Herschel to Hertzsprung*, 1984.)
30. See Ch. 9, "Galileo's *Dialogue Concerning the Two Chief World Systems (Dialogo)*"

Once underway, the arguments for repositioning the center of the world **Bruno** culminated in the disappearance of the center of everything. Giordano Bruno (1548-1600), taking a radical leap beyond Copernicus to the notion of an isotropically lumpy universe that has no center, demoted the sun to the status of an ordinary star in one of the lumps. Like Copernicus, Bruno did not offer physical proof. And like Democritos, he coupled intuitive insight with undauntable conviction in his infinity of worlds. Added to his "crimes" against theocracy, that conviction made him a casualty of the Inquisition: he was murdered at Rome in 1600 in a "Christian" ritual known as burning at the stake. Bruno's thinking was eclectic, blending Copernicus' breakaway from anthropocentrism with the infinity of worlds held by Nicholas of Cusa and by Thomas Digges. The result was an almost 20th century view of the universe. If Bruno saw farther than others, his foresight may owe more to his weaving than to the elements that he wove together.

Founder of symbolic algebra, Franciscus Vieta (François Viète,[31] 1540- **Vieta** 1603) created the modern language of mathematics by transforming algebra into a form by which we recognize it today. This achievement Dantzig [32] ranks as one of "the two epoch-making events" in the history of mathematics; namely, "the principle of *deductive reasoning* inaugurated by Thales and the *symbolic algebra* of Vieta". He adds: "... it not only endowed mathematics with a language, but armed it with such powerful weapons as *paraphrase, analogy, generalization.* Thus did Vieta turn a tongue-tied thinker into a fluent and convincing speaker, and, at the same time, immensely enriched the thinker's creative and critical faculties." He also created a generalized analytic approach to trigonometry.[33] With the possible exception of Galileo's failure to take note of Kepler's Laws of Planetary Motion, perhaps the greatest tragedy, and at the same time the most amazing fact, of Galileo's professional life was his non-use of Vieta's literal algebra. That Galileo burdened himself with verbal mathematical arguments and intricate geometric proofs such as those of his *Discorsi*, and then achieved as much as he did, is amazing. Though his *Principia* is largely still in the Euclidean form, Newton took full advantage of Vietan algebra in transforming Galileo's and Kepler's laws into his system of the world.

As a contemporary of his own achievement, Galileo could not see a **Benedetti** line between "medieval" and "modern" science, an abstraction that marks his position in history from the hindsight of centuries later. From a modern perspective, we can see him breaking through to the other side of that line as he replaced the concept of impetus pioneered by Giambattista Benedetti (1530-1590),

31. I refer to him as Vieta, the short Latin form of his name (Vietaeus) under which he published.
32. Tobias Dantzig, *Bequest of the Greeks*, pp. 184-185.
33. Cf. Carl Boyer, *A History of Mathematics*, p. 307.

student of Tartaglia and mathematician to the Duke of Savoy. Benedetti brought the medieval theory of free fall to the refined form in which Galileo inherited it. To do this, he had to correct "Aristotelian" doctrine. Appealing to a principle found in Tartaglia's 1551 translation of Archimedes' first book on hydrostatics, he showed that the speed of a falling object is constant for a given medium. Benedetti retained the notion of impetus that was a hallmark of medieval kinematics, and that Galileo had to overcome. Unlike the modern concept of inertia, impetus was, for Benedetti, a self-exhausting force that would temporarily sustain motion produced in the moving object by the mover, not a natural property inherent in motion. And though the impetus of a body released from circular motion is rectilinear along a tangent line, an important characteristic of Galilean *impeto*, it was lacking in the quality of permanence that characterizes the latter.[34]

Tycho   It seems to me that Bruno had climbed to the peak of theory outdistancing the data available to support it. His contemporary in this later Renaissance period, Tycho Brahe (1546-1601), son of a Danish nobleman, turned the tide of this imbalance with a great information harvest. Tycho compiled massive observational data on the orbit of Mars, so accurate that in the deluge of numbers Kepler found a secret of planetary motion that neatly complemented Galileo's evidence for the Copernican system. The combined effect relieved astronomers from having to go on refining theory to "save the appearances" of planetary motions. Tycho advanced a hypothesis of planetary motion, compromising between Ptolemy and Copernicus. He placed the wanderers we see in the night sky in orbit around the sun, and the sun with its entourage of planets in orbit around Earth. Here was another incorrect hypothesis, which Galileo, who never saw any use for compromise in science, rejected.

Kepler   Johannes Kepler (1571-1630) was the most important contemporary of Galileo (1564-1642) among the developers of the solar system model with which we are familiar today. While Galileo showed that Copernicus had been right to associate the center of planetary motions with the sun, not Earth, he held onto the assumed circularity of planetary orbits: adherence to this notion required the epicycles of Apollonios, Hipparchos, and Ptolemy to bridge the gap between the ideal geometry and the real observations. It took the genius of Kepler to banish forever these circles and epicycles to the closet of discarded scientific relics. His achievement, the Laws of Planetary Motion, became a cornerstone of the grand synthesis that was Newton's system of the world. To conform to all three laws, the planets must revolve in elliptical paths around the sun, so that their orbital radius vectors sweep out equal areas in equal times, and so that the squares of their periods are proportional to the cubes of their mean distances. And this is how modern astronomers report that the

---

34. See Gillespie, *Dictionary of Scientific Biography*. See also Ch. 5 "Galilean Impetus".

sun's planets, and even distant stars in gravitational interlock, are seen to behave.

Isaac Newton built his system of the world on a foundation that included those Keplerian laws of planetary motion and Galilean kinematics: his laws of fall and principles of motion. Kepler's laws can only be mentioned here. That other principal cornerstone of Newton's system, Galilean kinematics, will require much of Chapters 4, 5, 9, 11, and 12 to examine in detail. In Chapter 4, we will find Professor Galilei dropping weights from the tower of Pisa to convince his students and colleagues that Aristotle's laws of fall left much to be understood. In the remaining chapters we will see how he achieved that understanding until (by the end of Chapter 12) we can see in our minds' eyes a vision that Galileo made possible but apparently never enjoyed: his weights falling from the Tower of Pisa related to the sun, the planets, and their moons falling around their common center of mass located in the sun.

Chipping Away at the Rust (1587-1589)

Let's return to our story of the young Galileo. In the hours between his tutoring appointments, he continues to chip away at the rust that had coated and hidden away the glorious edifice of Archimedean/Aristotelian physics over the preceding thousand years of intellectual bad weather in the dark valley that stretches between two towering thrusts of civilization. As he chips and polishes, the gleaming of reflected sunlight from this edifice catches the eye of influential scholars.

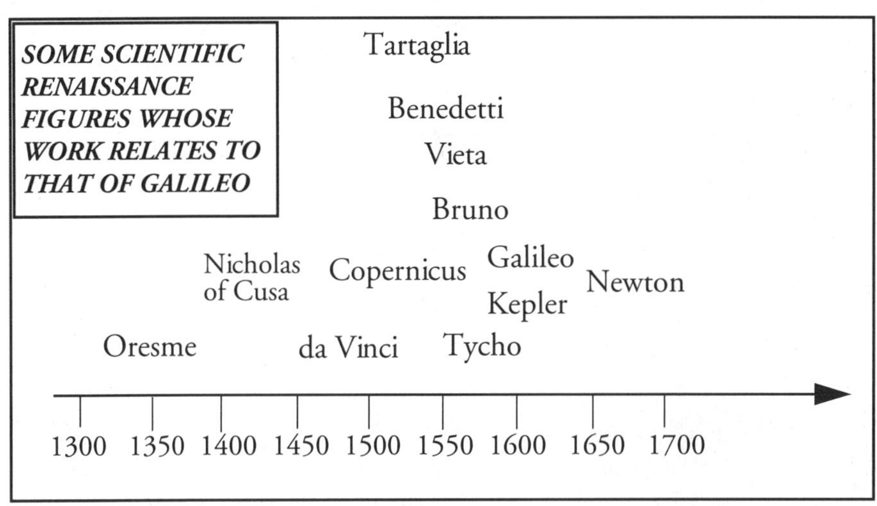

Monument to Leonardo of Pisa (Fibonacci)
(photo by author)

Nicolaus Copernicus, The Decentralizer,
Bust at his *Alma Mater,*
the University of Padova
(photo by author)

Copernicus on U.S., Rwandan
and Liberian Stamps Created by
Unspecified Artisans

Note: The Liberian stamp identifies Eudoxos as a contributor to the Copernican system. None of them identifies Aristarchos, Apollonios or Hipparchos. Copernicus combined Aristarchos' heliocentric arrangement of the planets with circular orbits. To reconcile the circular orbits with observation required the epicycles of Eudoxos as refined by Apollonios and brought to perfection by Hipparchos. The system called "Ptolemaic" by Galileo in his *Dialogo* is fundamentally the system of Hipparchos as recorded by Ptolemy; hence, both systems contrasted by Galileo contain Hipparchan theory; it was the Aristarchan heliocentrism in the Copernican system that caused the controversy.

Chapter 4. The Spiral Staircase

Professor Galilei at Pisa (1589-1592)

*"...Or let my lamp at midnight hour
Be seen in some high lonely tower,
Where I may oft outwatch the Bear,
With thrice great Hermes, or unsphere
The spirit of Plato to unfold
What worlds or what vast regions hold
The immortal mind that hath forsook
Her mansion in this fleshly nook;..."*

... John Milton, from *"Il Penseroso"* [1]

*"He (Galileo) soon noticed that when the lamp swung through a wide arc the
time it took to perform one swing seemed to be the same as when it swung
through a narrow arc."*

... Morris Kline, from *Mathematics in Western Culture*

Those bright rays — the new hydrostatic balance, the Archimedean extensions[2] and
the lectures at the Florentine Academy — catch the attention of some leading
European mathematicians. One of them, Guidobaldo del Monte, recommends
Galileo when the Chair of Mathematics opens up in 1589 at Pisa. In England,
Shakespeare is introducing *King Henry VI* on his stage. At Pisa, Galileo plays to a
different kind of stage: the lecture hall in which he will act out a growing inner
conflict between accepted teaching and his novel thoughts. At first, the new
Professor lectures on the Ptolemaic world system, a cornerstone of the ancient
cosmology that still permeates university teaching. He also prepares to publish.

**The Pisan
*De Motu***

Before his famous demonstration at the Leaning Tower, Galileo has
begun to build a tower of knowledge, his legacy to the world.[3] The
tenets of the Aristotelian natural philosophy that he is teaching now

---

1. from *The Portable Milton*, Douglas Bush ed., The Viking Press, NY, 1949.
2. In 1587, his extensions to Archimedean mechanics on centers of gravity of solids came to the attention of
   del Monte, Moletti at Padova, and the Belgian mathematician Coignet, from whom Galileo received his
   first recognition abroad. Though some of Galileo's theorems, particularly about parabolic solids, were
   advanced, he based them on his simple but powerful proof (see p. 31) that the center of gravity of weights
   in arithmetic progression equally spaced along a balance arm divide the arm in the ratio 2:1, a result from
   which del Monte had to retract his initial objection. (cf. Drake, *Galileo at Work*, Ch. 1)
3. Here I contrast the vertical approach of mathematical physics, which builds on earlier premises, with the
   horizontal way in which Aristotelian knowledge was apprehended by the followers of Aristotle.

include the concept that the universe is arranged onion-like in layers of heavenly spheres with Earth at the center. Yet even in his first manuscript, the *De Motu* of 1592, his readiness to theorize independently of Aristotle, while invoking the mathematical physics of Archimedes, is already manifest.[4] He attacks the question how bodies move in this layered universe by developing the science of kinematics, the study of the motion of bodies without regard to the cause of that motion.[5]

Aristotle's Law of Fall

If you are a reader who cringes at mathematical notation, bear with me for the next two paragraphs: I am providing the underpinnings to an important point. First I want to say that the following formulae are entirely anachronistic. They are presented here only for the modern reader's ease of understanding. All these thinkers expressed their notions in words only. None of them had any concept of "instantaneous" before Galileo; indeed, he must invent "velocity", too. Without algebra, this is particularly difficult. Without algebra, even to understand what he invented is difficult; therefore, I am going to "cheat" and use algebraic notation that he would not have used to explain it.

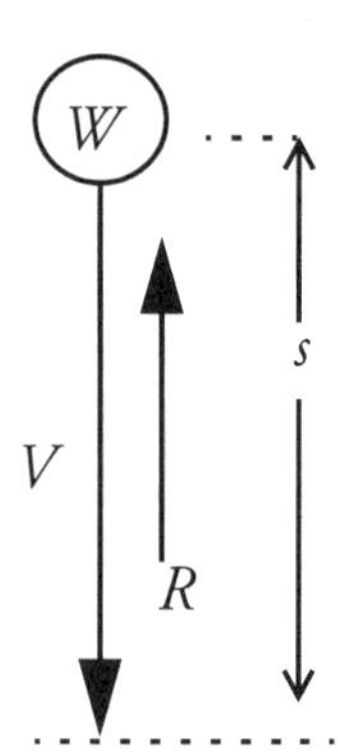

Galileo's predecessors in kinematics were curious how the speed (V) of a falling object relates to certain other variables, especially its weight (W), the distance through which it falls (s), and the resistance (R) of the medium[6] through which it falls. The law attributed to Aristotle can be rendered in modern notation as:

$$V = \left(\frac{W}{R}\right) \cdot f(t)$$

where f(t) is some unspecified function of time. This so-called "law" leaves a lot to be desired: it isn't computable because the function f is not specified, nor does it work in a vacuum (as that would involve division by zero), nor is it consistent with observation. It means to Galileo that heavier objects are supposed to fall faster, an effect he does not see or cannot measure in his tests.

---

4.   Cf. Drabkin and Drake, *Galileo Galilei on Motion and on Mechanics*, U. of Wisconsin, 1960: "Aristotle, as in practically everything that he wrote about locomotion, wrote the opposite of the truth on this question [by what agency projectiles are moved] too. And surely this is not strange. For who can arrive at true conclusions from false assumptions?", trans. by Drabkin from *De Motu*, 1590-1592.

5.   The equations defining acceleration a in terms of speed v and initial speed v_0 as $v = v_0 + at$ and defining average speed v over distance s and time t as $(v_0 + v) \cdot t = 2s$ are kinematic equations of motion. Other kinematic equations, relating distance and speed to time, have a dynamic interpretation in the context of falling bodies, and will be presented later in that context. Galileo's kinematics was adequate for solving problems that only required descriptions of motion; i.e., causes could be ignored if knowing them did not add anything to one's solution (jm).

6.   In non-falling situations, Aristotle viewed the medium as assisting motion.

The Concept of Instantaneous Velocity
Between Aristotle and Galileo, reformulations were attempted; some commentators implied that $V = k \cdot s$. Avempace held that $V = k \cdot (W - R)$; Bradwardine, that $V = k \cdot \log(W/R)$, where k is some proportionality constant. Ultimately, Galileo could not accept any of these. The latter two do not jibe with the observed nonuniformity of the speed of fall. The relation $V = k \cdot s$ hints at acceleration, but does not predict what can be measured with a water clock[7] while a metal ball rolls varying distances down an inclined plane, a favorite test of Galileo's. And for his precursors, instantaneous speed was incomprehensible and unmeasurable.[8]

It must be hard for us in this century who drive our cars to work, ever wary of the speedometer and that cop lurking behind bush #57, to imagine the strangeness of instantaneous speed to a medieval mindset in which time itself is little more than the integral number of chimes of a tower bell. Imagine how you would have been able to monitor instantaneous speed without being able to read the instrument of your car as you coast without braking to a lower speed, so as to fool the cop behind you into thinking you have all day to get to your next appointment. Cars and speedometers weren't developed yet, nor even the concepts of air resistance and tire friction. Galileo could only measure average speed over an interval of time. To recognize that variables like rolling friction are extraneous to the phenomenon under study, to remove them from his results, to collapse time intervals to infinitesimal instants, and then to arrive at instantaneous speed and direction required his imagination and ingenious application of Greek mathematics. In the medieval philosophy with which the Roman Church opposed him, and to which it still clings today, there is much metaphysical babbling about the essences of things, but it was the "heretic" Galileo who showed the world how to strip away the accidental properties of the objects and phenomena of our ordinary experience. Let's return to Pisa in 1590 as he begins to attack this problem.

The Continuous Abscissa
Those countable chimes of the tower bell seem to endow time with an accidental property of discreteness. But recognizing that free fall entails a continuous change of distance and speed, Galileo seeks to calculate distance and speed for every instant of fall.[9] To do this, he

---

7. Galileo describes his water clock: "As to the measure of time, we had a large pail filled with water and fastened from above, which had a slender tube affixed to its bottom, through which a narrow thread of water ran; this was received in a little beaker during the entire time that the ball descended along the channel or parts of it. The little amounts of water collected in this way were weighed from time to time on a delicate balance, the differences and ratios of the weights giving us the differences and ratios of the times, and with such precision that, as I have said, these operations repeated time and again never differed by any notable amount."; Galileo's *Two New Sciences*, Drake trans. (see Bibliography), p. 170.
8. Instantaneous change was not rigorously defined until the 19th century limit concept of Cauchy.
9. See Ch. 5, "Experimental Derivation of the Law of Falling Bodies".

must clear a mathematical hurdle, indeed an infinity of hurdles. The relationship of distance fallen to time over arbitrarily small intervals of fall contains infinitely many magnitudes that cannot be expressed as ratios of whole numbers. If he were to omit the Pythagorean "irrationals", his law would fail to model time as a continuum. Here, Eudoxos comes to the rescue. Ostilio Ricci had used Tartaglia's translation of Euclid's *Elements*, unique among texts at the time in its clear and correct treatment of Eudoxian proportions between continuous magnitudes.[10] Applying Eudoxian technique, Galileo defines equality of ratio so that the magnitudes need not be rational — and thus bridges over these gaps in rational numbers.[11]

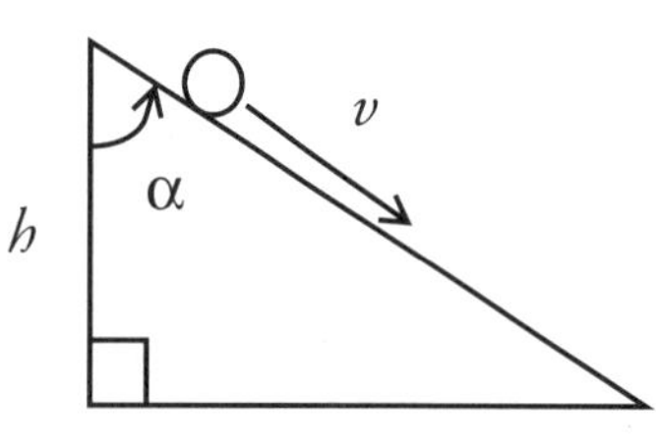

Besides this mathematical hurdle, there is a physical one: how to observe the behavior of bodies in free fall. To meet it, Galileo creates a method. He seizes on the inclined plane as a way to control the phenomenon and render it measurable. In the *De Motu*, he says that he has gone beyond Archimedes and all earlier philosophers to discuss ratios of motions of a body over a variety of inclined planes. First, he takes it as observable that the speed of descent would be greater as the angle of inclination to the horizontal ($90°-α$) is greater. Next, he adopts the postulate that a heavy body descends with the same force with which it resists rising when lifted. Just as cargo being loaded on a ship can be carried up an inclined plane with less effort than if lifted vertically, the change in rate of descent with inclination implies a difference of "weight" on various inclined planes. Following this line of thought, he conjectures that as the "weight" of the cargo is inversely proportional to the ratio that the length of the inclined plane bears to its vertical height, so the speed (v) of descent on an incline is to the speed of descent along the vertical as the ratio of the vertical height (h) to the length of the incline (i.e., the cosine of the angle $α$ of inclination to the vertical).

Applying this conjecture to compare ratios of speed acquired by the same body on two differently inclined planes of the same height leads him to the incorrect result that *speeds attained in fixed time intervals vary inversely in proportion to the lengths of the oblique paths.*[12] Though he will have to correct this early attempt at a mathematical law of falling bodies, an important thing to note here is that he employs a concept of weight or heaviness that is closer to what we mean today by "mass" and, being a forceless concept, is consistent with a kinematic treatment of motion.

---

10. See Ch. 3, "Tartaglia".

11. Cf. Euclid Book V and Galileo's *Two New Sciences*, Drake trans., p.114. Concerning strength of materials Galileo says: "But the ratio of GN to X is compounded from the ratio of GN to NC and that of NC to X, which is [that of] FB to BO. Hence the weight A has to the power that sustains it at G, the ratio compounded from [that of] GN to NC and [that of] FB to BO; ..." This statement applies Euclid V Prop. 4 to a Galilean theorem on statics; hence, a Galilean application of Eudoxian proportions.

Falling and Floating Trying to put myself in his place, I cannot help thinking how neat it would be if Archimedean hydrostatics, the laws of floating bodies with which Galileo is already familiar, would turn out isomorphic (or at least applicable in some way) to gravity.[13] He begins with a bold stroke: rejecting acceleration altogether, except initially. This allows him to treat free fall as if it results from the medium, like the pond in which the duck is floating. The duck is buoyed upward by the fluid, an effect that is uniform over time. Might a falling object be "buoyed" by its medium? Might one expect the effect on the object to be uniform, once any contrary forces imparted to it initially have decayed to zero?

You might object that a falling object seems to pick up speed. Fine. Have you watched it fall long enough? Will it reach a uniform state? Its initial increase of speed could be the result of a decaying resistance to motion imparted prior to release.[14] It turns out that this conjecture, Galileo's initial hunch about how bodies would fall in a resisting medium, made him right about the loss of acceleration in fall through a medium, but for a wrong reason. The real breakthrough had to wait until he considered the limiting case of free fall as the resistance of the medium tends to infinity: while the speed increases, what is the effect of the resisting medium (the increasing "relative wind" in the case of air) on acceleration? But that is getting ahead of our story.[15]

In the *De Motu,* Galileo denies the continuation of acceleration in free fall, but not on account of the medium. He assumes a contrary force impressed upon the object by initial conditions, an extension of Aristotle's idea that motion depends on a quality imparted to it by the mover. Building on Aristotelian physics, he first assigns the cause of natural motion, upward or downward, to the *density* of the moving body. Aristotle's idea that "heavier" objects seek lower positions in the universe is convenient to this argument. The speeds of objects falling through different media, Galileo concludes, are proportional to the differences between the density of each body and that of the medium through which it falls. While his base is Aristotelian, his extensions are Archimedean: he treats bodies in natural motion analogously to the weights of a balance. The analogy enables him to say that an object less dense than water cannot be completely submerged, just as the heavier cannot be raised by a lighter object equidistant from the fulcrum.

---

12. We will see later how Galileo corrects it (Ch. 5, "Inclined Planes and Pendula"). In practice, the extrapolation from rolling down an inclined plane to vertical free fall also requires taking into account the conversion of linear to rotational acceleration in the inclined plane that is not present in free fall, but this consideration involves Galilean physics beyond the *De Motu* and is ignored here (jm). Later, we will see how he calculated acceleration thus lost to fall (Ch. 5).

13. This is not necessarily Galileo's thought, but with the advantage of hindsight, I conjecture that it was.

14. similar in principle but opposite in effect to the quality called "virtus" that earlier theorists had suggested a mover communicates to the object being moved.

15. Cf. Chapter 12, "Falling Through the Air".

Can anything transit a vacuum? "No", said Aristotle, unknowingly denying the future possibility of space travel. His conclusion that speed of fall varies inversely with the density of the medium would preclude motion in a void, because the speed would have to be infinite. Galileo counters by converting a continuous increase of speed to an exercise in discrete arithmetic, showing that a speed can continue increasing indefinitely without reaching a finite ceiling.

The Pisan *De Motu* shows the first winding in the spiral staircase of his growing knowledge. Beyond the *De Motu*, Galileo will ascend one more winding before abandoning the concept that motion requires an external mover.

The *Mechanics*

In less than five years, an exciting thing will happen. Chronologically, it belongs to the next chapter of our story — but I want you to know about it while *De Motu* is fresh in your mind. Resisting blind faith in Aristotle, Galileo quietly reconciles the two approaches to handling the problems of mechanics: Archimedes' statics and Aristole's dynamics. He does it in a new manuscript: the *Mechanics* of 1594.

How Machines Work In Galileo's treatment of capacity to do work, a starting point for the conservation principle of modern physics, he calls upon Aristotle,[16] who saw that if a force can move an object through a distance, it can move an object $(1/n)$th its weight through n times that distance. If the converse is true, that it can move an object n times that weight through $(1/n)$th the distance, would that not lead to the "absurdity" that a man could move a ship?[17] Galileo responds by denying the absurdity: the machines of Archimedes, he explains, enable humans to move superhuman weights by dividing the weights indefinitely according to increments of greater distance. The ability to move a weight is limited only by the distance through which one can apply a contrary force.

We see here also the influence of Thales' axiomatic methodology in Galileo's early definitions of concepts fundamental to his mechanics:[18]

---

16. Cf. Stillman Drake in *Galileo Galilei on Motion and on Mechanics,* op. cit., p. 141 ff.

17. Aristotle, *Physics*, VII, 5, 249b, 30 ff. as quoted by Drake, ibid.

18. trans. by Drake, ibid. Drake notes that (1) Galileo "generalizes the term *moment* to include forces in other directions", (2) besides weight and position Galileo considered velocity a factor in *moment*, and (3) the definition of center of gravity is due to Commandino, but "that the center of gravity might be treated as the seat of every received force or impulse seems to be new with Galileo".

Heaviness (gravita): "that tendency to move naturally downward which, in solid bodies, is found to be caused by the greater or lesser abundance of matter (materia) of which they are constituted."

Moment: "the tendency to move downward caused not so much by the heaviness of the movable body as by the arrangement which different heavy bodies have among themselves." [19]

Center of gravity: "that point in every heavy body around which parts of equal moments are arranged."

The Perpetual Screw

Even a cursory glance at the *Mechanics* reveals its Archimedean influence. Here the machines of antiquity are analyzed by one principle, that of the lever.[20] Galileo shows us the principles behind the steelyard, the pulley and the screw. Even when we see the geometry underlying Archimedes' amazing screw for raising water, we can still marvel at it.[21]

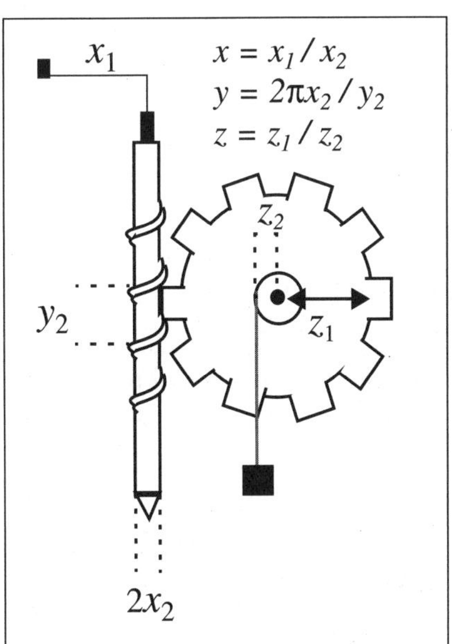

In his lecture notes, Galileo shows the capability of gear trains to multiply forces and demonstrates a screw mated to a load-bearing wheel with gear teeth so that the screw will turn the wheel perpetually, as long as a force exceeding a fraction $1/[c \cdot (xyz)]$ of the load L is applied to the screw. Here, L is the load, x is the radius of the rotary force applied to the screw divided by its radius, y is the ratio of the length of one revolution of the screw to the distance between the centers of the threads, z is the radius of the gear wheel divided by that of the axle to which the load is attached, and c is a proportionality constant.[22]

---

19. This 1594 definition can be compared with his 1612 definition of moment (see Ch. 8, "Floating Bodies").
20. A postulate from which Archimedes derived the law of the lever was: "Equal weights, suspended at equal distances from a center, will remain in equilibrium." Galileo later derived the same law in a different way than had Archimedes, using this postulate alone. Stillman Drake, *Cause, Experiment and Science,* op. cit., p. 212 and *Two New Sciences,* pp. 110-113.
21. Visitors to the Reuben H. Fleet Science Center, San Diego, have seen a model of Archimedes' water screw.
22. Each of x, y, and z is assumed greater than 1; sketch by Vincent Di Canzio modified in graphics imaging.

Conservation of Motion: an Inertial Concept

From his synthesis of Archimedean statics and Aristotelian dynamics, he will launch a science of motion built on his postulate of the indifference of an object to its state of motion. This postulate has one very non-Aristotelian consequence: an external mover is not necessary to produce motion. Speaking of a spherical ball resting on the surface of a frozen pond, Galileo enunciates his conservation-of-motion principle in the *Mechanics*:

> *"... on an exactly balanced surface the ball would remain indifferent and questioning between motion and rest, so that any the least force would be sufficient to move it, just as on the other hand any little resistance, such as that merely of the air that surrounds it, would be capable of holding it still. ... From this we may take the following conclusion as an indubitable axiom: That heavy bodies, all external and adventitious impediments being removed, can be moved in the plane of the horizon by any minimum force."*[23]

In more modern terms, the force required to produce motion of an object tends to vanish as the external impediments to motion, such as friction or deviation of its path from the horizontal, vanish. If we couple this idea with the axiom that motion can be a natural state of an object, then will not an object already in motion indefinitely continue in motion once the resistance of the medium is removed? While there is not an infinity of time to test this conclusion, we will see in the next Chapter how it emerges from Galileo's mature theory of which we see the seeds here in the *Mechanics*.

One Idealization, Two Principles of Nature

By a daring double stroke — the *De Motu* and the *Mechanics* — Galileo exercises a faculty of abstraction pushing human imagination to its limit. Bound up with technique improvised from the ancient Greek tool set, mathematical abstraction will (as we will see) ultimately enable him to wrest *two principles of nature*, a law of fall and a conservation-of-motion principle, out of extreme idealized ends, vertical and horizontal, of a continuum of experiments with inclined planes.

Mathematics and abstraction joined by Galileo will pave the way for Descartes and Newton to imagine infinite lines in idealized space. It is a combination that will be Galileo's signature and hallmark, to be indelibly stamped on his future work and on science that will come after him.

---

23. trans. by Drake. Drake cites a letter of 1607 from Castelli to Galileo attributing to Galileo the concept that to continue motion a mover is not necessary; see Drake in *Galileo Galilei on Motion and on Mechanics, op. cit.*, p. 171 note 26.

The Spiral Staircase

"Of the (ring) areas bounded by the turns of the spiral and the line segments on the initial line, the third is the double of the second, the fourth the triple, the fifth the quadruple, and thus each subsequent ring area will be the subsequent multiple of the second; the first, however, is one-sixth of the second." [24]

... Archimedes, *"On Spirals"*, *Proposition 27*

The *De Motu* of 1590-1592 is a bold and original theoretical model, a hypothesis which fails to square with Galileo's later observations of free fall. So much is this true that in the next turn of the spiral we find him embracing the law that speed is proportional to distance fallen, a law that, though incorrect, recognizes acceleration. Unsatisfied, he refrains from publishing the *De Motu*.

From our 20th-century perspective, watching him ascend the spiral staircase from the *De Motu* through the *Mechanics* of 1594 to his final work on physics, the *Discorsi* [25] of 1638, we can see modern science being born in the evolution of Galileo's thoughts. Despite the confusion of these early years at Pisa, one thing was clear to him then: the "Aristotelian" hypothesis assumes a speed proportional to the weight of the object, and that can be rather easily counter-demonstrated. To do the demonstration, and let nature persuade skeptics, it was convenient first to mount the spiral staircase of *la Torre Pendente*, the Leaning Tower. [26]

The Leaning Tower of Pisa

While its bell sounded the death knell of authoritarian physics, the tower at the Cathedral of Pisa became the birthplace of the modern science of motion. A pre-Galilean engineering project that went crazily but happily awry, it leans implausibly from the vertical. From a corner of the adjacent cathedral, the lean of the tower with respect to the wall of the cathedral can be measured directly; it is about $6°$, and until recently stabilized by intervention, increased a fraction of a degree each year at a decreasing rate of increase. [27] Evidently, the tower was built not only on soft ground, but on a slight slope in the direction of the lean. For the year

---

24. from *Archimedes*, E. J. Dijksterhuis (see Bibliography). The Archimedean spiral is generated by a radius vector r increasing uniformly with the vectorial angle, θ (i.e., $r = a\theta$). An analogous theorem for a logarithmic spiral (one whose radius vector increases exponentially with the vectorial angle) would yield the cross-sectional area progression for the chambers of the nautilus. In modern terms, if R_n = the area of the nth turn of an Archimedean spiral bounded by the initial line, then $R_n = 6(n-1)R_1$. My restatement of it illustrates how indebted we are to Vieta for compactness and clarity of mathematical expression.

25. *Discorsi e Dimostrazioni Matematiche, intorno a due nuove Scienze Attenenti alla Mecanica & i Movimenti Locali*, Elzevir, 1638 (abbreviated English title: *Discourses on Two New Sciences*) .

26. The Leaning Tower staircase is an extension of the two-dimensional Archimedean spiral to a third dimension, with a constant rate of increase in the third dimension and a constant radius of curvature. (Actually, the Leaning Tower staircase models an Archimedean screw for lifting water.)

1987, the increase was recorded as about 8 seconds of arc. In 1984, I was permitted to climb the spiral staircase to the top of the tower and take the two photographs that are the final ones of this Chapter and of Chapter 2. In 1993, I found the tower closed to visitors while a network of cabling, 400 tons of steel ingots, and surgically cautious earth-moving procedures were applied to stabilize the lean of the 15,000 ton marble structure.[28] Incredibly, the tower was able to stand on its own with no assistance for over 800 years. Let's turn the clock back to 1580, enter the cathedral here with the 16-year-old Galileo, replay one of the earliest legends about him, and then consider the evidence for the legend.

The Cathedral Lamp

While attending Mass he observes a cathedral lamp being lit, a process that requires drawing the lamp to one side, lighting it, and then releasing it. It then swings freely in a dampening series of simple harmonic oscillations. As the oscillations die down, the lamp moves through a shorter and shorter arc. If you are there with Galileo, and become curious how much time is required for each swing of the lamp, would it not be a reasonable starting assumption that the longer oscillations require more time than the shorter ones? After all, the lamp must move through a greater distance to complete a longer arc. Whatever may be his own guess, Galileo decides to settle the matter by timing the swing of the lamp. Wrist clocks with second hands haven't been invented yet. How does he do it? This story is based on a conversation between Galileo and Viviani, but the measurement technique is not part of Viviani's account. Whether as some say, by using his pulse, or by tapping out a beat, Galileo concludes that *the period of the oscillation is constant and independent of the amplitude*. The point of this legend is that despite experimental error [29] he is able to model the real phenomenon more accurately than if he were to assume that the period of an oscillation is proportional to its amplitude. His inattentiveness to the religious ceremony has enabled him to learn something about nature.

---

27. according to a graph of the tilt angle of this tower over the time since its construction, as shown in Levy and Salvadori, *Why Buildings Fall Down*, W. W. Norton & Co., NY, pp. 153-160. The top of the tower is about 53 meters above ground; the "drop point" for gravity demonstrations is about 48 meters above ground and leans out about 5 meters. At its summit, the tilt of the tower increased an average of more than a millimeter per year in the half century preceding stabilization.

28. This is perhaps unfortunate for the St. Vincent de Paul Society, whose collection box at the top of the tower solicited contributions from survivors of the climb.

29. That a simple pendulum is not exactly isochronous with respect to amplitude has been noted as early as 1639 by Mersenne and confirmed recently by MacLachlan; see MacLachlan, James, "Galileo's Experiments with Pendulums: Real and Imaginary", *Annals of Science*, 33 (1976), pp. 173-185. Wide-swinging pendulums actually "tick" more slowly than shorter-swinging ones. However, the excess time over a single small vibration is only 0.75% for amplitude of 20° and less than 3% at 40°. Professor Houston has observed that the dimensions of the cathedral are such that if the lamp is pulled aside to be lit, the angle of the supporting cable would be less than 20°. Though, as these facts imply, the anisochronism could not have been detected by Galileo in this manner, it follows from his later experiments and theory.

Galileo's Law of the Pendulum

Because Galileo did not comment on this story, attested by his student Viviani and by his contemporary Magalotti,[30] it is not certain that the legend is historically accurate. Still, it is consistent with his ostensible belief in the isochronism of the pendulum.[31] Skeptics have attempted to dismiss the story with the argument that the "Lamp of Galileo" in the Pisa cathedral was installed after the time of Viviani's account, as if the old lamp somehow obeyed different rules of nature than its replacement.[32] By participating in his father's musical experiments on tensions of equal length strings,[33] Galileo would have been able to observe nearly isochronous [34] pendulum effects on weights suspended from the strings. Whether or not the legend is accurate, we know that in his laboratory he studied the pendulum and that this research enabled him to announce a law relating the time of the swing to the length of the suspension: *the period of oscillation of a simple pendulum is proportional to the square root of the suspension length.*[35]

Leaning Tower Demonstration: a Legend?

Nothing is said in Galileo's pendulum law about the weight of the pendulum. But if, at the same suspension length, a heavy pendulum swings at the same rate as a lighter one, how could Aristotle be right that a heavy object falls faster than a lighter one? [36] Perhaps it was just this consideration that emboldened Galileo to demonstrate at the Leaning Tower the incorrectness of the law of falling bodies attributed to Aristotle. Despite conflicting speculation, the circumstantial evidence is compelling. Galileo was a native of Pisa, had lived about 18 of his 28 years of life in Pisa, and taught at Pisa. This is the tallest structure in Pisa, and stands adjacent to the University. It is leaning! What are the chances that a flamboyant

---

30. Filippo Magalotti, a prominent Florentine, only 6 years older than Galileo.
31. Galileo's law of falling bodies implies the anisochronism of the pendulum. Despite contrary assertions in *Two New Sciences*, Galileo's working papers bear evidence that he had measured and calculated deviations dependent on angular deflection in the oscillations of a pendulum but failed to publish them. See David K. Hill, *Pendulums and Planes: What Galileo Didn't Publish*, Nuncius Anno IX, 1994, fasc. 2.
32. as noted by Finocchiaro (see Bibliography).
33. ... in Vincenzio's victorious dispute with Zarlino about notes produced under certain tensions; See p. 23 above and Drake, *Galileo at Work,* pp. 16-17.
34. MacLachlan (op. cit., p.181) distinguishes three types of isochronism of the simple pendulum:
    "(1) that each successive small oscillation occupies exactly the same time; (2) that the period of oscillation of the pendulum is independent of the amplitude of the oscillations; and (3) that the period of oscillation of a pendulum is independent of all physical factors except length (and the acceleration of gravity), and in particular is independent of the weight of the bob." Galileo asserted that the simple pendulum is very nearly isochronous in sense (2), and by default of mentioning the weight of the bob, his pendulum law implicitly asserts isochronism in sense (3). However, Galileo knew more than he asserted; cf. Hill, op. cit.
35. For a derivation of Galileo's Pendulum Law from dimensional analysis, see Pólya's *How to Solve It*, Princeton U. Press, 2nd ed., 1971, p. 204.
36. Galileo did not, to my knowledge, raise this question in his writings, but since his pendulum discovery immediately suggests it to me, I conjecture that it did to him also.

young professor who had every motivation to perform gravity demonstrations for his disbelieving colleagues would not have used this tower for this purpose? Viviani affirmed that Galileo demonstrated more than once to the assembled faculty the incorrectness of the notion that heavier objects fall faster. Engraved on the inside wall of the Leaning Tower is a Latin inscription which I interpret that Galileo had here overturned the previous "law" of falling bodies. Yet Simon Stevin had done this in 1586 and Jerome Borri [37] not later than 1576. Galileo's original achievement was not the disproof of an incorrect law, but the proof of the correct law, which came later. No doubt he really gave this demonstration in order to convince others of what he, Borri and Stevin had already established. So let's throw historical caution to the winds and recount this (probably true) legend, complete with contrived details.

The Scattering of Loose Dust

Imagine the crowd of professors and students, all confident that the heavy weight would leave behind the lighter weight with an ever increasing lag. That has to happen, because they believe that the great Aristotle has declared it so. And who is this upstart young Professor examining nature in order to question, even to doubt, all the eminent authority that has enshrined Aristotle in a static body of doctrine? We can imagine the daring Galileo emerging from the spiral staircase near the top of the tower, probably with a referee that he would have brought along to make certain that he releases the weights at the same time.[38] The crowd looks on in a breathless hush, as Galileo extends his arms, suspending the weights over the railing at the maximum point of lean. His fingers part. The contest begins. The silence of suspense is severed by the nearly simultaneous [39] thud of the two weights upon the ground, and the scattering of loose dust that falls back to the earth, symbolically burying the nearly 2,000 year old "law" of falling bodies.[40] The shock of that collision at the base of the tower must be felt by the "Aristotelians", setting up a frantic search for reasons why, in this particular case, the weights seem to fall together. For, if they should acknowledge that Galileo is correct,

---

37. Settle, Thomas B., "Galileo's Use of Experiment as a Tool of Investigation", *Galileo: Man of Science,* McMullin ed. (see Bib.) p. 325. Settle cites Drabkin's translation of Galileo's *De Motu*, p. 106 and note.
38. There is no historical data known to me that would let us be sure what objects Galileo dropped, but it is probable that they were not meatballs and certain that they were not car keys.
39. Galileo does not claim, as is sometimes stated, that different weights falling from the height of the tower strike the ground simultaneously. What he says is that, on impact of the first weight, the second weight lags about two fingers' breadth from 100 cubits (roughly the Leaning Tower's height). Citing the example that when gold, the densest substance, is beaten into a thin leaf, it floats in the air, he attributes the lag to the difference in buoyancy of the objects. Galileo evidently assumed that air is a fluid and would follow laws isomorphic to those of objects buoyed by water, with the observed separation being consistent with the laws and the variables.
40. Later, we shall see that Galileo's Leaning Tower demonstration has been successfully repeated on the moon, and from even as far away as the supernova of 1987, when allowance is made for the variability of acceleration in a force field that is not constant (cf. Chapter 12 "Modern Corroborations of Galileo's Law of Falling Bodies".

how many more of their Master's doctrines might become suspect? And so, rather than congratulate Galileo, some of them hiss at him in his later public lectures.

That is the legend, but however it really happened, there is not much doubt that Galileo made the test, as he told Viviani (in response to a letter of 13 March 1641 from Vincenzio Renieri, Galileo's successor at Pisa) that he had dropped two objects of the same materials but of different weights from the tower.[41] What's the probability he had no doubting observers?

Against the Wearing of the Toga

Preferring issues that are philosophical rather than sartorial, Galileo chafes at the stuffy pomposity of an administration that would prescribe what type of academic robe is required of his position. He reacts by writing a ribald poem *Contro Il Portare La Toga*. Among his arguments "against the wearing of the toga" is that it puts wearers of heavy academic gowns at a disadvantage when visiting loose women.[42]

Conflicts with Pisan Philosophers and Royalty

Galileo's championing of Archimedean science has clashed with the Aristotelian dogmatism of the faculty at Pisa. His manner of dealing with philosophical adversaries has alienated them. With the same wit that attracts friends, he creates enemies by cutting down opponents in debate with laser-like logic and searing sarcasm. With any new idea, he is painstakingly cautious. But having convinced himself, he advances it with force and daring. Too candidly criticizing a machine invented by Don Giovanni de' Medici for dredging the harbor at Livorno, Galileo angers an influential member of the royal family.[43] The incident is alarming to Guidobaldo. Soon, everyone [44] seems so hostile to him that the young professor expects his contract not to be renewed; he begins to look elsewhere for employment.

Promotion for the "Troublemaker"

While Galileo is looking around, the University of Padova,[45] where Copernicus studied, has been recruiting for a vacant chair of Mathematics. A worthy successor to the post has long been sought, and when Galileo hears of the opportunity from Guidobaldo, he at once applies for it. Though scorned by the "philosophy" faculty at Pisa, Galileo has continued to garner a reputation among mathematicians for original thinking and for being correct. These same two attributes of Galileo, liabilities among

---

41. Stillman Drake, *Galileo at Work,* pp. 414-415.
42. Cf. Ludovico Geymonat, *Galileo Galilei,* McGraw-Hill, NY, 1965, Chapter 1 footnote 13.
43. Allan-Olney, op. cit., p. 12.
44. except Jacopo Mazzoni, a newly hired philosophy professor with enlightened views.
45. Strengthening its mathematical faculty would presumably have helped the University of Padova to recover from the embarrassment of having opposed the use of "Arabic" numeration as late as the mid-fourteenth century (cf. Rouse Ball, op. cit.)

the group whose method of validating knowledge is to consult authority, become assets among the mathematicians, who are more impressed by originality and correctness of results. With the help of Guidobaldo, who had secured his position at Pisa, as well as of Valori [46] and of Pinelli ,[47] he obtains the appointment at Padova, where free expression of thought and international communication among scholars are encouraged.

Looking from high above Galileo's Toscana, independent but influenced by the Papal States, across that fertile plain flanked by the Ligurian Sea on the west, the Appenines that wrap around Pisa from the north toward Florence and smaller hills to the south, one sees the Papal States that had expanded far north of the Vatican. Beyond the horizon and the northern extension of the Papal States lies the Republic of Venice, commercial gateway to the Adriatic.

Will the fruits of Galileo's inventiveness emerge as he moves to a region of relatively free commerce, where originality in thought or in production is relatively unhindered by an authoritative social institution? Those of us who use binoculars, thermometers, pendulum clocks, or computing machines are affected by events this story is about to reveal as it resumes on a happier trend on the road from Pisa to Padova. As Galileo departs his ancestral homeland for Padova, he begins a happy period of his life, enduring but not permanent. It is the summer of 1592, a moment in history when Shakespeare is making famous the opening line:

> *"Now is the winter of our discontent*
> *Made glorious summer by this sun of York;*
> *And all the clouds that lour'd upon our house*
> *In the deep bosom of the ocean buried."* [48]

In stark contrast to that villain whose homicidal character Shakespeare portrays, Galileo is about to convert the intellectually barren winter of medieval philosophy into a life-enhancing and durable summer of civilization, warmed by the radiance of his coming discoveries and inventions.

---

46. Consul of the Florentine Academy in 1588, and doubtless well aware of Galileo's assistance and presentation concerning Dante's *Inferno* to that eminent body.
47. a cultural and literary leader who enjoyed strong ties to scholars and churchmen, and who would later become one of Galileo's strongest supporters.
48. Opening lines of *King Richard III*, 1592.

Bell Tower of Pisa,
Death Knell
   of Authoritarian Physics
   (photo by author)

Vector indicates path of
drop; angle of tower lean
can be measured from the
parallelogram.

"Lamp of Galileo" at
Pisa Cathedral
(photo by author)

Inscription, dated MDCCCXXXVIIII, Inside Base of Leaning Tower,
"GALILEUS GALILEIUS EXPERIMENTIS E SUMMA HAC TURRE
SUPER GRAVIUM CORPORUM LAPSU ..."
Refers to Galileo's Counterdemonstration of Mechanical Laws of Motion
"through experiments on the fall of heavy bodies from the top of this tower"
(photo by author)

Leaning Tower Bell, Facing the
Point of Release of the Weights
(photo by author)

Chapter 5. The Royal Road to Geometry
Professor Galilei at Padova (1592 - 1610)

" ... The place is stored with great variety of sextants, quadrants, telescopes, astrolabes, and other astronomical instruments. But the greatest curiosity, upon which the fate of the island depends, is a load-stone of a prodigious size, in shape resembling a weaver's shuttle... By means of this load-stone, the island is made to rise and fall, and move from one place to another. ... This load-stone is under the care of certain astronomers, who from time to time give it such positions as the monarch directs. ... They have ... discovered two ... satellites, which revolve about Mars; ... so that the squares of their periodical times, are very near in the same proportion with the cubes of their distance from the center of Mars; ..." [1]

... Jonathan Swift, *Gulliver's Travels, Part III Chapter III*

In the summer of 1592, the wheels of a coach rolling through the three hundred kilometer slice of countryside between Pisa and Padova, bearing in their visible turning motion the man who would reveal invisible turning motions of the Earth, finally come to rest at the gates of one of the most renowned universities in all of Europe. Here, too, it is soon to be that this traveller on the public road, having undertaken a new professorship of mathematics, will develop and announce his "royal road to geometry". As he alights, eager to teach at the same university where Copernicus studied,[2] he can scarcely imagine the chain of events that will bring him enduring fame. At Pisa, he has developed a seasoned and captivating lecturing style. On 7 December 1592, he delivers his inaugural lecture to a large audience in the Great Hall, a room which will still exist in the 20th century. Attracting students from all over Europe to attend his daily classes at three o'clock in the afternoon, here he toils the next eighteen years, a fruitful period of teaching and discovery. During this time also, he invents revolutionary things and sends the shock waves of his telescopic discoveries in all directions from this sleepy little suburb of the Venetian commercial center.

But first he finishes a task left over from Pisa.

---

1. Swift here notes that the moons of Mars follow Kepler's 3rd Law, implying that the behavior of objects under gravity, including Galileo's and Kepler's laws, is universal. *Gulliver's Travels* was published in 1727, the year that Newton died.
2. Here too, William Harvey, best known today for his discovery of blood circulation, will study under Fabricius, a colleague of Galileo and pioneer sunspot researcher.

Galileo's Padovan Science

It is one thing to ascend the Leaning Tower and disprove notions about objects in free fall. It is quite another to discover and prove the correct law that governs their motion, as Galileo will do at Padova. In order to build a bridge to his new mechanics on mutually supporting members of theory and experimental results, he will measure what no one before him has measured: the relationship among variables that describe free fall. He has consistently observed that light and heavy weights fall together, but — how can this be? To answer the question, it may be helpful to relate this phenomenon to the behavior of an Archimedean machine.

Levers, Inclined Planes, and Falling Bodies
While embarking on a new science of kinematics, Galileo continues to lecture on the old astronomy. In 1598, he abruptly suspends his astronomy lectures in order to lecture on "Questions in Mechanics", an outgrowth of his 1594 papers. When a machine, such as an Archimedean lever, is used to do work, there is a price paid for the increase in mass that can be moved by a given applied force. The force must be extended over a greater distance.[3] In modern terms, the end result is that work done is the same in this case as in direct application of a force not extended through a distance. Galileo generalizes that one cannot cheat nature with a machine. Here we catch an early glimpse of a physics beyond his time, an insight that is part of the conceptual backdrop of the modern conservation laws of physics. With a fresh perspective on experimentation gleaned from his analysis of machines, he returns to his study of accelerated motion and its application to free fall.

Relationships among force, distance and capacity to do work are preserved by alterations in leverage. In the case of a falling body, would those between acceleration and distance fallen be preserved by an inclined plane? Galileo sees in the inclined plane the experimental method of completing the theory of falling bodies. Sometimes it is instructive to work backward from a result. Let's first describe the law of freely falling bodies that follows from Galileo's experiments, and then, with the destination clearly in sight, look at how he got there. We begin this retrograde path here and will finish it in Chapter 11 after discussing his mature work. For ease of understanding, we can express Galileo's law of falling bodies in modern terms that he would not recognize, using the notation of Leibniz,[4] in which the symbols ds/dt and d^2s/dt^2 represent

---

3. By Drake's rendering (*Galileo at Work*, p. 56, see Bibliography), "the price paid is a loss in speed equivalent to the gain in power", an unfortunate mistake that may mislead students of Galileo. Power is the time rate of doing work, and does not realize a gain from a loss in the speed of the operation.

4. Leibniz, co-developer of the calculus, tried to found an Academy of Science at Vienna, but was prevented by Jesuits. A.D. White (cf. Bibliography) says of this: "the priests ... would not allow him the privilege of aiding his fellow men to ascertain God's truths revealed in nature".

instantaneous speed and acceleration, respectively. For objects falling short distances near Earth's surface: [5]

The distance fallen is proportional to the square of the time: $\quad s = (g/2)t^2$

Instantaneous velocity is proportional only to elapsed time: $\quad \dfrac{ds}{dt} = g \cdot t$

Objects near Earth's surface fall with constant acceleration: $\quad \dfrac{d^2 s}{dt^2} = g$

... from which it follows that: $\quad v = \dfrac{ds}{dt} = \pm\sqrt{2gs}$.

These are equivalent statements. Before Newton and Leibniz show their mathematical equivalence, Galileo experimentally verifies the first and shows the physical equivalence of the second and third with the first. To demonstrate the second statement, he generalizes the medieval 1:3 relationship between speed of fall in the first and second time intervals, to the "odd-number rule" for spaces traversed: *Over equal time intervals, the speed of free fall increases in arithmetic progression of the odd-number sequence beginning with unity.* Since the rule generates the distance sequence of squares 1, 4, 9, 16, ... , it is clear that the differences in spaces traversed for successive time intervals is just a multiple of the odd-number sequence. Galileo's theory differs from that of his precursors in that it contains the first coherent general formulation and experimental verification for this family of laws under various initial conditions. He does not stop there.

Propping up the mystery of light and heavy weights falling together is that other pillar of academic dogma, the ancient concept of an object's natural state of motion, summed up by *quod movetur, ab alio movetur.* [6] For the ancients, it was easy to assume that the Earth requires no external mover because it does not move.

Quod Movetur, ab Alio Movetur

Had not Galileo's teacher, Archimedes, declared: "Give me a place to stand, and I will move the Earth"? Conceptually, Galileo is converging on nothing less than setting the whole Earth in motion. Under the old concept of Aristotle, rest is a natural state of being motionless. A force is required to produce motion. There is a special velocity state, the zero velocity. Zero is natural, every other velocity is unnatural and must be produced by an energy expenditure. Is this not analogous to

---

5. Recalling that t = elapsed time of fall; s = distance of falling object from its initial release point, at which t = s = ds/dt = 0; g = acceleration due to gravity, commensurate with s and t . In cgs units, g = 980.665 cm per sec per sec is accepted as standard for Earth's surface (sea level at 45° latitude), while the measured value undergoes slight positional variations.
6. "Whatever is moved, is moved by another": the ancient concept of motion.

the idea that there is a special place, the center of the universe, which happens to be where Earth is? The idea that whatever moves is moved by another continued right up through the Parisian school that preceded Galileo, in Buridan's theory of *impetus*. An impetus impressed in a falling object would endure unless opposed by another force. The object's speed, acquired initially by a downward motion due to heaviness, adds impetus to the body's natural motion. Buridan's impetus replaced (by an impressed force) the push a medium was supposed to exert on a thrown object, but it still required an external agent to impress the impetus.

What Galileo does now is to invert this concept. Assume first that Aristotle is right. Then, motion *ad infinitum* ought to be impossible. But as Galileo watches his bronze ball roll down an inclined plane, across a horizontal beam, and then onto another plane inclined upward, he sees and recognizes that the substitution of a retarding force for the original downward acceleration does not immediately bring the rolling object to its "natural" state of rest. He asks himself: If Aristotle were right, why didn't the mere removal of moving force directed along the path of

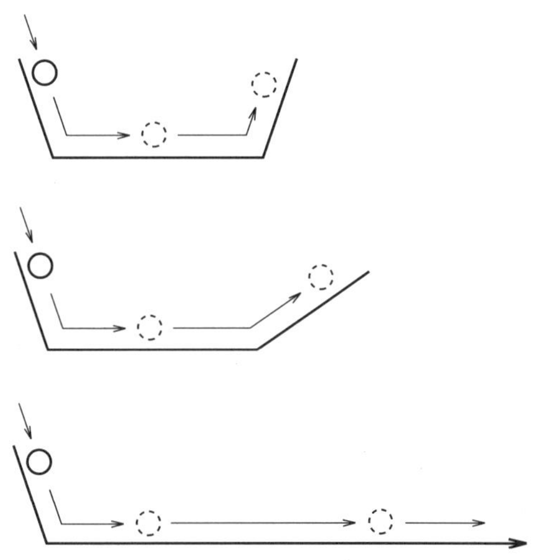

motion bring the object to rest? Even while moving against the direction of downward acceleration, the ball continues to roll. It reaches nearly the same height from which it started. What if the ball is deflected to roll onto a more shallow plane? He sees now that in pursuit of the height from which it was originally released, the ball rolls farther along a plane as that plane approaches the horizontal. A merely astute experimenter might have noted these data in his logbook, stopped there, and congratulated himself about the questions his experiment has raised against the ancient accepted idea that a force is required to sustain motion. Or perhaps he would have tried new experiments, with different variables, noting in each case that the object never regains its original height and does, after all, finally attain its "natural" state of rest (as Aristotle said it would) after struggling against the redirected force.

But not Galileo. He does not try to resolve a paradox with more experimentation. A physical paradox is an event for him like a contradiction in geometry. It implies an error in a starting premise and requires a new theoretical framework. The experiment had been designed with Aristotle's starting premise in mind: that rest is a natural state of motion. It is this premise and not the original experiment that he rejects. While retaining that experiment, he pushes it to limits that observation cannot reach, extending it into a new dimension: the dimension of imagination. In Galileo's fertile imagination, an infinity of repetitions of the experiment can be tried, stretching the variables beyond values measurable in his workshop.

One of those variables is the difference between rolling descent and ideal (frictionless sliding) descent or free fall. Recall from Chapter 4 Galileo's assertion in the *De Motu*:

Inclined Planes and Pendula

(1) Speed of descent on the incline is to speed of descent along the vertical as the cosine of the angle to the vertical.

In his tests at Padova he recognizes (1) as mistaken in that it relates speeds rather than times of fall to the length and height of the incline. Correcting (1) is this formulation, equivalent to what he discovered by 1602:

(2a) "*The ratio of times of* [ideal] *descent over planes differing in incline and length, and of unequal heights, is compounded from the ratio of lengths of those planes and from the inverse ratio of the square roots of their heights*",[7]

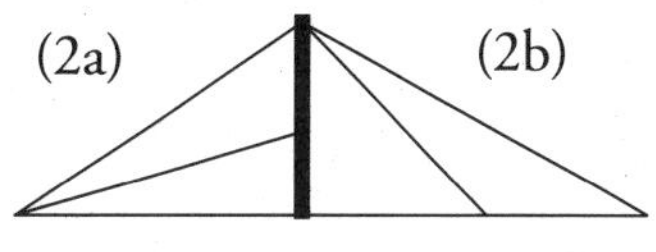

a simple consequence of which is:

(2b) Descent times are proportional to the lengths of the inclined planes if their heights are equal.

He proceeds to discover his "rule of chords":[8]

(3) "*If, from the highest or lowest point of a vertical circle, any inclined planes whatever are drawn to its circumference, the times of* [ideal] *descent through these will be equal.*"

An ordinary experimenter might be tempted to extrapolate, using laws (2) and (3), from inclined plane experiments to inferences about free fall along the vertical. But these laws express ideal cases in which no factors affect descent on an incline other than those which affect it in free fall. Confronted with a choice of time metrics, Galileo makes a decision that will have a consequence beyond the ease with which accidental effects on real inclines can be separated from ideal descent. He chooses to measure descent times by the "ticks" of a pendulum. Having been released at time T, the bob would audibly strike a surface at time T+T', so that T' measures one "tick" of time. How do the times of

---

7. published later in Galileo's *Two New Sciences*. See Drake's trans., p. 177. Algebraically, $t_1/t_2 = (l_1/l_2) \cdot \sqrt{h_2/h_1}$.

8. Ibid., Theorem VI, p. 178.

descent along the chords of law (3) compare to those through the circular path of a pendulum bob?

Remember Archimedes' proof of the area of a circle? He inscribed a polygon in a circle and increased the number of sides to infinity. Galileo recognizes that the arc through which a pendulum bob falls in one tick is the limit of a sequence of conjugate chords representing a polygon inscribed in that arc and increasing in number of sides. He uses his measurements of times of descent on inclined planes to 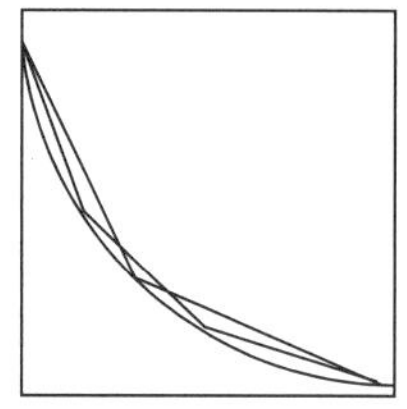 deduce the time of descent through an arc of a pendulum bob as a series of inclined plane traversals. His results show that the descent time for the chord traversing a quarter circle is a multiple about 1.3 times that for the arc.[9]

Through a small arc, the descent time to the vertical of his pendulum bob becomes $\pi/(2\sqrt{2})$ times the time of free fall through a distance equal to the length of the pendulum.[10] In the data of this relationship he will ultimately recognize the time-squared law of fall — not surprisingly to us who with hindsight can derive this same quantity by equating the acceleration due to gravity in his pendulum law and in his law of falling bodies.[11] Combining experiment with Eudoxian/Archimedean technique, he thus arrives at time as a function of distance through a unit radius.

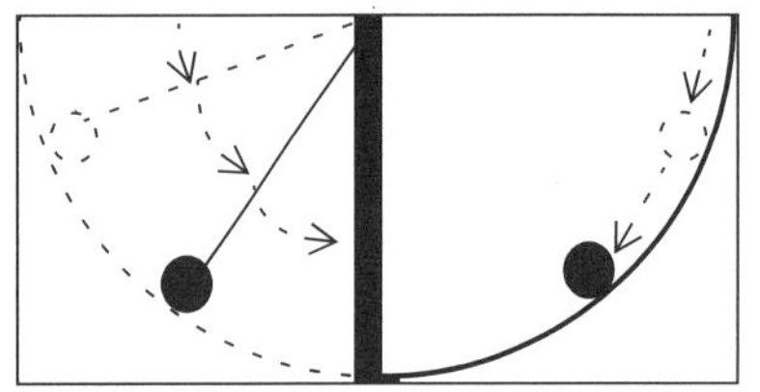

Rolling Balls Gather No Moss But Leak Acceleration

Having arrived at this relationship between the tick of a pendulum and time of free fall, an experimenter might assume that it applies equally to timings of bronze balls rolling through the arc traversed by the pendulum. But not Galileo. His working papers contain results implying that, to test for other variables, he constructs a track in the shape of a circular arc and times rolling descent with a pendulum of the same radius,[12] obtaining a ratio of about 6/5. In modern terms, this means that the spinning of the ball converts gravitational acceleration to rotational acceleration so as to retard the descent speed by nearly 20% relative to the speed of the pendulum bob. By choosing pendulum motion as a baseline, he has tricked nature into yielding a result equivalent to that which he would have gotten had it been possible for him to construct a frictionless inclined plane down which the ball would slide without rolling.

---

9. See David K. Hill, op. cit., p. 507.
10. Cf. Stillman Drake, *Galileo: Pioneer Scientist*, op. cit., p. 21.
11. By Galileo's pendulum law $T = (\pi/2)\sqrt{L/g}$ and his law of fall $t = \sqrt{2 \cdot s/g}$, setting the distance fallen s equal to the pendulum length L yields the ratio $T/t = \pi/(2\sqrt{2})$.
12. See David K. Hill, op. cit., pp. 510 ff.

Another variable retarding speed of fall is the acceleration that is not converted to rotation but is dissipated as friction. In the real experiment, more highly polished bronze balls and smoother tracks **Greasing the Skids** could have been tried, up to a point. In the ideal experiment, friction can be reduced toward the vanishing point. But what if the air acts as an inhibitor of motion, rather than a sustaining medium, as Aristotle believes? In the ideal experiment, the resistance of the air can be made to vanish too. Here he sees, in his mind's eye, that in the absence of resisting media, the ball indeed attains its original height.

Having removed friction as an extraneous variable, he pushes another variable toward its limit: the inclination of the second plane! As the ball rolls onto ever more shallow planes, it must travel farther to reach its original height, and in the case of a horizontal plane, it can never regain its original height and will roll on eternally, even though the initial force that produced motion has been completely removed. In order to remain forever horizontal on a spherical Earth, the track must curve into a closed ring circumnavigating the entire planet.[13]

Are we to believe with Aristotle that such a ball, rolling relentlessly forward on a frictionless horizontal surface, is being pushed by air? If **Galilean Impetus** so, then we must be prepared to believe that a mover can exert a moving force eternally. From a modern viewpoint, an eternal energy expenditure would deny the constancy of total energy. From the viewpoint of Galileo, who developed the more fundamental concept of conserved mechanical work in his *Mechanics* circa 1604, the air cannot exert a force on the rolling object greater than that object's resistance to motion (zero in the absence of a resisting medium). Hence the medium through which an object moves does not sustain its motion, as Aristotle implied, but acts as an interference. His new viewpoint implies that continued motion requires no more force than suffices to overcome the interfering factors. Whatever uniform velocity a body may have, be it zero (rest) or any other number, is a natural state. In summary:

> *A force is required, not to produce motion,*
> *but to produce any change in motion.* [14]

---

13. In an ideal experiment, one can easily construct a plane of infinite dimension tangent to Earth at the track's junction with the inclined plane. In order to remove any cause of retardation due to gravity, Galileo here signifies in the phrase "horizontal plane" the arc of the Earth, locus of tangency for the infinite sequence of horizon planes through which the ball rolls as its "geocentric speed" is conserved.

14. I do not know the actual chain of Galileo's thoughts as he led himself to this great discovery. Only the end result exists in his two dialogues. The preceding paragraphs are my attempt to reconstruct the approximate direction of Galileo's thinking from the starting point of his precursors to that end result based on inductive inference concerning his methods.

He will wait until 1612 to announce the conservation of motion. In the first public statement of it, he says that a body on a horizontal surface will conserve its state of motion or of rest; viz., a ship "having once received some *impetus* through the tranquil sea, would move continually around our globe without ever stopping; and placed at rest it would perpetually remain at rest, if in the first case all extrinsic impediments could be removed, and in the second case no external cause of motion were added." [15] Galileo's principle of impetus can be restated:

> *A body will retain its state of rest or of uniform motion indefinitely unless acted upon by an external force.*

In 1644, René Descartes will speak of the tendency of a body when once moved always to continue in the same state of motion. Later, Newton and Einstein each will formulate equivalent alternative statements of the inertia principle, generalized to all objects in the universe:

> *"Every body continues in its state of rest, or of uniform motion in a right line, unless it is compelled to change that state by forces impressed upon it."* [16]

> *"A body removed sufficiently far from other bodies continues in a state of rest or of uniform motion in a straight line."* [17]

> *"There exists a coordinate system (or coordinate systems) with respect to which all bodies not subjected to forces are unaccelerated."* [18]

With this peek ahead at the evolution of a concept known today as inertia, we see that Galilean impetus is a step toward it. Galileo has achieved the critical breakthrough, the

---

15. Emphasis mine. See Galileo's 14 August 1612 letter to Mark Welser, Drake's *Discoveries and Opinions*, pp. 113-114, see Bib. There is a debate in the literature about whether the indefinite and uniform continuance of motion along a horizontal plane refers to a single tangent plane (rectilinear inertia) or an infinity of planes tangent to the local horizon as the body moves (circular inertia). A modern interpretation of Galileo's statement that velocity is permanently maintained would rule out circular inertia, since that would imply continual change of direction. Drake has stated that in "applying the principle to physical problems, however, he [Galileo] included the more important case of bodies moving uniformly along straight lines, neglecting the force of gravitation"; ibid. p. 113 note 8. The debate is irrelevant to my point: Galileo pioneered the concept that a body will retain its state of rest or of uniform motion indefinitely in the absence of an external force. Here we find something different from Buridan's "impetus", a quality of a projected body which is an indefinitely lingering cause of both natural and forced (accelerated) motion produced by the mover and varying directly with speed.
16. Newton's statement in his *Mathematical Principles of Natural Philosophy* (from the Motte/Cajori translation in *Great Books of the Western World*, Encyclopaedia Brittanica, Inc., 1952)
17. Einstein's original statement from *Relativity*, Crown Publishers, NY, 1951. Einstein refers to this as "the fundamental law of mechanics of Galilei-Newton, which is known as the *law of inertia*".
18. A more precise version of the Einstein statement, quoted from Peter Gabriel Bergmann, *An Introduction to the Theory of Relativity*, Prentice-Hall, NY, 1946. Einstein collaborated with Bergmann on this book and said of its treatment that it is "systematic and logically complete"; therefore, I presume that he either inspired or would concur with this restatement of the inertia law.

principle of conserved motion, that enables the modern concept of inertia. But he is not the author of the latter concept. Indeed, he holds at least initially a postulate (conservation of geocentric speed) with which it is incompatible. This incompatibility is a theoretical difficulty. And with Newton's and Einstein's statements of the inertial principle, quoted above, there are observational difficulties. Can we observe purely inertial motion in a Newtonian universe where there is no hiding place from gravitation — what body is not acted upon by an external force? And how can stable rectilinear motion exist in a universe fabricated of Einstein's gravitationally warped spacetime — where is there a straight line geodesic? Is there any body "sufficiently far removed from other bodies"? In Chapter 11, we will revisit this topic when we go "beer sliding at the lunar pub".

Galilean impetus, as a conserver of motion, will figure into Newton's reasoning about inertial mass. If a force is measured by the acceleration it produces, then the force required to **Impetus and Inertial Mass** overcome impetus — either in the case of rest or of rectilinear motion — and produce a change in velocity is proportional to the rate of that change. We can test the mass of an object relative to another by determining the resistance of each object to acceleration under an applied force. The constant of proportionality between an applied force and the resulting acceleration gives us a way to characterize the mass of an object. Though he extended Galileo's formulation, Newton will generously credit Galileo with using the law of proportionality between change of motion and applied force to discover the laws of fall.[19]

When a body falls, a gravitational force overcomes its tendency to conserve motion. We call the resulting change in velocity its "acceleration due to gravity". Acceleration **Impetus and Gravitational Mass** due to gravity decreases in proportion to *inertial mass* and increases in proportion to *gravitational mass*. In one object, are these not equivalent? Binding light and heavy objects together in a thought experiment,[20] Galileo finds that acceleration is independent of the mass of a falling object. Newton will arrive at the same result by equating the gravitational and inertial mass.[21] In Newton's thought, Galilean impetus conceptually links inertial mass to the numerically equivalent gravitational mass.

---

19. In Newton's formulation, this is his second Law of Motion: "The change of motion is proportional to the motive force impressed; and is made in the direction of the right line in which that force is impressed" (*Mathematical Principles of Natural Philosophy*).
20. Galileo tied together a light and heavy object, deriving from the Aristotelian assumption (that the heavier weight would fall faster) the contradiction that the still-heavier pair would be retarded by the lighter weight and thus fall more slowly; cf. *Two New Sciences*, Drake trans. p. 67.

Leaning Tower Revisited This new relationship of mass to acceleration will enable Newton to combine the gravitational view of the weight of an object with the common notion of mass as a quantity of matter to yield three characterizations of the mass of an object:

Table 1: Proportionality Characterizations of Mass $a \propto b$ denotes "a is proportional to b"

Quantitative Mass	$Mq \propto V \cdot D$	volume times density
Gravitational Mass	$M_g \propto F_g$	Force of attraction to a given gravitational attractor (e.g., the Earth)
Inertial Mass	$M_i \propto F_i$	Force that accelerates the object

Equating gravitational mass to inertial mass, one finds that a more massive body supplies a proportionately greater component of the mutual force of attraction required to overcome its greater resistance to acceleration, so that acceleration will be constant.[22] And since two bodies starting from rest and accelerating at a constant rate will, in the absence of interference, fall together, we have here an explanation of the Leaning Tower demonstration.[23] Conservation of motion, the Galilean concept that a force is required to produce not a motion but an acceleration, dispenses with the Aristotelian requirement for a medium through which an object moves. After Galileo and Newton, the medium is an interference instead of a requisite for motion. Gone is the dependence on falling speed of the weight of a body. And gone is the mysterious function of time.[24]

In thought experiments Galileo often represented natural circular motions without reference to forces. Newton explained these motions by forces operating at a distance. Later, Einstein replaced the mysterious Newtonian forces by spacetime curvature. In the meantime, Newton postulated the reciprocality of gravitational force, added the parameter of distance between attractors, showed the equivalence of a spherical mass to a point mass at its center, and extended the result to all masses in the universe. But that is getting quite a bit ahead of our story. Suffice it at this point to say that Galileo's results at Padova enabled Newton to determine that the Earth attracts a falling object proportionally to the amount of matter it contains.

---

21. Unable experimentally to alter the Earth's gravity so as to treat it as a variable, Galileo treated it as a constant. By combining the law of proportionality between change of motion and applied force, which Newton credited to Galileo, with his own law of universal gravitation, Newton found that the acceleration due to gravity increases with the mass of the attractor (Earth) and decreases with the square of the distance between the attractor and the falling object.
22. For an algebraic demonstration, see Chapter 11 "Falling Bodies Revisited".
23. Newton's Universal Gravitation Law: see Ch. 8 for a statement of the Law.
24. from the Aristotle law, as I represented it algebraically (cf. Chapter 4).

In his early years at Padova, Galileo idealized a bronze ball rolling so that changes in its acceleration vector with the slope of the

Experimental Derivation of the Law of Falling Bodies

inclined plane are pushed to a limit. He doubtless realized his own innovation in method, as he later occasioned Salviati to remark:

> *"So that we may say the door is now opened, for the first time, to a new method fraught with numerous and wonderful results which in future years will command the attention of other minds."* [25]

Remember the "mean speed theorem" developed by Paris and Oxford physicists for general cases of acceleration?[26] For Galileo, here is a mathematical construct that, if falling bodies accelerate uniformly, ought to apply to them. He can use it to derive his law of fall as a theoretical consequence, a derivation that he later will publish in his final work on physics. But no physical theory can rest solely on theoretical derivations, and though the imaginative visualization of idealized experiments informs Galileo's theoretical vision, its end product is a conjecture supported by the interplay of observation and prior theory, still to be empirically tested. In his early tests at Padova,[27] he repeatedly rolls a bronze ball about 19 mm in diameter down a groove (8 to 9 mm wide) of an inclined plane about 2 meters long and tilted about 1.7° from the horizontal. Probably by singing a marching tune of half-second beats and tying gut-string markers to the inclined plane to mark the position of the rolling ball at each beat, he can adjust the marker positions until the thump of the ball striking each marker synchronizes with the notes of the song.

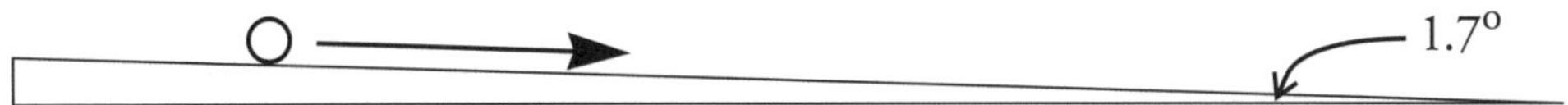

Denoting positions at equal times, the marker separations give distance as a function of time. Galileo notices that the separations increase as the odd number sequence, implying that the distance fallen is proportional to the time squared. In later experiments, instead of equalizing the time intervals, he will measure them with a water clock whose flow rate will be nearly 3 fluid-ounces per second, or with his pulse. In more than a hundred samplings, he will never find the time of fall to vary by more than a tenth of a pulsebeat from that predicted by the time-squared relationship.[28]

The method of calculating speed of fall came as the culmination of many theorems

---

25. Galilei's *Discorsi*, near end of the Third Day. Translation by Stillman Drake.
26. If not, see Ch. 3 "Oresme, DaVinci".
27. Details of these tests are based on Drake's research; cf. *History of Free Fall*, see Bibliography.

Galileo had written out describing in terms of his experimental results how to calculate times along straight or curved inclines from any higher point to any lower point. In 1607 this research culminated in Galileo's note 2 on folio 164:

> "The moments of velocità of things falling from a height are to one another as the *square roots* of the distances traversed [from rest]" [29]

Having inferred his principle of impetus from one end of his inclined plane experiments, Galileo has arrived at his law of falling bodies from the other. Later, Newton will see the law relating acceleration to force and mass in the synthesis of the two ends of Galileo's experiments. We will resume the topic of how Galileo developed his law of falling bodies in Chapter 11 ("Falling Bodies Revisited").

In hindsight we can see Galileo's role in the understanding of mechanical motion from Archimedes to Newton. Yet his significance did not go unnoticed among the best of his contemporaries. After corresponding with Galileo, Luca Valerio (a mathematician who had applied refined Archimedean methods of proof to theorems on centers of gravity) said of Galileo's genius that it was at least equivalent to that of Archimedes. Ironically, today Valerio is remembered because Galileo generously referred to him as "the Archimedes of our age" in his *Two New Sciences*.

Galileo's Inventions

What if, instead of continuing along the path of reality as history records it, Galileo were transported by visiting space aliens at a speed close to that of light? He might then return to Earth only slightly aged in our own century, and, obtaining contemporary garb, quietly join a group touring the halls of this same University where he has taught while decrypting these secrets of nature. Here, in the anteroom of a hall where he lectured, he finds a bust with the name (his own) of the University's most famous faculty member, standing adjacent to a bust of its most famous student, Copernicus. In a corner of the chamber behind the hall, there is his podium, cordoned off, decrepit, untouchable and, in stark contrast to its immediate surroundings, still looking like the 16th century object that it is. A spiral staircase on the far side of the familiar 13th century courtyard outside this room leads to a small museum. Behind a glass case in a dimly lit corner there, he finds a posthumous

---

28. Drake, ibid., p. 39, footnote 58. In 1961, Dr. Thomas B. Settle repeated Galileo's experiment described here. Besides obtaining results that corroborated those of Galileo with double the precision, he demolished the old myth that Galileo's experiments were largely imaginary; cf. ibid. and Rom Harre, *Great Scientific Experiments*, Oxford U. Press, 1981, p. 72. See also Settle's "Galileo's Use of Experiment as a Tool of Investigation", *Galileo: Man of Science,* McMullin ed., op. cit.
29. Stillman Drake, *Galileo: Pioneer Scientist*, U. of Toronto Press, 1990, p. 105. Drake explains that in modern terms, "moment of velocità" is the speed component of the velocity vector.

diploma bearing his name. Two hundred fifty years after his death, the University honored itself by awarding Galileo his first and (as far as I have determined) his only diploma. Probably, he would get a good laugh out of that!

Let's now cancel our fantasy trip, so that Galileo remains on the real time-skein, as we remain in the 20th century removed 220 kilometers from Padova to the city of Firenze. Outside a museum, the radiance of the mid-day sun reflects diffusely from the ripples of the Arno River. A man rows his canoe beneath the Ponte Vecchio in a visible interaction of the laws of inertia, acceleration, and floating bodies. Galileo's instruments, invented and developed at Padova, are here.[30]

In the lifespan of an individual, Galileo created and inspired enough inventions to fill a museum. His writings and correspondence (by, to and about him) fill an encyclopedia — Favaro's National Edition. Walking between the stacks of books, we see all around us works exploring the consequences of his discoveries, and evidence of a swelling tide of knowledge through and for the ages. Let's exit the museum's library and climb to an exhibit area, home to Galileo's instruments.

As we climb the stairs, not a smooth ramp but a sequence of levels through which our vertical movement progresses and pauses momentarily at the end of each step up, we can ponder the ascent of humans up the staircase of knowledge. While a close range view of discovery shows discontinuities in which the potential for its application takes sudden leaps, scientific knowledge can appear to be smoothly progressive from a remote viewpoint. Like the process of discovery, the process of inventing is the introduction of a minor discontinuity in a larger-scale apparent continuity of development. From that large-scale perspective, an airplane propeller, for example, is a form of Archimedean screw and not something fundamentally different. And from that perspective, it would be difficult to find an invention in history without precedent of some kind; the instruments with which Galileo worked and either invented or re-invented, and which we will see on this 20th century museum tour, are no exceptions. His inventiveness became evident early on at Padova, when in 1594 the Venetian senate awarded him a patent on a one horsepower water pump — approximately one horsepower, that is, because the pump was driven by exactly one horse.

We come now to a large room on the west side of the second floor, where rays of the setting sun cast a pinkish pallor on the wall opposite a display case in which two of the most famous
Galileo's Telescopes
telescopes of all time are on display. It is well established that the first Italian-made

---

30. During the 1966 flood, according to my conversation with the research librarian here, the curator who managed this museum, Maria Luisa Righini Bonelli, carried Galileo's instruments up to the roof and across to the higher roof of the Uffizi Gallery, as flood waters came roaring up the staircase.

telescopes were those made by Galileo. In the next chapter, I will use Galileo's own writings to show why he is rightly considered an inventor of the telescope,[31] not merely a designer of improvements. The following specifications of Galileo's early telescopes can be inferred by correlating the data of Henry C. King's *History of the Telescope* with those of the museum's booklet.[32]

Property	Telescope #1	Telescope #2
Primary Aperture	2.6 cm	3.7 cm
Stopped to	1.1 cm	1.6 cm
Magnification	14	21
Resolution	15 arc seconds	10 arc seconds
Obj. Focal Length	132 cm	98 cm
Field of View	17 arc minutes	17 arc minutes
Objective	biconvex	biconvex
Eyepiece	planoconcave	biconcave
With it you can see	moons of Jupiter	lunar surface detail
Tube Composition	wood, paper cover	red leather, gold decor

A Galilean telescope, commonly found in the "opera glass" binoculars, gives an erect and enlarged upright image with a magnification equal to the ratio of the focal length of the objective to that of the eyepiece.[33] Galileo constricted the aperture or "stopped" the objective lens in order to get more image definition at the expense of brightness. In the table, the primary aperture is the diameter of the unstopped lens; "stopped to" refers to its diameter after constriction. Telescope #2 has a fainter image than #1, but with it you can see more detail on the moon.

Opera Glasses and the Telescopic Helmet

The Galilean refractor telescope, utilizing a concave eyepiece that does not invert the image as does the Keplerian refractor, is the same lens

---

31. Galileo invented the first convex/concave refractor sufficiently powerful to be useful in astronomy. Kepler later (1611) invented the convex/convex refractor. The reflecting astronomical telescope was invented by Newton in 1668; cf. Henry C. King, *The History of the Telescope*, (Dover, 1955). The word "telescope", meaning literally "remote watchman", was devised by Giovanni Demisiani, mathematician to Cardinal Gonzaga (see Ch. 8 "Triumphal Visit to Rome", 1611) to describe Galileo's instrument; cf. Willy Ley, *Watchers of the Skies*, Viking Press, NY, 1966 (My copy was a gift from my mother and father in 1967.)

32. Bonelli and Settle, *The Antique Instruments at the Museum of the History of Science in Florence*, Arnaud, undated.

33. The Galilean telescope has a converging convex objective lens and a diverging concave eyepiece lens. The objective lens forms at its focus an inverted image of a distant object (it is also reversed between right and left). Rays that converge as they enter the concave eyepiece are made to diverge as they leave, so as to restore the orientation of the original object.

system used in opera glasses. Galileo applied it to nautical uses by adapting it for use on shipboard. He designed a helmet mounting for a telescopic lens system[34] in order to produce a steadier image than if the instrument were hand-held, so that, for example, a sailor positioned high on the main mast could more easily spot approaching land or other ships that were over the horizon as seen from the deck. While the telescopic helmet does not appear to have met with commercial success, its potential for extension and for applications other than nautical ones remains to be considered by future inventors.

Galileo's Handmade Lenses

Returning now to our 20th-century museum tour, we find the lens that Professor Galilei used in discovering the four moons of Jupiter. The lens fracture shown in the photograph at the end of this chapter is a consequence of mechanical laws including his law of falling bodies: he dropped it.[35] Viviani inherited it and in 1677 presented it to Prince Leopold, who commissioned Vittorio Croster to mount the lens in an ivory frame. Here at the museum, the lens is on display in a glass case next to Galileo's telescopes. And, to make it all eerily complete, here is a fingerbone, probably from the hand that dropped the lens.[36] Some of his lenses, obtained and preserved by the Medici family, were tested by Giuseppe Molesini and colleagues at the Italian National Optical Institute at Florence, using laser-based instruments. They found that Galileo's objective lenses are ground and polished with such accuracy of curvature as to "deviate from perfection by less than one quarter of the wavelength of green light".[37]

Suppose that as we tour the museum in the 20th century, we look back at 17th century Padova with an imaginary spacetime telescope. Every real telescope is, of course, a spacetime telescope. If light travels about a foot per nanosecond, then when we watch the Professor writing on the board 20 feet away, we see at each instant the

---

34. Galileo called his invention a *celatone*. Whether this lens system is a simple or binocular telescope is disputed and is a topic worthy of historical research. See Albert Van Helden, "The Invention of the Telescope", *Transactions of the American Philosophical Society*, vol. 67 part 4, 1977, p. 20 footnote 6.

35. This lens has a focal length of 169 cm, and in the test by Abetti and Ronchi, showed lunar craters as small as $10''$ to $15''$; cf., Henry C. King, op. cit., p.43. Thanks to the discovery of Astronomer Royal William Rutter Dawes, published (ibid. p. 272) in 1867, that the minimum angle a telescope will resolve (the Dawes Limit) equals 1.1 radians times the wavelength of light divided by the telescope aperture, we can calculate the effective aperture of Galileo's telescope. Taking the Dawes limit for this telescope to be $10''$, and wavelength of visible light to be approximately 5000 angstrom units, this implies that Galileo had stopped the aperture to about $[2.269 \cdot 10^5 (5 \cdot 10^{-5})]/10''$ arcsec(cm)/arcsec = 1.135 cm.

36. i.e., if he carried the lens in his right hand. Anton Francesco Gori detached the middle finger of Galileo's right hand when in 1737 Galileo's remains were relocated within the Cathedral of Santa Croce. Source: http://galileo.imss.firenze.it/museo/4/eiv10.html. So many bones, including several fingers and the 5th lumbar vertebra, have been removed by Galileo cultists that one wonders what remains of his remains.

37. *Washington Post*, July 13, 1992. This article by "B.R." further states that "the lenses he [Galileo] used were 'optically perfect' even by today's standards."

position of his hand 20 nanoseconds ago, for it took the information that long to reach our eyes. From a position 386 light years from Earth, simultaneous to the year 1996 on Earth, is it theoretically possible to look back at Earth, and with a sufficiently powerful telescope, watch Galileo at work in 1610? As we tour the earthbound museum, it will suffice to invoke the telescope of our imagination and envision Galileo turning out one invention after another, like his successors Edison and Tesla.

The Jupiter Satellite Ephemeris and Analog Computer

Hunched over a brass plate on which he has inscribed a table of numbers, Galileo is at his workbench connecting a large wheel with a long slotted flange to a dial pointer. What is this thing? It is a prototype of an invention that seems to be in hot demand in the early 17th century. This is a time when ships are coming back from the New World laden with gold. To pirates rampant on the high seas, sailors who lose their bearings can be easy prey. The ability to pinpoint one's position at sea raises the probability of delivering the gold. European states are offering sizable prizes to anyone who can solve this problem.

In the northern hemisphere in this millenium [38] we are fortunate that there exists a star, Polaris, near the north celestial pole. This pole marks an intersection of Earth's axis with the "sky" or celestial sphere, an assumed sphere of infinite radius concentric with the Earth and whose equator is coplanar with that of the Earth. In clear weather, local latitude can be determined simply as the altitude of Polaris [39] (the angle measured vertically to the star from the horizon), or at local noon, as a function of the sun's zenith distance and declination. [40] Determining longitude, conceptually equivalent to local time, [41] is more challenging. Accurate clocks that can operate on nautical vessels do not exist in the 17th century. How does Galileo measure the time of a sighting before shipborne chronometers, before radio, and before stopwatches exist? [42] How does he convert the sighting to positional data before navigational almanacs exist, and before Hadley's sextant, more than a century

---

38. Hipparchos' determination of precession tells us that in several millenia Polaris will no longer be the north star (and will probably be renamed).

39. plus or minus a small correction for the deviation of Polaris from the north celestial pole.

40. The zenith distance of an object is the complement of altitude; i.e., the vertical angle from the "overhead" point exactly above the observer to the object. Declination is its (positive or negative) angular distance (north or south) from the celestial equator measured along the great circle (the "hour circle") passing between the object and the poles. (If the sun is considered at infinity, its rays directed at the observer and at Earth's center can be considered parallel for practical purposes.) Local noon occurs when the sun's zenith distance is at a minimum for that day.

41. The difference in the hour angle of the sun between the reckoning position and longitude zero can be converted to local apparent time at the rate of 15° per hour; hereafter, I refer to the longitude zero hour angle of the sun converted to time as the reference time. (Hour angle of a celestial object is the angular distance on the celestial sphere measured westward along the celestial equator from the local meridian to the great circle of the celestial sphere that passes through the poles and the object.)

in the future? He addresses both problems with a combination astrolabe and analog computer to read the accurate celestial clockwork he has discovered: the Medicean satellites of Jupiter. He calls it "*giovilabio*".

Because these satellites transit the planet regularly, their cycle of positions forms a visible sky clock. The difference of "sky clock" time and the local time obtained from sighting the altitude of a reference object, like the moon or a fixed star, can be converted to local longitude. Neither the original *giovilabio*, nor Galileo's two letters describing it will survive. Favaro and others will believe it to have involved some form of micrometer.[43] A copy of a non-micrometric "*giovilabio*"[44] will survive to our time. To measure distances of Jupiter's satellites, what better unit of measurement than Jupiter's own radius can there be? Concentric rings for the orbits of each satellite are inscribed on a disk divided into 24 parallel chords equally spaced so as to represent the radius of Jupiter. A long arm for sighting and alignment with the zodiac is attached to such a disk revolving in a fixed ring that can move 12° to the left or right. This arm is hinged to a dial pointer, and time tables for satellite positions corrected for Earth's annual motion are inscribed on the brass plate that serves as a mounting for the disk and sighting arm. The instrument bypasses trigonometric calculations; only training in the use of the device, not in the underlying mathematics, is required.

Although the Medicean satellites orbit Jupiter and transit unseen behind its disk like clockwork, they do not pop in and out of sight with quite the same regularity. After considerable trouble reconciling his predictions and observations, Galileo finds the cause of the irregularities. He reasons that each satellite is eclipsed by the cone of the giant planet's shadow, whose direction in space varies with the relative position of Earth, sun and Jupiter. To this discovery he adds yet another and more surprising factor: the distance of each satellite from the parent body varies.[45] About this time, and despite these and other complications,[46] he breaks new ground with the

---

42. A small-boat owner in the 20th century, before the advent of the Global Positioning System (GPS), might have equipped himself with a sextant, a stopwatch, a radio tunable to WWV, and an almanac relating date and dead-reckoning position to local horizon coordinates of various celestial objects. ("Dead reckoning" is a phonetic abbreviation of "deduced reckoning", an estimate of position deduced from estimated mean velocity and time elapsed since a previous reckoning.) He could then correct his dead-reckoning position by taking a sextant sighting at a reference time expressed as the sum of the most recent WWV minute signal and the stopwatch reading at the instant of the sighting.

43. Cf. Silvio A. Bedini, "The Instruments of Galileo Galilei" in *Galileo: Man of Science*, McMullin ed., op. cit. Bedini himself disputes the micrometer hypothesis but offers no alternative explanation.

44. Cf. Bonelli and Settle, op. cit., p. 24.

45. This fact must have clashed with his notion of circular orbits, yet the conflict did not stop him from noticing it, or from announcing it in a letter he wrote to Prince Cesi on 15 February 1613.

46. Another difficulty for Galileo was that "near the bright disk of Jupiter it was hard for him to see one [satellite] less than one arc-minute from the center of the planet"; Drake, *Galileo: Pioneer Scientist*, p. 146 note 2. One arc minute is about 1.5 times Jupiter's usual visual diameter.

giovilabio: he discovers a method of determining whether the satellites' orbits are parallel to the ecliptic.[47] He announces to Kepler, via the Tuscan ambassador at Prague, that he has compiled the orbital periods of the four satellites and constructed tables for determining their past and future positions to an accuracy of one second of time. The announcement takes Kepler completely by surprise.

Galileo's Micrometer

As an outgrowth of his project to track the Medicean satellites, Galileo invents a telescope-mounted astrometric device. So accurate are the positional data from the use of this micrometer that he stuns Kepler with tables of satellite positions and nearly discovers the planet Neptune, invisible to the naked eye, more than two centuries before its actual discovery.[48] A grid-bearing disk mounted to the telescope tube on a sliding ring is viewed with one eye while the observer looks through the eyepiece with the other. The optical effect is to superimpose the telescopic image on the grid. The grid can then be rotated to align one axis with the plane of the Medicean satellites and moved toward or away from the eye until the target spans two of its spacings. If the target is Jupiter, any interval can be measured in units of Jovian radii by counting the grid lines in that interval. As Jupiter's angular diameter varies roughly from 30 to 50 arc-seconds, the grid spacings are of such resolution that angles down to about 15 seconds of arc can be conveniently measured. Considering Galileo's invention and his use of it to compile accurate tables of satellite positions and to measure changes in the angular diameters of planets, he is a father, if not *the* father, of astrometry, the measurement of celestial positions. To determine motion from these positions requires voluminous calculations, and this problem leads Galileo to analyze the available calculating devices and to improve on them.

A Revolutionary General-purpose Computer

One reason why Galileo is a founder of computer science is a calculating device described in his publication of 1606, *Operations of the Geometric and Military Compass*. Galileo's compass is a combination of two inventions: the proportional compass invented by Federico Commandino prior to 1568, and the sector, an instrument with calculating scales that Galileo endows with general-purpose computing capabilities. The cover art on his manual is of his own hand and his own design. The title includes the word "military" to catch the attention of princes: the instrument is rather general and by no means limited to military uses.

---

47. Each of the four Medicean satellites orbits in a plane inclined a half degree or less to the plane of Jupiter's orbit, which itself is inclined about 1.3° to the ecliptic; cf., *The Astronomical Almanac*, issued jointly by the United States Naval Observatory and the Royal Greenwich Observatory, Tables E5 and F2.
48. See Chapter 12, "Galileo's Neptune Sightings and Their Future Implications" and the article "Galileo's Sighting of Neptune", Stillman Drake and Charles T. Kowal, *Scientific American*, Dec. 1980.

The instrument has four principal parts: two arms with scales, a quadrant, and a leg that facilitates vertical placement and extension of one of the arms. On the front of the compass are engraved an arithmetic scale, useful for dividing lines into proportional parts, and a geometric scale useful for scaling polygons and extracting square roots. On the back are lines that perform mappings between circles and polygons. The scales give the compass practical uses that would require no special mathematical training on the part of the operator, including interest rate computations (the gain on 140 scudi in 5 years at the annually compounded rate of 6%), and currency conversions (240 ducats = 186 scudi). In this respect, it has no precedent as a calculating instrument. Introducing his English translation of this book, Professor Drake has this to say of Galileo's calculator:

> "Something of the importance to society of such an invention as Galileo's can be grasped from the modern introduction of the pocket electronic computer ... that suddenly made it possible for everyone to deal effectively with almost any problem arising in practical matters by following rather simple instructions. The effects on society are already noticeable after only a few years." [49]

How does he make such a device, using 16th century hardware? First, he must decide what it will do, then puzzle out how to calibrate lines to various scales adapted to this astonishing variety of arithmetic and geometric operations. Galileo describes these operations under the following headings: [50]

* Division of a line
* How in any given line you can take as many parts as may be required
* How these same lines give you infinitely many scales for altering one map into another, larger or smaller
* Solving a proportionality by a compass and these same arithmetic lines
* Rule for Monetary Exchange
* Rule for Compound Interest
* Use of the geometric lines to find ratios of similar polygons or to alter their sizes
* How you can construct a plane figure similar and equal to several others
* Given two unequal similar figures, to find a third, similar to these and equal to their difference
* Square root and cube root extraction
* Rule for arranging armies with unequal fronts and flanks
* Discovery of the mean proportional by means of these same lines
* Use of the stereometric lines [to find ratios of similar solids or to alter their sizes]
* Given two similar solids, to find the ratio between them
* Given any number of similar solids, to find one equal to all of them together.
* Discovery of two mean proportionals
* How every parallelepiped can be reduced to a cube by stereometric lines
* Ratios and difference of specific gravity and of weight between metals: gold, lead,

---

49. from Drake's English trans. of *The Geometric and Military Compass,* 1606, by Galileo Galilei.
50. Headings translated by Stillman Drake, ibid., in some cases abridged or paraphrased by me.

   silver, copper, iron, tin, marble, and stone
* Gunner's calibration for cannonballs of any material and weight
* Calculating the measures of a body of specified material and weight based on
   measurements of a small scale model made of any material
* Drawing of regular polygons
* Division of a circle into as many parts as is desired
* Transforming any regular figure into any other regular figure of equal area (a
   special case of which is squaring the circle or any other regular figure)
* Constructing a regular figure equal to the sum of dissimilar regular figures
* Constructing any regular figure equal to any given irregular rectilinear figure
* Quadrature of segments of a circle
* Measurement of vertical heights by sighting

Among clients whom he trains in the use of the instrument are Archduke Ferdinand of Austria; Phillip, Landgrave of Hessia and Count of Nidda; Johann Friedrich, Prince of Holsace and Count of Oldenburg; Vincenzo Gonzaga, Duke of Mantua[51].

What is Galileo's own view of his proportional compass? Legend has it that the King of Syracuse, being tutored by Archimedes [52] in geometry but impatient with the tedious logic and wanting quick results, asked Archimedes for a simpler approach. Archimedes is said to have replied: "Your majesty, there is no royal road to geometry". Hear Galileo now, in his own words,[53] describe his invention:

"The opportunity of dealing with many great gentlemen in this most noble University of Padua, introducing them to the mathematical sciences, has by long experience taught me that not entirely improper was the request of that royal pupil who sought from Archimedes, as his teacher of geometry, an easier and more open road that would lead him to its possession; for even in our age very few can patiently travel the steep and thorny paths along which one must first pass before acquiring the precious fruits of this science; frightened by the long rough road, and not seeing or being able to imagine how those dark and unfamiliar paths can lead them to the desired goal, they falter halfway there and abandon the undertaking... I fell to trying to open this truly royal road — for with the aid of my Compass I do that in a few days, teaching everything derived from geometry and arithmetic for civil and military use that is ordinarily received only by very long studies. ..."

Archimedean/Galilean Stamps on Mechanical Devices

His proportional compass helped motivate the development of computing machines by illustrating and advertising the concept that

---

51. Not to be confused with Ferdinando Gonzaga, a Cardinal, or Aloysius Gonzaga, a famous Jesuit Saint.

52. Proclus had also attributed this story to Euclid as tutor to Ptolemy I, and W. W. Rouse Ball to Menaechmos, as tutor to Alexander "the Great".

53. from *The Geometric and Military Compass*, 1606, op. cit.

algorithms embodied in machines can be of quite general purpose.[54] Today one finds Galileo's visible stamp on a wide variety of mechanical devices, like the hardware components of a computing machine. If I remove the casing from the disk drive on the machine with which I am producing the manuscript of this book, for example, I see a 6.5 cm diameter flywheel mounted on one side of the integrated circuit card containing the logic that controls the drive operation. This is an application of Galileo's inertial principle to stabilizing the rotational rate of the disk. And on the other side of the card from the Galilean flywheel is an Archimedean screw, attached to a motor shaft. Its function is to convert the torque of the motor shaft into a linear force that "seeks" the drive heads to a track on the rotating disk where data are to be stored or retrieved. All the modern technology in which these simple parts are encased has not done away with the need for Archimedes' or Galileo's contributions; it has been developed on them as a base. In this and other ways, the man about whom this book is written has even assisted in its production.

Returning to our museum tour, we come to a room housing optical instruments other than Galileo's telescopes. What's this? Did he make microscopes, too? Let's focus our spacetime telescope now on his Padovan laboratory in 1623.

In the candlelight we see the master hunched over his workbench, mounting a lens that he has toiled far into the night to polish. On the table by the flickering candle lies a parchment

Compound Microscope

bristling with hastily but accurately drawn geometrical lines: the "blueprint" for a kind of telescope in reverse that will open up the world of the very near and very small, too small to be seen without this instrument. What he is doing tonight that no one has done before him will be debated by historians.[55] Despite the confusion, Galileo will achieve recognition as an inventor of the compound microscope. What is not in dispute is that he made magnifying instruments of his own design and gave them to various noblemen, including the Duke of Bavaria. One of the lucky recipients of an instrument handmade by Galileo will say, in a letter of 1624, that the microscope made a fly appear as large as a hen. The shock of seeing the world of the familiar suddenly become unfamiliar generates widespread excitement. The medical botanist Johann Faber deserves to be remembered for his remark, after having seen a fly under Galileo's *occhialino*, that Galileo was "a kind of creator, having exhibited something no one before had known to have been created".[56]

---

54. See also Chapter 13 "Galileo, Computer Science and System Methodology".
55. See Albert Van Helden, "The Invention of the Telescope", op. cit., p. 7, footnote 32 and p. 20 footnote 6.
56. English paraphrase by Stillman Drake, *Galileo at Work*, Ch. 16.

In the box to the right, I list all Galilean microscopes that will be known to have survived 300 years:

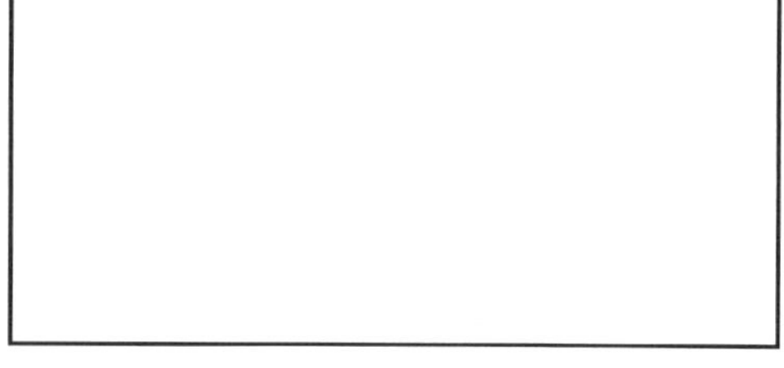

That's right, there is none. Early Galilean microscopes will be lost, and only post-Galilean specimens will survive from the collection of the Accademia del Cimento.[57] The construction of Galileo's instruments, which earlier generations could inspect, is unknown to us.

Thermometer

Returning to our tour of his artifacts, we pass to a display case that houses the earliest surviving thermometers. With our spacetime telescope, we can watch his original measurement of temperature.

Here he is overturning a flask of pre-heated air into a dish of fluid. He is testing a prototype of the thermometer. It will be called a thermoscope [58] (actually it is a thermo-baroscopic instrument). As the air in the tube cools toward room temperature, it contracts and sucks the fluid up into the tube. After that, the fluid seeks a level in reaction to the room temperature. A problem with this procedure is that the thermometric substance is the air. Air is subject to random pressure fluctuations, of which Galileo is apparently unaware; later he will assign a question arising from siphoning experiments to his student Torricelli, who will discover the existence and variability of air pressure.

Another student and friend, Giovanfrancesco Sagredo, whose name Galileo will immortalize in the *Dialogo*, improves the sensitivity of Galileo's thermoscope so that it reacts to the proximity of a human body. This instrument is adapted to medical uses by a colleague of Galileo. Santorre Santorio conceives of using it for the examination of feverish patients and, inspired by the writings of Heron of Alexandria, alters its design so as to direct fluid displaced upward by air, expanding as the air temperature rises, to register on an inscribed scale of degrees. A similar instrument is apparently independently invented by Robert Fludd in England about this time, but Galileo's is the first of these new developments. Torricelli will convert Galileo's thermoscope to a thermometer by isolating expansion/contraction effects from air pressure.[59]

Astrolabe

Is Galileo turning out from his laboratory a combined sighting device and analog calculator, like the one in the museum called

---

57. founded by Leopold de' Medici to pursue Galilean science. (See Ch. 11 "The Academy of Experiment").

58. (See photo.) We know from Sagredo's letters to Galileo, as well as Viviani's biography, that Galileo is the original inventor of the thermoscope, though Heron of Alexandria noted the principle about A.D. 100.

"The Astrolabe of Galileo"? Designed for finding the time of local sunrise and sunset, the altitude and azimuth of a star, local solar time, and what stars are visible at a given time, this astrolabe is set up for use near Galileo's villa at Arcetri. Using the imaginary line through Polaris ("The North Star") and the stars of Ursa Major ("The Big Dipper") that point to Polaris as an hour hand of a 24 hour clock in the sky, the astrolabe can read the hour hand with fine enough angular resolution to act as a sidereal clock.[60] Is it marketing hype, or did Galileo make this instrument? The image in our spacetime telescope turns to the mark of a question awaiting some ambitious historical researcher.

Galileo is attracted to the work of William Gilbert, whose textbook on magnets appears in 1600. Dr. Gilbert will have the honor of being repelled by an Inquisition Consultant [61] as

Magnetic Lift

a perverse heretic and as a quarrelsome and sophistical defender of Copernicanism. A distinguished physician and philosopher, he will be better remembered for pioneering research in magnetism. Inspired by his work, Galileo amplifies a natural magnet, reinforcing a lodestone with an armature of iron connected by brass strips to the poles of the magnet. In this way, he strengthens a magnetic lift by a factor of 80, so that it holds up 26 times its own weight. A small reinforced magnet lifting a heavy iron casket accompanies instruments that he designs, makes, and leaves to posterity. He has thus converted a lodestone to a "load-stone", as Jonathan Swift will call it. Swift's description (quoted at the heading of this Chapter) of the astronomers' cave on the island of Laputa appears to be directly or indirectly motivated by this lodestone and the other Galilean inventions described here.[62]

Looking ahead a century beyond Galileo, one finds the German physicist Leibniz, an inventor of the calculus, deploring the loss of a reinforced magnet which Galileo had given to Cosimo de' Medici. Leibniz apparently thinks that his Italian predecessor was on a promising track in his investigations of magnetism and that his work in this field ought to be studied and resumed. But Galileo's exciting and controversial discoveries in astronomy lured him from magnetic research into a

---

59. A thermometric device being sold in the late 20th century (as I am writing this book) and inaccurately advertised as the "Thermometer of Galileo" is actually modelled on the post-Galilean specimens called "*infingardi*" or "lazy ones" that are a product of the Accademia del Cimento. The *infingardi* thermometers were named for the slow motion of crystal spheres constructed with varying propensity to drop in alcohol as the temperature of the fluid rises. Galileo inspired them as he did all thermometers, but these do not use air as the thermometric substance and thus are not of his design.

60. There is no doubt that Galileo actually used such an instrument; the altazimuth conversion dial ("*timpano*") of the one in the museum is suggestively adapted for 43° 40′ N. latitude, which would position it for use in the vicinity of Galileo's villa at Arcetri.

61. Melchior Inchofer, a Jesuit who vehemently attacked Galileo's *Dialogo* in his report to the Inquisition.

62. If motivated by the Royal Society of Swift's time, this satire illustrates how closely the Royal Society was modelled on the Accademia del Cimento.

protracted mission as principal defender of the new world view of Copernicus.

Medical Innovations Galileo may have done the medical field more good by staying out of it. His impact on it will have far reaching effects, indirectly in theory and directly in instrumentation. Galileo's mechanics will inspire and equip his disciple Gian Alfonso Borelli to undertake the complex analysis of the human body as a machine.[63] At least two Galilean inventions, the compound microscope and the thermometer, will be used in medical diagnosis for centuries after Galileo.

In our spacetime telescope we can see him at Padova in 1602 discussing his pendulum experiments and his thermoscope with that same physician friend Santorio, who will reduce Galileo's findings to practical medical applications by improving the thermoscope and by designing a pendulum-based instrument to measure the pulse. Santorio's invention, the "pulsilogium", makes use of the quasi-isochronism of the pendulum.[64] The pulsilogium allows a physician to alter the length of a pendulum and, having synchronized it to the heartbeat of a patient, to read a pulse rate from the pointer of a dial. The physicians are surprised and delighted, and the oldest hospital in Europe (at Siena) puts the device into use. In our own century, innovations in medical technology still flow from Galileo's concept of implementing the field of medicine with measuring instruments.

Other Pendulum Devices: Clock and Vibration Counter In later life, he returns to the pendulum, designing vibration counters and regulators for clockwork. By 1637 he has designed a pendulum-driven clock in which a crown wheel advances one tooth per oscillation. Slowed by his blindness, then stopped by his death, his project to build a more sophisticated clock to his design awaits a successor. His son Vincenzio records that design and produces an incomplete model interrupted by Vincenzio's own death. That is the model of which Viviani produces an engineering sketch still on display. [65]

As we exit the museum housing remnants of Galileo's inventions, we have seen that he recognized and tapped the potential of scientific instruments, many of which had existed in some rudimentary form, and that he improved and found new

---

63. Gino Loria, *Galileo Galilei*, p. 135, cites Borelli's *De Motu Animalium*, 1680.
64. Besides having inspired the pulsilogium through his 1602 conversation with Santorio (see Drake's *Galileo at Work*, p. 69 and p. 466), Galileo may have built a prototype vintage 1582 when he was a medical student. Citing Viviani's letter to Prince Leopold de'Medici dated 20 Aug 1659, Silvio Bedini awards the invention to Galileo, saying that it was "further exploited by Santorio" (see *Galileo: Man of Science*, McMullin ed., p. 257). It is undisputed that Galileo is at least a conceptual contributor to the pulsilogium and that Santorio is at least a practical inventor of this instrument, which he also developed further.
65. A pendulum-regulated clock will finally be built and patented by Huygens circa 1659.

applications for them. Silvio Bedini has put into perspective all the controversy about whether he was first with this or that invention:

> "Much useless effort has been expended by scholars in controversies about whether Galileo ... merely improved or developed (scientific instruments) .... The historical origin of the instruments attributed to Galileo is not, in the final analysis, of crucial importance in assessing Galileo's own achievement in their regard. The important, and apparent fact, is that it was Galileo who developed and improved them and first applied them scientifically to make observations that were crucial in their effects on the scientific revolution he helped to shape." [66]

Attacks on Invention

In our spacetime telescope, we can look back to 1607 and see a student plagiarizing Galileo's instruction manual on the calculating instrument.

Galileo's Defense Against a Plagiarizer

He even claims priority for the invention. To clear his name, Galileo brings an action before the University's board of governors. A recipient of an early prototype of Galileo's instrument certifies under oath that this student had borrowed the instrument and the manual from him. Galileo shows that he has been making and selling his instrument five years prior to this student's starting his studies in mathematics. The student is expelled from the university.[67] Later, Galileo will encode certain astronomical discoveries in anagrams in order to protect his priority.

Despite this frustration, Galileo flourishes in the intellectual climate of Padova. Here he is befriended by Vincenzio Pinelli, who maintains an extensive

Rewards and Risks of the Renaissance

personal library and regularly invites local and foreign scholars to meet in his home. One of Galileo's best friends at Padova, an articulate diplomat with a well developed interest in science, is that same Sagredo who improved the thermoscope, and whose name Galileo will immortalize in his dialogues.

Another friend, Fra Paolo Sarpi,[68] a Servite theologian, has been a professor of philosophy and a scientist who discovered the contractility of the iris. Called in last year (1606) by the Venetian senate as a counsellor in canon law, he advised resistance to Pope Paul V who, acting under the advice of Cardinal Bellarmino, demanded jurisdiction over civil matters affecting the clergy and placed all of

---

66. Silvio A. Bedini, "The Instruments of Galileo Galilei" in *Galileo: Man of Science*, op. cit.

67. A faculty member (Simon Mayr) who seems to have put him up to it, then leaves Italy to take up a foreign post, where he will later publish a claim, easily shown to be false, of priority in the discovery of the first four satellites of Jupiter. cf. Galileo's *Assayer*, Drake trans., p. 233.

68. The source of information here related about Sarpi is Will and Ariel Durant, *The Story of Civilization*, VII: *The Age of Reason Begins*, Simon and Schuster, New York, 1961.

Venice under ecclesiastical interdict — religious services were banned. Galileo, who then had his contract renewal up for negotiation, had to wait in Venice for a hearing while the battle between the Doge and the Pope culminated in expulsion of the Jesuits for defying an order to ignore the interdict. An advocate of complete separation of church and state, and of the view that papal power extends only to spiritual matters, Sarpi is about to have his own problems with the Inquisition. In 1607, assassins will attack him with daggers and leave him for dead. He will recover and remark, in one of the classic puns of history:

"Agnosco stilum Curia Romanae". [69]

Later, he will write a massive and eloquent indictment of the Council of Trent, exposing its culpability in the failure to reconcile the Protestant schism. [70]

In our imaginary spacetime travels, we have followed Galileo to trace the flow of his inventions. Let's return to the year 1604, when Galileo is still deeply immersed in terrestrial physics. He has not yet invented his telescope nor even heard of a principle by which lenses can be combined to make remote objects appear close. On August 10 of that year, he begins to translate a Homeric verse on the battle of rats and frogs. Exactly two months later, he sets aside all such recreations to take up the challenge of a new and unexpected event in the heavens. That event sparks an interest in astronomy that will set the stage for what is to happen five years later, when he will turn his telescope on the night sky.

ABRIDGED CHRONOLOGY OF GALILEO'S INVENTIONS

1586	hydrostatic balance		1609	astronomical telescope
1590	use of inclined plane in gravity test		1611	ephemeris calculator (*giovilabio*)
1593	water pump		1612	telescopic micrometer
1597	geometric proportional compass		1617	telescopic helmet
1602-8	magnetic lift		1620	doubly convex microscope
1604-6	baroscopic thermometer		1637	pendulum-driven clock

---

69. This can be translated "I recognize the pointed style of the Roman court". Cf. Will and Ariel Durant, ibid., p. 230. The pun is on the word "stilum": a style of communication or a dagger.

70. The Council of Trent had decided, by a vote taken 18 years before Galileo's birth, which books of the Old and New Testaments are "canonical", meaning that the Church has established by its authority that they are the Word of God. This event will pave the way for the indictment of Galileo by the Inquisition and the censoring of the work of Copernicus.

Edifice, University of Padova, Where Galileo Taught for Eighteen Years
(photo by author; courtesy, University of Padova)

Courtyard, University of Padova,
Outside Hall Where Galileo Lectured
(photo by author; courtesy, University of Padova)

Galileo's Original Podium
(photo by author;
courtesy, University of Padova)

Galileo's Posthumous Diploma
(photo by author,
courtesy, University of Padova)

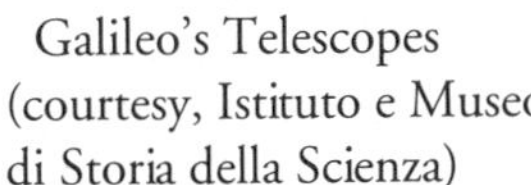

Galileo's Telescopes
(courtesy, Istituto e Museo
di Storia della Scienza)

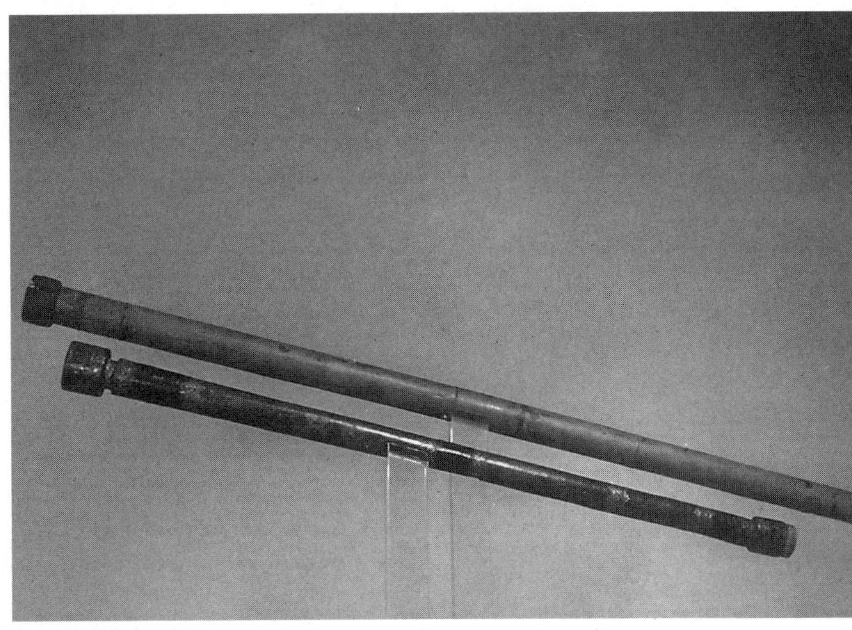

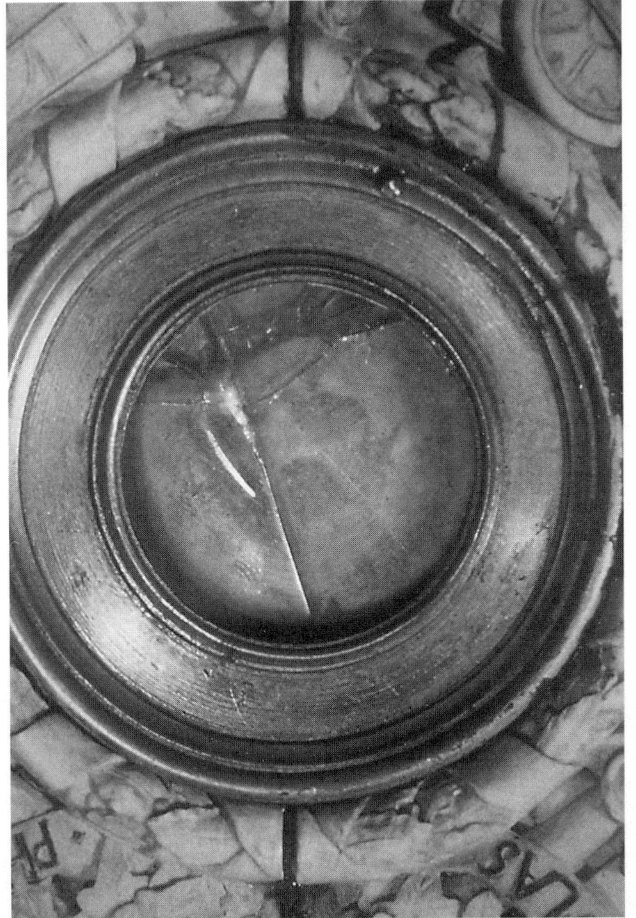

Lens Used in Discovery of
Jupiter's Moons,
Dropped and Broken by Galileo
(courtesy, Istituto e Museo di Storia
della Scienza)

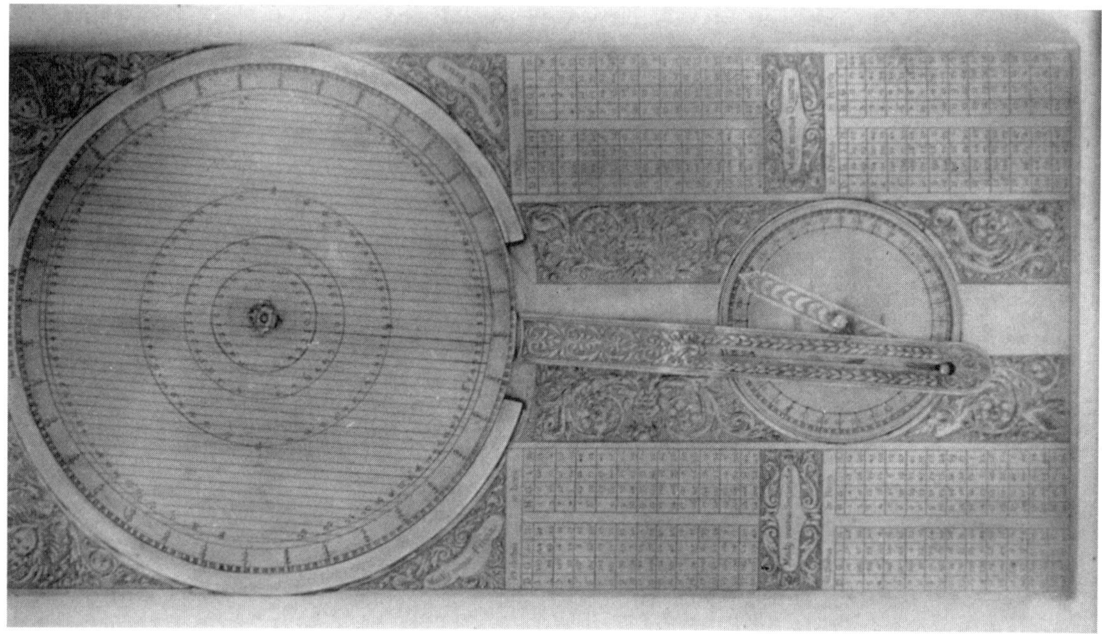

Giovilabio, Based on Galileo's Design
(courtesy, Istituto e Museo di Storia della Scienza)

Magnetic Lift Using Armored Lodestone,
Made in Galileo's Workshop
(courtesy, Istituto e Museo di Storia della Scienza)

About this device Benedetto Castelli said
that Galileo had given it to Grand Duke
Ferdinand after arming it so that a six ounce
lodestone holds a 15 pound iron casket.

Cover Illustration Drawn by Galileo
for the Instruction Manual
on His Proportional Compass
(courtesy, Istituto e Museo di Storia della
Scienza)

Galileo's Proportional (Geometrical and Military) Compass
(courtesy, Istituto e Museo di Storia della Scienza)

Sliding Tube Compound Microscope
(courtesy,
Istituto e Museo di Storia della Scienza)

Model of Galilean Thermoscope,
an Early Thermometer
(courtesy, Istituto e Museo di Storia della Scienza)

Chapter 6. The Celestial Beacon

Galileo Ventures into Astronomy (1604 - 1610)

"... Then felt I like some watcher of the skies
When a new planet swims into his ken;
Or like stout Cortez when with eagle eyes
He star'd at the Pacific — and all his men
Look'd at each other with a wild surmise —
Silent, upon a peak in Darien."

... John Keats
from *"On First Looking into Chapman's Homer"*

"With what admiration the reading of excellent poets fills anyone
who attentively studies the invention and interpretation of concepts!" [1]

... Galileo Galilei, *Dialogo, End of First Day*

It is not supposed to happen. In October 1604, the blazing light of a stellar explosion ruptures the serenity of the sky — and of Aristotle's followers. They have been proclaiming that any change in the heavens would violate their Master's doctrine. Galileo rivets his attention on the sky. Unlike Tycho before him, whose valiant vision swept through dim spaces, Galileo will open a bright window on the universe with his invention of optically aided astronomy. Little more than five years after the supernova, he will use this invention to launch a stunning series of revelations. These revelations will catapult an obscure 40-year-old mathematics professor, still unknown outside academia, into a place in history harder to hide than the tomb sequestered in a dark closet of a cathedral where the Church will consign his body for a century after he dies.

That superlatively rare and cataclysmic natural event, the supernova, can be witnessed only by a lucky minority of astronomers. Centuries normally come and go between observable supernovae. Only seven [2] have been recorded from the time of Ptolemy, and their nature as celestial objects was mysterious from antiquity; they

The Supernova of 1604

---

1. Galileo was especially fond of the poetry of Tasso, Ariosto, Petrarch, and Dante.
2. Probable supernovae were recorded by pre-telescopic astronomers in the years 185, 393, 1006, 1054, 1181, 1408 and 1572. Of these, only three (the Lupus supernova in 1006, the Crab Nebula Explosion in 1054 and Tycho's star in 1572) became brighter than second magnitude. Tycho's star and the 1604 supernova seen by Kepler and Galileo were stunning events, brighter than all the fixed stars. (Cf. Hartmann, *Astronomy: the Cosmic Journey*, Wadsworth, Belmont Calif., 1978, p. 294 ff.)

were known as "new stars" by their appearance. Now in 1604, as in 1572, once again the processes of the heavens randomly synchronize with new watchers of the skies, observers primed to recognize irregular events and to measure their distance. In this space of only 32 years, two supernovae are visible from Earth, and each occurs while at least one outstanding astronomer is at the prime of his career: first Tycho Brahe, then Galileo and Kepler.

The doctrine of the unchangeability of the heavens begins to come undone in 1572, as Tycho proclaims the "new star" of that year. But in the immutable skies of Aristotle there is no place for a new star: only the terrestrial elements of Empedocles (earth, air, fire, and water) may change. To fend off this challenge, the "Aristotelians" say that the new star is atmospheric and therefore terrestrial.

Countering, Tycho shows that the object must lie beyond the moon. His proof uses the parallax method of Hipparchos. Even objects as near as the sun, if their separation angle from nearby "fixed" stars is accurately measured, exhibit detectable parallax — the difference in apparent direction of an object seen by observers in different places. Though the new star shows no parallax, Tycho's analysis falls on deaf ears. When the keeper of new truth is in sufficient minority, the majority can try to pretend that they are right — first by ignoring, later by suppressing, still later by subtly expropriating, the new truth. Here the "ignore-ance" [3] ploy is a natural consequence of a philosophy having been converted to institutional dogma.

New Columbus of the Heavens What Tycho has done is to make a correct finding with a correct method. What he has not done is to persuade clerics and academics. Just as men before Columbus had found the New World, Tycho had found change in the heavens. Soon after the 28th of October, it is Galileo who makes his debut as the new Columbus of the heavens when he recognizes the new star of 1604. Like Tycho, he makes careful parallax measurements. Unlike Tycho, he announces his finding in three public lectures in which he focusses finely honed oratory on Hipparchan trigonometry and its consequences. This proof, Galileo's first contribution to observational astronomy,[4] will set a precedent for his design of a parallax experiment which Herschel will use in discovering the planet Uranus in 1781, and another which Bessel will use to obtain a real image of Earth's orbit in the 1830s, two centuries after Galileo. Thus it happens that prior to his development of the telescope, Galileo has begun to shake the tottering edifice of "Aristotelian" cosmology.

---

3. I accent the second syllable (ig-NORE-ance) to distinguish this act of "looking away" from ordinary ignorance, which implies a passive deficiency of observation or knowledge.
4. Kepler independently developed a similar proof, published in 1606.

Galileo's contributions to astronomy, and to Newton's system of the world, can best be understood by considering his special relationship to Kepler. Each of them will play a necessary role in consigning to oblivion the Aristotelian duality of worlds — the heavens and Earth — through a unified physics that explains terrestrial and celestial phenomena according to the same set of laws. From 1604, let's step back one nova, just 32 years, and step over an Alpine range to the north.

In the village of Weil der Stadt in the Swabian region of Germany, Johannes Kepler is born when Galileo is almost eight years old. Kepler's childhood is burdened

Johannes Kepler, Cosmic Theologian

by frailty, frequent illness, and quarrelsome parents. A sensitive and precocious boy, he is required (at the age of nine, for two years) to do hard labor instead of attending school. While attending universities in the Würtemburg area, he leaves behind the harshness of his youth to take up earnest studies, to which he applies a keen intellect and unusual mathematical prowess. Intending to train for the priesthood, he enrolls at the University of Tübingen. Shortly thereafter, a temporary teaching assignment diverts his career path; then a vision transforms the diversion to a lifelong quest.

The vision emerges from a question: Why are there only six planets? [5] On the 9th of July, 1595, as he is sketching geometrical figures for his class at the Protestant school in Graz, Austria, an answer comes to him suddenly. For a moment, he stands in stunned silence as he sees in his mind's eye the five regular solids of nature [6] fit neatly between the orbits of the six currently known planets. Here, in the number and spacing of the planets, is evidence of a mathematical harmony of the world and of a Creator who designed and constructed the universe around geometric forms, using as his building blocks the five regular polyhedra (discovered by Pythagoreans) of solid geometry. Boldly and proudly, after repeated calculation and checking, he announces this arrangement in the *Mysterium Cosmographicum*. He sends a copy to Tycho and two copies to a correspondent who forwards them to Galileo. [7] Tycho and Galileo respond with encouragement and Tycho, though unconvinced, is so impressed with Kepler's

---

5. Six planets, including Earth, were known from antiquity, and even today only these six are visible to humans without optical aid. Uranus, Neptune and Pluto had to be found with telescopes, and long after Kepler. In the time of Kepler, there was no detectible hint in nature that others might be found; the knowledge required to detect such hints grew out of Newton's system of the world.

6. The tetrahedron (pyramid), hexahedron (cube), octahedron, dodecahedron and icosahedron. These are known as the Platonic solids, since it was probably Theaetetos, a friend of Plato, who discovered the remarkable fact that five and only five regular polyhedra are possible to construct. (see Carl B. Boyer, *A History of Mathematics*, p. 85).

7. The intermediary is Paul Hamberger. Galileo's reply of 4 Aug 1597 indicates that he has agreed with the Copernican hypothesis for some years. Cf. Drake, *Galileo at Work*, p. 41.

creativity that he suggests a meeting, after which Kepler becomes his assistant.

Tycho's carefully cloistered cache of data on the orbit of Mars will, after his demise, allow Kepler to convert his mathematical harmony into science. Nothing less than determination bordering on the surreal lets Kepler achieve this conversion. While across the Alps, Galileo is discovering kinematic laws and is re-uniting geometry and physics, here in Germany the music, mystery,[8] and mathematics of the cosmos are fusing into a unified trinity under Kepler's synoptic vision and scrupulous introspection.[9] Tempered in a crucible of self-scrutiny, the fusing elements ignite and merge after Kepler perseveres to re-conceive his conceptions more times than men less than fanatic in the pursuit of truth can manage.

About the first element of his vision, that the universe is governed by mathematical laws, he is right. About the second, the geometry of the planetary orbits, the *Mysterium Cosmographicum* is, by his later admission, wrong. But accompanying that admission are the first two of three correct laws of planetary motion, laws that today are known to be obeyed throughout the universe — by planets, by distant binary stars, and even by giant galaxies locked in gravitational embrace.[10]

And though the conclusion of the *Mysterium Cosmographicum* is wrong, the data it contains will be useful — to Galileo. In 1601, Galileo will combine his data from inclined plane experiments with Kepler's data in order to investigate the relationship of orbital speed to the radius vector of a planet.[11] Later still, Kepler will discover a generalization of this relationship, his Third Law of Planetary Motion, on the 8th of May, 1619. This law establishes the proportionality of the squares of the sidereal periods of any two planets to the cubes of their mean distances from the sun. The remarkable story of Johannes Kepler is the story of a lone scholar struggling against formidable obstacles to wrest out of his mathematical vision of nature certain of its most elusive secrets. That is a story we will leave to others,[12] for we must now return to Galileo, and to the year 1604.

Trials, Tribulations, and Telescopes

By the time of the 1604 supernova we find Galileo still having to supplement a rather meagre salary despite the increase attendant on his move to Padova. Professors in other subjects, especially those hired from abroad, are being paid multiples of Galileo's salary. Following the death of his father in 1591, he

---

8. For my sense of "mystery", see Ch. 2, footnote to "Science in Toscana before Galileo".

9. Cf. Max Caspar, *Kepler*, Abelard-Schuman Limited, London 1959, p. 39.

10. Kepler announced Laws I and II in *Astronomia Nova* (1609) and Law III in *Harmonices Mundi* (1619).

11. Stillman Drake, *History of Free Fall*, op. cit., pp. 64 and 83. See also Chapter 11, "From Parabolic Fall to Planetary Periods".

12. For example, see *Kepler* by Max Caspar, op. cit.

assumed responsibility for the family, including a large dowry for his sister. His brother, Michelangelo, did not contribute; instead he "borrowed" from Galileo, apparently with an infinite payback period. Since 1598, Galileo has been manufacturing and selling his compass, but by now, the time of the supernova, there are new burdens. A Venetian woman, with whom he lives, bears him two daughters and a son between 1600 and 1606. Galileo and Marina Gamba will separate permanently upon Galileo's return to Florence in 1610. Afflicted with arthritis and chronic colic, he has to mix philosophical ponderings with physical pain. Suddenly, in 1609, Galileo's troubles are overshadowed by a new and exciting development emerging from Holland: the telescope.

Late in the summer of 1609, Galileo's friend, Fra Sarpi, hears of a Dutch invention [13] combining lenses so as to magnify distant objects. Sarpi's news

The First Astronomical Telescope

electrifies Galileo. Having no specification for the shape or arrangement of the lenses, Galileo extracts from his knowledge of geometry a lens system to produce such an effect. By his own account, he produces the design for a telescope in one night. Two days later, he has manufactured the instrument by his own hand. His first documented use of it is to demonstrate it to the Venetian authorities.

On 21 August, 1609, Galileo unveils his far-sighted invention at Piazza San Marco to a gathering of educated but shorter-sighted men whose interest in

Demonstration of the Telescope

what he has to show is mainly political.[14] He has brought them to the top of the bell tower where one can look out over the Adriatic. According to an eyewitness, Antonio Priuli, he uses a 9-power metal-plated telescope, covered with a variety of crimson cotton material, of diameter 42 mm and length 0.6 meter. More than 20 km away, the bell tower and facade of Santa Giustina of Padova can be seen, looking perhaps 2 km away in the telescope. And here are the sails of approaching ships still two hours from port. To the Venetians, this means that if the city were under attack, they would have two hours warning. The Doge of Venice is so impressed that he grants Galileo a salary increase to 1,000 florins per year, coupled with a lifetime professorship, an offer from which Galileo will soon walk away.

---

13. Three Dutchmen apparently independently made original working models: Hans Lippershey, Zacharias Janssen, and Jacobus Metius. We will recount this story in detail later in this Chapter (see the section titled "Invention of the Telescope"). Sarpi advised the Venetian government in August 1609 against buying a spyglass at a large sum from a foreigner because Galileo assured him that he could make a more powerful and useful instrument, which in fact he proceeded to do (cf. Stillman Drake, *The Unsung Journalist and the Origin of the Telescope*, Zeitlin & Ver Brugge, Los Angeles, 1976, pp. 3-4).

14. (rdc)

While Galileo is preparing to turn out more than a hundred telescopes in his own workshop, he asks himself: If he were to turn his telescope on the night sky, would not visions of things heretofore unseen swim into his ken?

While the Slow Moon Climbs

We can well imagine this,[15] though we don't know the exact day — it is late in the year 1609. Excitement, perhaps mixed with some apprehension of the unknown, quickens his pulse as he mounts the narrow stairs to his observatory, the upper balcony of his Padovan villa. What might confront his first gaze upon the magnified celestial vault? Pausing momentarily, he braces himself for a steady view. In a confident sweep of the arm, like that of his teacher Ricci as he had traced out the imaginary arc of a circle, he raises the instrument skyward. How long he stands there, looking through the tube and lost in awe, we will never know. We do know that the moon, as if smiling on this grand adventure, was probably out that night, for Galileo will mention it in his first accounts of what he sees. As dusk sets in and the brightest stars have not yet cut through the waning diffusion of twilight, it is a natural first target. Lights begin to twinkle from the sky, the long day wanes, and the slow moon climbs. There loom the dark, broad seas of the moon.[16] Our little sibling planet is not at all the perfectly spherical orb that it is supposed to be. Dappled with mountains [17] and giant rayed craters, its rough topography is interspersed with large smooth areas which might be seas. Following this thought, he calls them "maria",[18] and though he will later conjecture the absence of rain on the moon from the absence of clouds, the term will stick long after astronomers learn that they are dark plains.

After moonset, the night sky darkens and fainter stars shine from the blackness of the firmament. Open clusters, dense star fields that are close in angular separation and brightness, reveal new pockets of light under magnification. Our galaxy, the Via Lactea, far from appearing milky, shows countless clusters of individual stars toward whatever part of it he aims his instrument. If the cool breeze were playing havoc with Galileo's arthritis, this moment might have left him oblivious.[19] It is only the beginning. New adventures await him in the unexplored sky.

Jupiter's Companions

As the year 1610 opens and he prepares to announce these discoveries, Galileo stands on the

---

15. Here for two paragraphs I supply imaginative detail in the style of a historical novel.
16. Yes, this is an intentional twist on Tennyson's *Ulysses*.
17. In the coming months, he will calculate the height of these mountains so accurately that astronomers centuries later do not dispute his findings.
18. The singular form "mare" is used in current names of major lunar plains; e.g., "Mare Spumans", Foaming Sea. There is even an "Oceanus Procellarum", Ocean of Storms.

threshold of yet another. On January 7 he observes Jupiter, one of the first objects to shine through the twilight. Nearly centered between the full moon, only one constellation eastward along the ecliptic, and the bright red-giant star Aldebaran, the Eye of the Bull, less than 10° westward, Jupiter transits the meridian at 9:45 p.m., a convenient target more than 60° above the horizon. On that night, the yet undiscovered Uranus [20] is near conjunction with Jupiter, less than three degrees away, but well outside his instrument's field of view, twelve minutes of arc. Like an unfaceted jewel in a setting, Jupiter has three splendid little "stars" flanking it, never before seen by the eyes of man. But they shine with a steadier light than ordinary stars with their diffraction spikes; they are "very rotund ... without rays". What are these Jovian jewels? Galileo will identify them within eight days, in an episode that we will visit in the next chapter of our story.

Invention of the Telescope

Let us return to our 20th-century vantage point and to the question: Did Galileo invent the telescope? The first thing to be noted about this question is that the telescope is not a single invention. It had to be invented conceptually and then mechanically. Then, because of special circumstances surrounding its mechanical invention, it had to be re-invented for practical use. Let's briefly recount the story. [21]

The conceptual invention began with the question: How might one make an instrument that would extend human vision by making faraway objects appear nearby? Suppose you had lived in a time before telescopes or any mechanical aids to vision were invented. Imagine how radical-sounding this question would have seemed. Yet more than 23 centuries ago, Aristotle recommended the use of long tubes without lenses for seeing distant objects clearly. But if telescopy began with that ancient advice, its practical realization depended on the later concept that human vision can be improved by the bending of light passed through lenses.

That concept underlies a spectacular 13th century invention: eyeglasses. Within a century after convex lenses were applied to the correction of presbyopia, concave

---

19. The renowned astronomer Sir John Herschel, quoted by Col. G. F. Young in *The Medici,* p. 653 (see Bib.) says of this moment: "It is difficult to conceive what Galileo must have felt when, having constructed his telescope, he for the first time turned it to the heavens, and saw the mountains and valleys in the moon. — Then the moon was another earth; the earth another planet; and all were subject to the same laws. What an evidence of the simplicity and magnificence of nature! But at length he turned it again, still directing it upwards, and again he was lost: for he was now among the fixed stars; and if not magnified as he expected them to be, they were multiplied beyond measure. What a moment of exultation for such a mind as his!"
20. Uranus is a planet that will remain undiscovered for more than a century after Galileo; yet its discovery will be due in part to him (see Ch. 12, "Galileo's Role in the Discovery of Uranus"). Data on the moon, Uranus and Aldebaran at Padova, Italy on 7 Jan 1610 are as calculated by Astronomer™ 8420 software (a product of Expert Software, Inc.).
21. Much of this story is based on the research of Cornelis de Waard as summarized by Van Helden, op. cit.

lenses began to be used to correct myopia. The first telescopes evolved conceptually out of a combination of these two types of lenses. Here is yet another riddle of history within the larger riddle surrounding the inventor(s) of the telescope — who invented corrective lenses? We do not know by whose divine artifice we are saved from the effects of grotesque visual defects, and that unsung genius is also among the conceptual inventors of telescopy.

Still, the telescope required the intent to extend human vision, not just correct it. Inspired by the 13th century writing of Roger Bacon, who had understood light refraction and visualized an invention with which "from an incredible distance we might read the smallest letter and number grains of dust and sand",[22] in 1589 the magus Giovanbaptista Della Porta wrote of the potential to recognize our friends "some miles off" by using lenses. He described a combination of concave and convex lenses with the property that "if you know how to fit them both together, you shall see both things afar off, and things neer hand, both greater and clearly."[23] It is evident that he tested the magnifying effect of such a combination. In 1610, he joined the Lincean Academy (see p. 146) and by that time had met Galileo. Raffael Gualterotti, a Florentine, wrote to Galileo in 1610 mentioning a similar lens combination that he had made in 1598 for a cavalry soldier as an aid in battle. However, the lenses available at that time magnified little and could not achieve Porta's goal of recognizing people miles away. Neither man's effort extended human vision to great distances, and neither Porta nor Gualterotti attempted to secure a patent. Roger Bacon's conceptualization of the telescope thus inspired real (but weak) prototypes of it in Italy in the last decade of the 16th century.

In October 1608, at least two spectacle-makers of Middelburg in Holland, Hans Lippershey and Jacobus Metius,[24] in that order, independently applied to the governing States-General for a patent on a magnifying lens combination, and a third claimant to the invention, probably Zacharias Janssen, was interviewed by the States-General. Metius and Lippershey were paid by the Dutch government for prototypes. Unable to decide on the priority and independence of the rival claimants, the States-General rejected their patent applications. A patent was never issued. Despite the Dutch government's commendable stewardship of relevant documents, the continuing difficulty for historians of resolving these questions among the rivals has been complicated by the Nazi Luftwaffe's bombing in 1940 of the municipal archives of Middelburg and the consequent destruction of such documents. Happily, modern technology offers historians the option to distribute

---

22. from Bacon's *Opus majus*, trans. by Robert Belle Burke, as quoted by Van Helden, op. cit., p. 12.
23. Van Helden, op. cit.., p. 15; quotation is from the English translation by Edward Rosen.
24. Lippershey's name has been variously written as Lipperhey, Laprey, and Lippersein. Jacob Metius, an instrument maker of Alkmaar, was also known as Jacob Adriaenszoon.

information on network file-server nodes and not to act as if ignorant and uncaring political despots would treat historical archives with the respect they merit.

Circumstances of the lens-manufacturing industry in Holland and the unguarded testimony of the son of a rival claimant provide clues pointing back to the earlier developments. Italian glass-making techniques had been introduced in Holland at the glass factory in Middelburg established by Italian immigrants and managed by an Italian, Antonio Miotto, from 1605. During a lesson, Janssen's son told Isaac Beeckman, rector of a Latin school and friend of Descartes, "that the first telescope in the Netherlands had been made in 1604 by his father, after the model of an instrument in the possession of an Italian", which bore the date 1590.[25] In magnification, the lens systems produced by the Dutch claimants were no great advance over the earlier Italian prototypes. Indeed, when Porta learned of the developments in Holland, they were (to him) so unremarkable that he denounced them as a hoax. What was all the fuss about? Had he not made an instrument about as good as these, years earlier, and found it inadequate to Bacon's vision? Unlike Porta, Lippershey had created a stir with his patent filing, then Metius and Janssen had added to the excitement. The Dutch developments would turn out to have great historical significance. For the man who would ultimately convert this early effort into a scientific instrument — Galileo — did not know about the earlier developments in his own country but first heard about magnifying lens systems in a report from Holland relayed by his friend Sarpi. For the knowledgeable Paolo Sarpi, who had met Porta in 1580, excitedly related the news from Holland to Galileo, who would convert these early efforts into a powerful scientific instrument.[26]

Recently, the Lick Observatory museum was selling a T-shirt bearing a portrait of Galileo captioned "Inventor of the Telescope", as well as the Lick Observatory logo. Consulting Galileo's account of the matter in the *Assayer*, I find the probable basis for crediting him as a mechanical inventor of the telescope:

"... we know that the Fleming who was first to invent the telescope was a simple maker of ordinary spectacles who, casually handling lenses of various sorts, happened to look through two at once, one convex and the other concave, and placed at different distances from the eye. In this way he observed the resulting effect and thus discovered the instrument. But I, incited by the news mentioned above, discovered the same thing by means of reasoning ...

My reasoning was this. The device needs either a single glass or more than one. It cannot consist of one glass alone, because the shape of this would have to be convex ... or concave ... or bounded by parallel surfaces. But the last-named does not alter visible objects in any way, either by enlarging or reducing them; the

---

25. Ibid., p. 8.
26. Long befriended by Federico Cesi, Porta would be elected to Cesi's Lincean Academy in 1610; cf. p. 131.

concave diminishes them; and the convex, though it does enlarge them, shows them indistinctly and confusedly. Passing then to two, and knowing as before that a glass with parallel faces alters nothing, I concluded that the effect would still not be achieved by combining such a glass with either of the other two. Hence I was restricted to discovering what would be done by a combination of the convex and the concave. You see how this gave me what I sought ; ...” [27]

Well, not quite. First, he had to derive geometrically the requisite curvature of the lenses (Galileo is uncharacteristically humble in this passage). Secondly, he passed over the choice he must have made between convex/convex and concave/concave, this last being negligible since a double concave might be expected to reinforce the diminutive effect of a single concave. The convex/convex combination, upon which he also defaulted comment, later became Kepler's choice. Galileo chose the convex/concave. This choice had a splendid side benefit. The upright image it gave permitted it to be shown easily to men of political influence who would pay him the money he needed to develop the instrument further. This he did, advancing the technology of lens grinding toward longer focal lengths and a telescope which magnified up to 30 times, as compared with the 3-power Dutch instrument. Turned skyward, it carried to the eye and mind of Galileo the information content of his message from the stars, *Sidereus Nuncius*.

Galileo, Kepler, and the Telescope

In April 1610, Kepler — who had solicited Galileo's comment on his book *Mysterium Cosmographicum* — received a copy of *Sidereus Nuncius* along with Galileo's request for comment. Kepler accepted Galileo's findings and published a letter of approval. Within six months, Kepler completed his *Dioptrice*, explaining the optics of lens systems, and announced his own convex/convex design. Meanwhile, a second letter from Galileo to Kepler revealed the gratitude that Galileo had felt for Kepler's endorsement:

> *"I thank you because you were the first one, and practically the only one, to have complete faith in my assertions."*

Kepler did not need faith in Galileo's assertions. He understood geometric optics thoroughly, and with a telescope of his design could validate Galileo's claims. But Kepler endorsed them even before he had his own telescope.

Galileo's independent manufacture of a powerful scientific telescope, on the basis of nothing more than the conceptual invention having been conveyed to him, is not the sole ground on which he merits the title on the T-shirt. Supremely skilled at promoting, capitalizing and applying the invention, he made the discoveries that

---

27. Drake, *Discoveries and Opinions of Galileo*, pp. 245-246

occasioned Van Helden to identify the telescope's birth cry with *Sidereus Nuncius* and its baptism with the Lincean dinner party of April, 1611.[28] Kepler, too, independently invented the telescope, having worked out a less experimentally evident lens combination by mathematical means. Galileo's respect for Kepler, though somewhat eclipsed by the scant attention he paid to Kepler's astronomical discoveries, came through clearly in his letters. On hearing of Kepler's having mistaken a sunspot for Mercury, he writes to his correspondent Welser that spots large enough to be confused with Mercury would be recognizable by remaining at conjunction for days, whereas the planet would cross the solar disk in fewer than seven hours. He then praises Kepler's honesty (and his own theory of sunspots):

> *"Kepler, as a true philosopher and not recalcitrant about manifest events, will no sooner hear of these observations and discourses of mine, than he will lend his assent to them."*

In a later letter to Welser, Galileo adds his finest compliment to Kepler:

> *"Kepler ... is a man of free and brilliant mind and more a friend to truth than to his own opinions."*

The Dutch telescopes were not well suited for astronomy. Galileo designed a more practical telescope by means of reasoning, and Kepler did so after him. Galileo re-invented the telescope, introducing the technique of stopping the lens — using only the center portion to increase its power — and using precisely ground and polished lenses of suitable curvature. He increased its magnification by a factor of ten beyond that of the Dutch prototypes, and he armed it with a micrometer. Telescopes designed for astronomy began with Galileo. With his telescope, one may ascend the staircase of the bell tower to sight ships at sea, or the heights of a majestic mountain that we call cosmology — formerly a pursuit only of philosophers and, after Galileo supplied vision to astronomy and method to epistemology, a pursuit principally of scientists.

Poets are wont to extol beauty that arises from random processes of nature. How much worthier of praise is the conquest of natural defects and limitations by the human intellect! The invention of the telescope is a spectacle of human intellect prevailing over a perversity of nature (myopia, for example) and in the process improving upon nature at its best.

---

28. Ibid. p. 20. See Chapter 8, "Triumphal Visit to Rome, 1611". In this same passage, Van Helden refers to Lippershey's patent application (the first filed) as the telescope's birth certificate. This is a good analogy, not that Lippershey was first to create a working prototype, but in the sense that he was the first of several Dutchmen who triggered patent examinations and promoted the invention effectively enough to bring word of it to the ear of one who would transform it into a practical instrument.

Campanile (Bell Tower) at
Piazza San Marco,
Site of Galileo's First Telescope
Demonstration, 21 August 1609
(photo by author)

Piazza San Marco, Seen from Grand Canal,
Site of Galileo's Demonstration of Newly Discovered Satellites of Jupiter
(photo by author)

Chapter 7. Message from the Stars

"... the broad circumference hung on his shoulders like the moon,
whose orb through optic glass the Tuscan artist views
At ev'ning, from the top of Fiesole, or in Valdarno, to descry new
lands, rivers, or mountains, in her spotty globe ...
... As when by night the glass of Galileo, less assured, observes
Imagined lands and regions in the moon..."

... John Milton, *"Paradise Lost"* I:286, V:261

As pinpoints of light come streaming through Galileo's spyglass from every direction in the night sky, injecting life into the grand theoretical visions of Democritos and Bruno, Galileo prepares to reveal his observations to all students of nature and to challenge each individual of the civilized world to verify his discoveries. This he does beginning in 1610, with his Message from the Stars, *Sidereus Nuncius,* and his Letters on Sunspots, *Lettere sulle Macchiae Solari.*[1] The discoveries announced in these two publications are almost beyond belief and would not enjoy public acceptance were they not verifiable with the telescope. Here are the first indications that:

> The earth's moon is cratered and mountainous.
> Jupiter has its own moons.
> The Milky Way is a vast conglomeration of individual stars.
> ... announced in *Sidereus Nuncius, 1610*

> Venus has phases, like those of earth's moon.
> The sun has blemishes crawling around like worms on its surface.
> ... announced in *Lettere sulle Macchiae Solari, 1612*

Galileo's other-worldly concerns might be supposed to have little effect on a mankind steeped in misery. The Black Death is stalking Italy. The Moriscos, who have contributed greatly to the economy and culture of Spain, are being expelled by a tyrant. A fanatic has just assassinated Henry IV in France. In the New World, the majority of colonists landed in Virginia have in the last several years died of a combination of starvation, disease and Indian attacks. In the midst of all this tragedy, a mathematician diverts popular attention away from dismal conditions that are local in space and time. The hint of a higher order of things, not the exclusive property of

---

1. An English translation by Stillman Drake of both works can be found in *Discoveries and Opinions of Galileo,* Doubleday & Co., New York, (Anchor Books), 1957, along with his letter to the Grand Duchess Cristina on the relationship of science and theology. *Sidereus Nuncius* was written in Latin, the official language of the scientific community, in order to be understood wherever the language of Cicero was understood; cf. Gino Loria, *Galileo Galilei* (Bib.). *Sidereus Nuncius* was published on 12 March 1610; all 550 copies in the first printing sold out within a week, one of which resold at auction in 1977 for $39,000 (cf. Derek Gjertsen, *The Classics of Science,* Barber Press, NY, 1984).

priests and mystics, but accessible to the man in the street, fires imaginations and raises the vision of the humans skyward, toward a new purpose and a new cosmology. In this glimpse of a universe that can be observed and comprehended, there is the promise of personal freedom attainable through individual action rather than through reliance on kings and priests.[2] Let's see what Galileo reveals in these two crucial publications.

Sidereus Nuncius (1610):[3] The First Round of Astronomical Revelations

> *"Something there is more immortal even than the stars, ...*
> *Something that shall endure longer even than the lustrous Jupiter,*
> *Longer than sun or any revolving satellite,*
> *Or the radiant sisters, the Pleiades."*[4]

Star Clusters and The Stellar Composition of "The Milky Way"

In this passage, the poet Whitman suggests[5] two Galilean discoveries: Jupiter's satellites, and the open star cluster containing the Seven Sisters, or the Pleiades. Only seven stars are visible without his telescope; the rest elude the naked eye. Galileo counts 36 of these newly visible stars;[6] he publishes hand drawings of this and other star groups. He adds over 500 stars to Orion, including 80 stars to its sword and belt.[7] The Milky Way, he discloses, contains countless clusters. The ancients saw it as some wispy, luminous substance in the sky; Parmenides said that a mixture of the dense and the rarefied produces its color.[8] But Galileo reveals a vast conglomerate of star colonies, each with innumerable stars. Stars are like distant suns, seemingly countless, and clustered. Elements of Bruno's imaginative vision — a mediocre sun, a stellar Milky Way and four non-geocentric planets — only a decade ago held to be an irreverent flight of fancy, present themselves now to Galileo's eyes.

---

2. The personal freedom to which I refer is the condition of an observer who can verify his relationship to the rest of the world directly rather than through reliance on an authority. It was a promise not fulfilled immediately, nor yet fully in our own time. [Even Kepler had to borrow the Emperor's telescope. (jm)]

3. Small in size but giant in effect, his little book entitled *Sidereus Nuncius* is traditionally, and incorrectly, rendered in English as *The Sidereal Messenger*, or *The Starry Messenger*. Galileo points out, in response to a critic who charges that he is representing himself as an ambassador from heaven, "Nuncius" can mean "Messenger" or "Message", and the latter is intended.

4. From "On the Beach at Night"; Walt Whitman, *Leaves of Grass*, Bantam Books, 1983.

5. I am not asserting that Whitman made this connection intentionally; I do not know whether he did or not.

6. Assuming a human-eye aperture of about 5mm, the visibility improvement ratio 36/7 suggested to me an objective lens stopped to about $[5^2 \cdot (36/7)]^{1/2} \cong 11$ mm. This calculation caused me to suspect that Galileo had used Telescope #1 in Table 1 of Ch. 5, until dr pointed out my unwarranted assumption of uniform density and luminosity distribution between the 7 star unamplified view and Galileo's 36 star field. Based on Drake's evidence that Galileo had used a 20 power instrument with a cardboard stop of medium aperture for observing the starlike moons of Jupiter at about this same time (*Galileo at Work*, p. 148), I now suppose that Telescope #2 or one like it was used in the Pleiades discovery.

At first sight, the Pleiades is a rather fuzzy-looking object. The poet Tennyson said that it glitters *"like a swarm of fire-flies tangled in a silver braid"*.[9] What is that silver braid? Galileo showed conclusively that a stellar aggregate can be cloudlike in appearance to the naked eye; those individual stars of the Pleiades that are invisible to the unaided eye emit a faint and cloudy glow of background light. Coincidentally, two and a half centuries later, a gas cloud in space, a nebula, will be found *inside the Pleiades* by the astronomers Mr. Tempel and Dr. Pope.[10]

Magnified Stellar Images

Stars viewed with unaided vision are "fringed with sparkling rays". In Galileo's telescope they appear merely to scintillate, unbounded by the clear circular periphery that frames the planets. Now conspicuous by their absence, the sparkling rays are gone. Galileo sees that the fixed stars are so remote as to appear hardly at all magnified by the same telescope that enlarges nearer objects like the moon, sun, and Jupiter in a fixed ratio. Due to their globular appearance, he sees them as distant suns.[11]

M44: Praesepe

Among fuzzy objects that men of bygone ages had seen in the sky is one that Hipparchos named "little cloud", a name that persists until Galileo finds it to be a stellar aggregate. With his telescope trained on it, the object loses its mantle of mystery. Galileo can resolve its individual stars and is first to recognize it as an open star cluster. Charles Messier will later enter it as the 44th object in his catalogue; today it is known as M44. At magnitude 4.5, with the central cluster (of which Galileo viewed at least 40 stars with his telescope) [12] spread out over a larger angular diameter than that of the full moon,

---

7. Galileo drew Orion as seen through his telescope. Comparing the facsimile drawing (e.g., in Drake's *Discoveries and Opinions of Galileo*, p. 48) with a photograph of Orion (see photo and diagram on p. 125), one sees an angle (α) in both images; its vertex is the rightmost star of the belt and it is subtended by the line segment between the left star of the belt and Eta (η) Orionis. Angle α looks like about about $92°$ in the photograph and $110°$ in the 1609 drawing, where η Orionis is closer to the belt. Yet the fastest angular motion of a star transverse to the line of sight (proper motion) is about ten arc seconds per year, or about $1°$ since Galileo's time. Assuming that his drawing is accurate and accurately copied, how can this be? Hint: If you have a sky atlas, try moving η Orionis one degree in a northerly direction keeping other stars fixed; this movement causes about $10°$ increase in angle α and places η Orionis about where Galileo did. But proper motions of all four stars as published in the 20th century individually do not exceed one second of arc per century. A contemporary "Aristotelian" might argue that the vagabond star is an omen from God. Or, that the camera is bewitched (rdc). Was Galileo's telescope also bewitched — or was it (jm) lens aberration?

8. Source: Sir Thomas Heath, *Aristarchus of Samos: The Ancient Copernicus*, Dover, 1981, ch. IX.

9. from "Locksley Hall"; see *The College Anthology of British and American Verse*, ed. by A. Kent Hieatt and William Park, pub. by Allyn and Bacon, Inc., Boston, 1969, p. 390.

10. Sir John Herschel, *Outlines of Astronomy*, American Home Library Company, NY, 1902, p. 864.

11. Galileo did not infer from the symmetry of the image to that of the underlying reality, but he did refer to stellar images as "simple globes". In this observation, he may have been aided by the spherical aberration of his objective lens, which would cause peripheral rays to focus closer to the objective lens than central rays. If his telescope could have achieved perfect focus, he would have seen no stellar disc at all.

M44 is an excellent target for binoculars. Flanked by γ and δ Cancri, each of which is not more than $2°$ from its center, and lying nearly on the ecliptic, the cluster can be located near the right angle vertex of a right triangle having the stars Castor and Procyon as the remaining vertices.[13] Galileo called M44 "Praesepe", Latin for "manger". Its two flanking stars have Latin names that can be translated as northern ass and southern ass.[14] Their eastern and western counterparts do not appear on star maps (or in the sky either), but may have existed as two self-styled "Aristotelians", who in mule-like stubborness refused to look through Galileo's telescope.

Lunar Features Viewed through the Tuscan artist's optic glass, a cratered and mountainous moon looms in stark and irreverent contrast to Aristotle's smooth and spherical mental vision of it. On hearing Galileo's description, one self-proclaimed Aristotelian declares it impossible and, invited by Galileo to use his telescope, declines the offer.[15] Were he to see an irregular disk, another claims,[16] then the instrument would be deceiving him. It might be coated, he suggests, with a smooth, transparent, crystalline substance through which he would be looking into some underlying irregularities. Galileo responds by suggesting transparent mountains and craters superimposed on that smooth spherical surface. Later,[17] he compares the Aristotelian moon to a spherical mirror, whose surface would be dark and largely invisible except in the line of sight of the sun's reflection. This is not what we see; instead we see a diffused brightness. *"Therefore the moon, by being a rough surface rather than smooth, sends the sun's light in all directions, and looks equally light to all observers."* [18]

In honor of Father Clavius, who (despite the news from Galileo) refuses to recognize the existence of craters or of any surface irregularities, the anti-Copernican Jesuit Father Riccioli will in 1651 assign the name of Clavius to one of the most prominent lunar craters, and the name will stick. Crater Clavius, shown in a photograph on p. 112, is 230 km in diameter and 4900 m deep,[19] which means

---

12. Also known as the "Beehive", the cluster contains over 350 stars down to 18th magnitude, at a distance of 525 light years: the light from M44 that today enters our eyes left the cluster about the time Copernicus was born. Over 100 stars in this cluster are absolutely brighter than the sun (cf. Burnham's *Celestial Handbook*, Vol. 1). The Beehive is a young neighbor of the sun, with an estimated age of only 650 million years (cf. John Sanford's *Observing the Constellations*, Simon and Schuster/Fireside, 1989, p. 35).
13. If you prefer coordinates, look around right ascension 8h 38m, declination $+20°$.
14. See Mallas and Kreimer, *The Messier Album*, Sky Publishing Corp., 1987.
15. Julio Libri, Philosophy Chairman at U. of Pisa. Of Libri's death in 1610, Galileo remarked "...since he would not look at the new discoveries while on earth, he would perhaps see them on his way to heaven." (paraphrase by Stillman Drake, *Galileo at Work*).
16. Johann Brengger, a medical doctor of Augsburg. This argument was apparently independently advanced by Ludovico delle Colombe, of the Florentine Academy.
17. Galileo's *Dialogo*, First Day.
18. *Dialogo*, trans. by Stillman Drake.

that Clavius is bigger in area than the state of Virginia and deep enough that if Pike's Peak were dropped inside, the peak would not clear the crater rim. That astronomers named a prominent lunar crater after him (much more prominent than the crater honoring the discoverer of lunar craters) is less bizarre than it may sound. Galileo has no need of a monumental crater. On the other hand, Father Christopher Clavius, S.J., who staunchly denied the reality of craters, is now perhaps mainly remembered on account of having his name attached to one.[20]

The "Aristotelians" were unconvinced. Galileo's telescope did not show irregularities on the moon's edge. Today, a photograph of the moon taken at prime

Edge of the Moon: A Mindwitness Account

focus of a 20 cm. Schmidt-Cassegrain telescope hints at irregularities, but at lesser resolution a circular disk would be apparent. This telescope amplifies more than Galileo's, which did not show roughness: Galileo admits that the entire periphery of the full moon looks as round as if drawn with a compass. Visualizing nature mathematically, he adds this "mindwitness" account to his eyewitness account:

"... if the protuberances and cavities in the lunar body existed only along the extreme edge of the circular periphery bounding the visible hemisphere, the moon might (indeed, would necessarily) look to us almost like a toothed wheel, terminated by a warty or wavy edge. Imagine, however, that there is not a single series of prominences arranged only along the very circumference, but a great many ranges and mountains together with their valleys and canyons disposed in ranks near the edge of the moon, and not only in the hemisphere visible to us but everywhere near the boundary line of the two hemispheres. Then an eye viewing them from afar will not be able to detect the separation of prominences by cavities, because the intervals between the mountains located in a given circle or a given chain will be hidden by the interposition of other heights situated in yet other ranges. ... Similarly in a rough sea the tops of the waves seem to lie in one plane, though between one high crest and another there are many gulfs and chasms of such depth as not only to hide the hulls but even the bulwarks, masts, and rigging of stately ships. ..."[21]

This passage suggests a statistical law of large numbers. Only at a close enough

---

19. Alfonso Fresa, *La Luna*, Hoepli Editore Milano, 1952, p. 425.

20. Evidently wishing to support Galileo without offending Church philosophy, Clavius wrote a politically ambiguous comment in the final edition of his *Sphaera* (1611): "... when the Moon is a crescent or half full, it *appears* so remarkably fractured and rough that I cannot marvel enough that there is such unevenness in the lunar body. Consult the reliable little book by Galileo Galilei ... ", James M. Lattis, *Between Copernicus and Galileo*, U. of Chicago Press, 1994, p. 198, emphasis mine. Clavius is taking cover behind the protective word "appears", for in responding to Bellarmino, he went out of his way to supplement the report with an opinion that the lunar surface is not irregular but only appears so.

21. *Sidereus Nuncius* trans. by Drake in *Discoveries and Opinions of Galileo*, op. cit. Galileo's use of terrestrial observations to justify deductions about celestial phenomena must have caused consternation among the "Aristotelians" who believed in essential differences between the two realms (rdc).

range to resolve individual mountains and crater rims do the eye and brain of man cease to integrate the superimposed variations of the lunar periphery into a smooth arc. Even with Galileo's telescope, an earthbound observer is too distant. Here, the *Message from the Stars* surpasses the perceptions of its readers. Sensing this, Galileo shifts their perspective through calculations on deviations in the lunar terminator, the boundary between the light and dark regions. From the existence of light spots on the dark side of the terminator, he infers a mountain range. Consider, for example, the largest visible deviation from the terminator of a particular spot of light on the dark side as equal in size to 1/20 the moon's diameter, his own estimate.[22] From this, he is able to establish the height of the inferred mountain.

Surveying Lunar Mountains

Galileo's solution begins by identifying an observed spot of sunlight with a mountain peak. It continues by shifting the observer's perspective to the surface of the moon. Suppose that the sun is rising (or setting), so that if we place ourselves as observers on the moon's surface near a point T on the terminator we observe rays of sunlight coming in along the horizontal plane nearly tangent to the ground at T. The rays of light that clear ground obstructions continue along the horizontal plane some distance until intercepted at the mountain peak. By our choice of T, this is the distance estimated at 1/20 the diameter of the moon.

In his translation of *Sidereus Nuncius*, Drake warns the reader: "*Free translations have been made, and mathematical sections have been omitted, with the interests of the general reader principally in mind.*"[23] Since Drake omitted some calculation, and altered the rest to fit English units, I will reconstruct Galileo's calculation using numbers that I suppose would have been available to him. It was presumably known to Galileo that Aristarchos estimated the moon's angular diameter as 0.5 degrees, and Hipparchos estimated the moon's distance to be 59 Earth radii.[24] Archimedes had estimated[25] the value of π by the inequality $(223/71) < \pi < (22/7)$.

The arithmetic mean of the Archimedean bounds on π to six significant digits is therefore 3.14185. As we know from the experiment of Eratosthenes, the distance of 5,000 stadia measured between Alexandria and Syene subtends a geocentric angle of 7°. Taking the value 157.5 meters as the length of the stadion based on Pliny's explanation,[26] the distance of 5,000 stadia is equivalent to 787.5 km.

---

22. Deviations visible to Galileo can be seen in 10-power binoculars with one objective lens covered.

23. Preface to *Discoveries and Opinions of Galileo*. Can the result of this populist maneuver on the part of Professor Drake be other than to frustrate precisely the best readers, who are left wondering what data and derivations have been suppressed? (rdc).

24. Per Ivor Thomas' *Greek Mathematical Works*, cf. Bibliography. Galileo might have used 56, a less good "refinement" of this estimate; cf. *Dialogo*, Second Day, Sosio ed., p. 272.

25. See Petr Beckmann's *A History of* π, or Tobias Dantzig's *Bequest of the Greeks*.

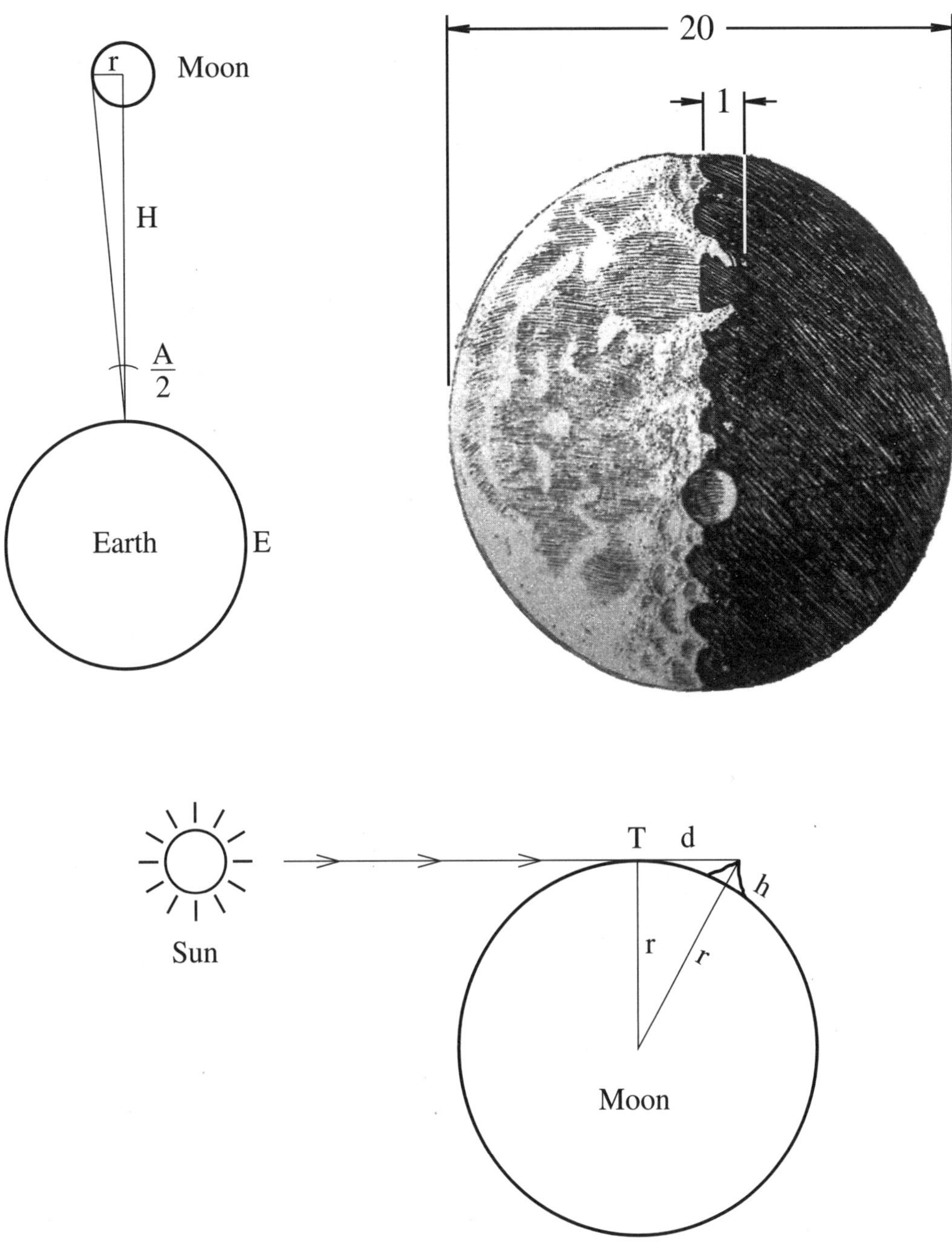

Galileo's Method of Determining the Height of Lunar Mountains
incorporating his hand drawing of the lunar terminator,
from Favaro's National Edition
(sketch by Don Rado)

---

26. trans. by Heath, cit. Thomas' *Greek Mathematical Works*, Vol. II Ch. XVIII (Eratosthenes), p. 273 note c.

Circular Lunar Periphery
Still Seen Through a Larger
Telescope Than Galileo's

(photo by author, 2000 mm
at prime focus)

Clavius,
the Great Lunar Crater,
Monument to a Jesuit
Who Denied the Existence
of Lunar Craters

Photo taken at the 5 Meter Hale
Telescope (see p. 298);
Courtesy,
U. S. Naval Observatory,
Dr. Edgar Woolard

Let r = lunar radius

d = Galileo's estimate of deviation = $0.05 \cdot (2r)$ = $0.1r$

E = Eratosthenes'[27] circumference of Earth

    = $(360/7)(787.5$ km$)$ = 40500 km

A = Aristarchos' estimate of moon's angular diameter= $0.5°$

π_a = Archimedes' mean estimate of π = 3.14185

H = Hipparchos'[28] estimate of moon's distance

    in units of $E/2\pi$ = 59

$$\text{Then } r = H(E/2\pi_a) \tan A/2 = 59 \ (6445.25) \tan 0.25 = 1659.25 \text{ km}$$

Let h be the unknown mountain height and apply the "Pythagorean" theorem:[29]

$$(h + r)^2 = r^2 + d^2 = (1.01) \cdot r^2 = (1667.53 \text{ km})^2$$

$$h = (h + r) - r = 1667.53 \text{ km} - 1659.25 \text{ km} = 8.28 \text{ km}$$

By comparison,[30] Mount Everest is about 8.85 km in height. What about Galileo's comment that the lunar mountains are at least four times as high as any on Earth? Did he not know about the Himalayas, nor high peaks in the Italian Alps, such as Monte Rosa? Perhaps not. But the point of this exercise is not to examine his knowledge of the earth's surface; about that of the moon he had discovered more than any man of his time or before him. A human who thinks he has knowledge of geography has at best only awareness of information about geography unless he has verified that information through observation. Galileo had knowledge of lunar topography that he created through his observations of the moon.

What, then, *is* the point of this exercise? It illustrates how a single seemingly simple calculation relies on insightful inference by at least six great thinkers: Thales, Aristarchos, Archimedes, Eratosthenes, Hipparchos, and of course, Galileo himself.

Modern measurements corroborate the accuracy of Galileo's lunar mountain survey. And his demonstration of a rough lunar surface explains the flashes, called "Baily's Beads",[31] at the endpoints of a total solar eclipse when rays of light come streaming through mountain passes. The result is a startling effect, like the one he will create when he sends elements of his stellar message streaming through the gaps in the cosmology of Aristotle and Ptolemy.

---

27. Per Ivor Thomas' *Greek Mathematical Works,* cf. Bibliography.

28. Per Arthur Berry's *A Short History of Astronomy,* p. 51

29. Intermediate results are carried to 8 digit precision, shown to 6 digit precision

30. If, as he may have, Galileo used 56 for the value H, the result would be 7.9 km.

31. Named for Francis Baily, who first described them after the eclipse of 1836. Professor Houston suggested to me the connection between Galileo's discovery and Baily's Beads.

Copernican Implications of the Sidereal Message

"It is a capital mistake to theorize before you have all the evidence. ... When a fact appears to be opposed to a long train of deductions, it invariably proves to be capable of bearing some other interpretation."

.. Sherlock Holmes (Arthur Conan Doyle), from *"A Study in Scarlet"*

While sleuthing about the sky, Galileo will uncover various clues opposing a long train of deductions in the intricate logical latticework of the Ptolemaic system, a massive but fragile theoretical construct that had depended on not having all the evidence. In January 1610, he finds the first of several key pieces of evidence that in the aggregate will be capable of bearing a new interpretation.

Non-geocentric Moons, By Jove!

In the last chapter, we left him on the brink of this discovery as he examines the lustrous banded disk of Jupiter shining against the black velvet backdrop of the night sky. Night after night, it is flanked, at first by three, then by four small "stars" of such nearly equal brightness and color as to frustrate any attempt to identify them individually. They follow Jupiter as a group. Like birds in flight they migrate together, but within the group they slowly move about, hour by hour. At first, they appear to be oscillating in a straight line. Each night, looking back at his drawings from previous nights, he observes and draws their configuration. A pattern begins to emerge. What if each object cycles through a distinct band of distance from Jupiter? "What if?" becomes a hypothesis borne out by further observation. Unable to tell them apart by appearance, he distinguishes them by their behavior relative to that hypothesis. Not only do they cycle through distinct bands of distance, but they are eclipsed periodically by the Jovian disk, as if revolving around the planet in the observer's line of sight.[32] Here are the first celestial objects recognized by an observer to be orbiting a planet other than the earth. He continues his observations into March, when he leaves for Venice to oversee the publication of *Sidereus Nuncius*. More calculations follow, in which he works out the motion, relative to Jupiter, of the satellites that he will call "Medicean".

The revelation of a non-terrestrial center of celestial motions deals a sudden blow to the geocentric system and to Church "philosophers" content with their imagined position at the center of everything. Galileo is quick to notice the contrary evidence:

---

32. Galileo drew his first observed (7 January 1610) configuration of moons (m) and planet (J) left to right as mmJm. He did not observe the fourth moon until six days later, when he recorded mJmmm. The intervening days (8-13 January) he recorded Jmmm, (clouded out), mmJ, mmJ, mmJm. Drake believes that he concluded on 15 January that the objects "m" were moons orbiting Jupiter.

"Here we have a fine and elegant argument for quieting the doubts of those who, while accepting with tranquil mind the revolutions of the planets about the sun in the Copernican system, are mightily disturbed to have the moon alone revolve about the earth and accompany it in an annual revolution about the sun. Some have believed that this structure of the universe should be rejected as impossible. But now we have not just one planet revolving about another while both run through a great orbit around the sun; our own eyes show us four stars which wander around Jupiter as does the moon around the earth, while all together trace out a grand revolution about the sun in the space of twelve years."[33]

It would be a mistake to suppose that Galileo rests his quarrel with Ptolemy on a mere analogy to some non-geocentric motions. His astrometric observations of Jupiter are to bring direct evidence against the Ptolemaic system, in which a planet circling the earth would remain equidistant and of constant diameter. In 1611 and 1612, Galileo finds Jupiter's angular diameter varying between 50 and 37 seconds of arc over a period less than a year.[34] He sees a 2-to-1 variation in brightness of its moons. He offers no explanation. He denies that this variation is caused by "terrestrial vapors", or that it can be due to circular or elliptical [35] motions in their orbits. His public statements are guarded. But in *Sidereus Nuncius* he has declared unequivocally, and almost poetically, that the sun is the center of planetary motions:

"Behold, then, four ... planets. Variously moving about most noble Jupiter as children of his own, they complete their orbits with marvellous velocity — at the same time executing with one harmonious accord mighty revolutions every dozen years about the center of the universe; that is, the sun."[36]

In the period from 1610 through 1612, Galileo's public declaration favoring Copernicus will appear to draw no reaction from the Church. The Church will remain officially neutral despite unofficial signs of opposition to Copernicus. How might a scientist convinced of a new truth, but still a devout Catholic, deal with a future declaration by the Church that such a belief is heretical? The outcome of this dilemma will emerge as our story unfolds.

While the "Aristotelians" reel from this blow to their geocentric view, Galileo comes up with a stronger one.

---

33. The translation is Stillman Drake's (*Discoveries and Opinions of Galileo*, p. 57), except that for consistency I have changed "rotation" to "revolution" and "rotating" to "revolving".
34. Drake, *Galileo: Pioneer Scientist*, p. 134 note 6.
35. It is interesting to note that here Galileo, almost certainly unaware of Kepler's great discovery of the previous year — that all planets move in elliptical orbits — raises the possibility that the Jovian moons have "oval" orbits and then dismisses it. Drake's translation (ibid., p. 58) of the phrase *ovalis vero motus* reads "oval motion". This leaves untranslated the adverb *vero* (literally, "truly"), which I believe refers to symmetry about the axes, so that the phrase may be better translated "elliptical". If Galileo had meant egg-shaped, he could have used the Latin *ovatus* instead of *ovalis vero*.
36. from *Discoveries and Opinions of Galileo*, trans. by Stillman Drake, op. cit.

Macchiae Solari (1612): The Second Round of Astronomical Revelations

Venus, the Imitator　　This next swing of the wrecking ball toward the crumbling edifice of the Ptolemaic world system comes with Galileo's observations of Venus. The first of these, in early November 1610, show a waning gibbous phase,[37] suggesting that Venus may cycle through phases. By 11 December Venus is nearing its maximum angular distance from the sun, is nearly in half phase, and since early November has grown in apparent diameter from 17″ to 23″. Reluctant to wait through a full cycle to fully verify his hypothesis, and thus risk another astronomer's prior announcement, he opts to declare cryptically that Venus will continue through a cycle of phases. To protect the priority of his discovery, he writes out an anagram to Kepler: *"Haec immatura, a me, jam frustra leguntur - oy."*[38] which can be unscrambled to read *"Cynthiae figuras aemulatur Mater Amorum."*[39]

For the rest of the month he watches the Venusian disc wane to a crescent. Then in a New Year's Day 1611 letter he reveals the whole story of his discovery:

> *"At first I saw Venus perfectly round* [he wrote to Julian de Medici] *neat and distinctly terminated, but very small; which figure she retained till she approached nearer to her greatest digression from the sun, increasing continually in apparent bulk. From that time her figure began to fail of its roundness on its eastern side which lay from the sun, and in a few days was reduced to a perfect semi-circle; and continued so without the least alteration, till she left the tangent to her orbit, and began to return towards the sun. At present the semi-circle becomes more and more hollow every day, its angles being changed into horns, which grow sharper and sharper, till they become so thin as to vanish with her passage into the beams of the sun."*[40]

On that same occasion he unscrambles the anagram he had sent to Kepler, and thus publicly announces his discovery of the phases of Venus.[41] How does this discovery affect the Ptolemaic/Copernican controversy?

---

37. This means that more than half (about 70% in this case, decreasing with time) of the Venusian disk is illuminated. By mid-November Galileo had noticed the gibbous phase; see Drake, *Galileo:Pioneer Scientist*, p. 137. Venus reached half phase about 17 December. Had he waited through a full cycle (a Venusian "synodic" year) he would have had a long wait — 584 days.
38. Headlining Galileo's letter to the Tuscan Ambassador at Prague, dated 11 Dec 1610. "One reads in vain what is not yet ready for disclosure."
39. "The Mother of Love imitates the phases of the moon". Castelli's 5 Dec 1610 letter to Galileo, asking if he had detected changes in the shape of Venus, may have arrived a day or so earlier. If so, it is a probable factor motivating Galileo to write out the anagram, but is not (as one myth-maker has loudly and falsely asserted — see Epilogue, Myth #4) temporally prior to Galileo's discovery.
40. Henry C. King, *The History of the Telescope*, Dover, NY, 1955, p. 38.
41. See Drake, *Galileo at Work*, p. 164.

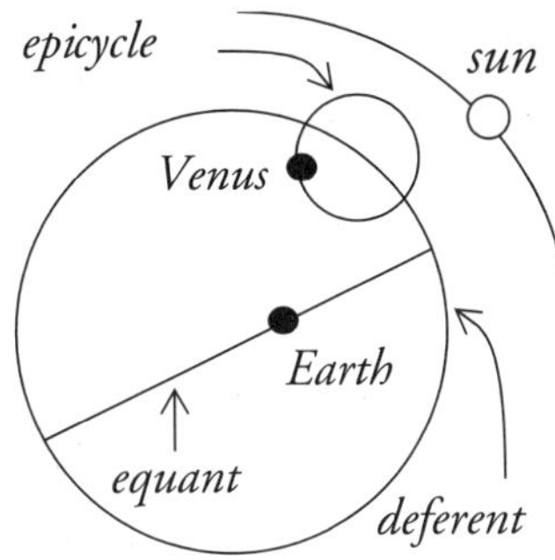

In the Ptolemaic system, Venus orbits Earth in an epicycle moving along a deferent at a uniform angular rate around its equant.[42] But such maneuvers are inconsistent with the cycle of phases and positional relationships to the sun observed by Galileo. Galileo's discovery of Venusian phases disproves the Ptolemaic system; it does not prove the Copernican one. And there is an alternative, due to Heraclides and Tycho, in which the other planets revolve around the sun, while the sun with its retinue of planets revolves around the Earth. Galileo rejects the Tychonic system, and rightly so, for it will eventually be disproved.

Saturn Devours Its Children

In a faint echo of Caesar's diary entry *"Omnis Gallia in tres partes divisa est"*,[43] Galileo tells of first seeing Saturn in 1610 as a trinity of mutually motionless planets that do not quite connect. What will later be seen (by astronomers better equipped) as space inside rings appears to him as a disjuncture, so that the overhang on each side of the planet is seen as a contiguous object. After three years, he looks again, and sees only the central body. By this time, unknown to Galileo, the plane of the rings has nearly aligned with an earthbound observer's line of sight, so that the formerly tripartite object looks like a single disk.[44] "Has Saturn devoured his children?" he asks himself. Later still, the "cup handles" re-appear as if attached to the disk of the planet; the rings, as usual, are not in a position relative to the line of sight for the gap to be seen in its entirety. It is surprising that his perception could be that good. His instrument's resolving power of 10 to 15 seconds of arc is probably marginal even for keen eyesight to resolve the gap between the planet and the rings, as the entire disk of the planet has a mean angular diameter of about 21 seconds.[45] Without declaring a structure separate from the planet, Galileo is first to recognize that Saturn is an oddly shaped planet with some appendages. It is left

---

42. See Ch. 1 "Apollonios" for explanation of deferent and epicycle. The equant is a point on the diameter of the Venusian orbit (deferent) whose distance from the center is equal to that of the earth.

43. The whole of Gaul is divided into three parts.

44. This observation by Galileo occurred in late November 1612, a few days prior to his December 1 letter to Mark Welser. Russell, Dugan and Stewart (*Astronomy,* op. cit.) show Barnard's drawing of Saturn with rings nearly edge on, slightly illuminated by sunlight passing through translucent portions of the ring structure, as seen through the Yerkes Observatory one-meter refractor on 12 December 1907. Ten Saturn years earlier, in late November 1612, the rings would have appeared similarly nearly edge on. Also it should be noted that Galileo did not have a one-meter refractor. I conclude that the rings were not positioned to exhibit themselves to him at this sighting.

45. "Mr. Flamsteed, by the micrometer, measured ... the diameter of Saturn's ring 50″ ... But the diameter of Saturn is to the diameter of the ring, according to Mr. *Huygens* and Dr. *Halley* as 4 is to 9; according to *Gallet*, as 4 is to 10; and according to *Hooke* ... , as 5 is to 12. And from the middle ratio, 5 to 12, the diameter of Saturn's body is inferred to be about 21″ "; Newton, *Principia*, Book III, Cajori trans., [15] .

to Huygens to observe and declare that Galileo's cup handles are detached rings. [46]

Sunspots and Annual Variations in Their Paths

Our sun, the source of light, heat, and life-giving energy, worshipped in some cultures as a god, has blemishes that come and go? These revelations fly against an authoritative belief system that has conditioned the imaginations of the public for two millenia. Had not the descendants of pagan converts to Christianity continued to cling to the notion of a sun that is perfect if not divine? Aristotle had said that celestial bodies are perfect and spherical, neither spotty like Galileo's sun nor pockmarked like Galileo's moon. And Aristotle had to be right because, well, he was Aristotle!

Of all the pronouncements of Aristotle and followers about the other-worldliness and perfection of the heavens, perhaps the most believable is that the sun, the steady light source of such apparent brilliance that the eyes must be shielded from direct vision of it, radiates with uniform and unabated splendor. Imagine then the disbelief early in 1611 when Galileo, one of the first solar observers to use a telescope, announces dark spots[47] on the surface of the sun. An anonymous contender to the discovery then publishes under a pseudonym "Apelles latens post tabulam", [48] claiming priority in the discovery of sunspots and maintaining that they are blacker than any of the spots on the moon. And small wonder he hides his identity, for Apelles is a Jesuit, Christopher Scheiner.[49] Galileo answers that the spots are brighter than the full moon, and provides a proof in his letter [50] to Mark Welser, a merchant of Augsburg, amateur of science, and correspondent with the Jesuits and with Fr. Scheiner. Soon Galileo will make a pivotal discovery: annual variations in the paths of sunspots that cannot be reconciled with the assumption of an absolutely stationary earth. With this discovery, he has in his possession evidence of the earth's motion that will place the system of Copernicus on firmer footing.

A protracted public debate breaks out in 1612 between Galileo and Scheiner, with Welser as intermediary. Scheiner holds that sunspots move east to west, whereas Galileo shows that they move west to east.[51] Perhaps to satisfy fellow churchmen

---

46. Despite the excellent optical quality of the Galilean lenses and relative lack of light pollution at Padova in 1610, it is surprising that he was able to see the extent of detail that he reported in the *Message from the Stars*. The richness of his reports suggests that Galileo had designed a stable telescope mounting. It would be an interesting challenge for an amateur astronomer and telescope maker to duplicate Galileo's 20 power telescope on a geosynchronous mounting with camera attachment and discover whether a photograph of Saturn's rings through such an instrument would appear to show solid protuberances.

47. The spots are only *relatively* dark, as Galileo proved through an ingenious argument. If a spot could be lifted off the sun into space, it would appear dazzlingly bright.

48. "Apelles hiding behind a public paper".

49. In order to protect the Jesuits from the ignominy of an un-Aristotelian argument, Father Scheiner's superiors prevented him from announcing his true identity — to no avail, for Galileo blew his cover.

50. written at Salviati's Villa delle Selve (near Signa) dated January 1612.

and other "Aristotelians" [52] who had to have an immaculate sun, Scheiner claims that the spots are unattached, like planets orbiting above its surface. And here the artistic dimension of Galileo's creative personality comes to the fore. Making splendid use of his early training in art, Galileo shows conclusively that, due to the effect of foreshortening[53] as the spots approach the limb of the sun, and to other effects,[54] they are surface phenomena. In his letter of 14 August 1612, the second to Welser, Galileo gives a more detailed report of their position and motion:

"... the dark spots seen in the solar disk by means of the telescope are not at all distant from its surface, but are either contiguous to it or separated by an interval so small as to be quite imperceptible. Nor are they stars or other permanent bodies, but some are always being produced and others dissolved. They vary in duration from one or two days to thirty or forty. For the most part they are of most irregular shape, and their shapes continually change, some quickly and violently, others more slowly and moderately. They also vary in darkness, appearing sometimes to condense and sometimes to spread out and rarefy. In addition to changing shape, some of them divide into three or four, and often several unite into one; this happens less near the edge of the sun's disk than in its central parts. Besides all these disordered movements they have in common a general uniform motion across the face of the sun in parallel lines. From special characteristics of this motion one may learn that the sun is absolutely spherical, that it rotates from west to east around its own center, carries the spots along with it in parallel circles, and completes an entire revolution in about one lunar month. Also worth noting is the fact that the spots always fall in one zone of the solar body, lying between the two circles which bound the declinations of the planets — that is, they fall within 28° or 29° of the sun's equator."[55]

Galileo is the first terrestrial to scrutinize the surface of a star and to link events on the surface with the orbital characteristics of planets bound to that star. The above quotation is a small sample of observational data about this difficult observational target,[56] the sun, including hand-drawn diagrams, that Galileo releases in his letters to Welser. In these he projects a strong sense of the capacity of the human mind to discover properties of nature against formidable obstacles to that knowledge. There

---

51. He also refers to a south-to-north motion, an apparent motion related to the 7° 15′ inclination of the solar equator to the ecliptic. This inclination was reported (as 7° 30′) in the *Rosa Ursina* by Scheiner, who criticized Galileo for defaulting mention of it.

52. Aristotle personally may have believed in an immaculate sun, but it was his followers the "Aristotelians" who turned celestial spotlessness into dogma.

53. Thanks to rdc for mentioning to me the origin in painting of the concept "foreshortening" and the significance of Galileo's early training for his later observations.

54. Galileo evidently finds that the spreading (toward the center of the apparent solar disk) and convergence (toward the edge of the solar disk) of sunspot clusters fits the assumption that the clusters are moving in an arc roughly coplanar with our line of sight and with radius about equal to that of the sun itself.

55. from Stillman Drake, *Discoveries and Opinions of Galileo*, p. 106.

56. Galileo used telescopic projection onto a sheet of white paper. Later solar astronomers will be equipped with heliosynchronous clock-driven mountings, but for Galileo, however stable his mounting, it is necessary continually to adjust the aim of the telescope to center the magnified image of the moving sun.

is no greater example than this of his perceptiveness, patience, perseverance, and general excellence as an observer and reporter of natural phenomena.

Father Scheiner, an innovative scientist,[57] probably found spots on the sun independently of Galileo.[58] Unfortunately, Scheiner spoiled his book on sunspots, *Rosa Ursina*, with a venomous diatribe against Galileo whom he imagined to have slighted him. Of two discoverers and investigators of sunspots, the one who observes with no pre-conception turns out to be correct about the major properties of sunspots — though Scheiner deserves to be remembered for his measurement of the inclination of the solar equator to the ecliptic. Galileo is a discoverer of sunspots who published a correct description of their position, motion, relative brightness, density, opacity, luminosity variations, birth-death cycle, dimensions, composition, configuration, and clustering. His is an accurate eyewitness description by a pioneer who has beheld and studied the magnified face of a star.

Comets: The Toy Planets

The "Aristotelians" do not have enough "appearance savers" like Scheiner to plug the leaks in their doctrine of celestial perfection as fast as Galileo can find them. After refuting Scheiner, who has tried in vain to lift the spots off the sun, Galileo takes a parting shot at the doctrine from yet another direction: his and Tycho's observations of comets. In closing his second letter to Mark Welser, he says:

"I should even think that in making the celestial material alterable, I contradict the doctrine of Aristotle much less than do those people who still want to keep the sky inalterable; for I am sure that he never took its inalterability to be as certain as the fact that all human reasoning must be placed second to direct experience. ... And as if to remove all doubt from our minds, a host of observations come to teach us that comets are generated in the celestial regions."[59]

To the ancients, the Milky Way looked like a streamer of alternately dense and rarefied material. Comets looked like "hairy" stars, and that is what the Greek word (transliterated) "cometes" implies.[60] Unlike the Milky Way, comets hold their tenuous connectedness in Galileo's telescope. Among Galileo's contemporaries,

---

57. Scheiner invented pantography, the reproduction of plans on any scale, and discovered from his own opthalmological experiments that the retina is the "seat of vision"; cf. Gillespie, op. cit.

58. The same can be said of Thomas Harriot, Johann Fabricius, and Domenico Passignano.

59. from Stillman Drake, *Discoveries and Opinions of Galileo*, p. 118-119.

60. Comets were not named after a popular kitchen/bathroom cleanser, but the other way around. Advertisers seem to have a penchant for astronomical words, possibly because they are Ptolemaic clingers who, despite Galileo's findings, live mentally in the Middle Ages and embrace the notion that all things in the heavens are perfect. Hence, "Milky Way" candy bars, "Quasar" television, "Mercury", "Saturn" and "Pulsar" cars, etc., ad nauseam. Even Galileo has suffered the inappropriate application of his name to a brand of salami. If this use of his name commemorates anything relevant, it must be the ignorance of certain meat vendors who used Galileo's original papers to wrap their product after he died (cf. Ch. 11).

their composition and even their location in or above the earth's atmosphere are intensely disputed. Aristotle suspected that comets are weather phenomena; Galileo places them in the celestial realm for reasons including parallax.[61] And in countering another opponent, the Jesuit Fr. Grassi (pseudonym "Sarsi") about the nature of comets, Galileo holds that cometary material is rarer and thinner than that of the aurora borealis. It will turn out that, with careful qualification, Galileo is correct about both claims. One qualifier is that the thin and rare material is a good description of what Galileo can see; this excludes the nucleus. In response to Grassi's opposing claim that the comets are solid bodies more like planets, and to his lack of substantiation, Galileo retorts:

"... I am not so sure that to make a comet a quasi-planet, and as such to deck it out in the attributes of other planets, it is sufficient for Sarsi or his teacher to regard it as one and so name it. If their opinions and their voices have the power of calling into existence the things they name, then I beg them to do me the favor of naming a lot of old hardware I have about my house, 'gold'. " [62]

Lacking the twin tools of astrometrics and spectroscopy that will facilitate 20th century knowledge about the density of comets, Galileo's description of comets is still surprisingly accurate:

I myself believe that the light of a comet may be so weak and its material so thin and rare that if anyone could get close enough to it he would completely lose it. ... Comets may be dissolved in a few days, and they are not of a circular and bounded shape, but confused and indistinct — indicating that their material is thinner and more tenuous than that of fog or smoke. In a word, a comet is more like a toy planet than the real thing." [63]

Galileo nevertheless believed that a comet is something more substantial than a rainbow. Regrettably, as Professor Boyer notes, he never applied his geometric approach to the analysis of the rainbow.[64] His description can be compared with a

---

61. The idea that comets are inchoate terrestial phenomena has been falsely imputed to Galileo: That they may sometimes visit terrestrial regions does not deny that they move in orbits beyond Venus, a conclusion of Tycho with which Galileo agreed. (Tycho had the great comet of 1577 crashing through multiple of Ptolemy's crystalline spheres — that must have rattled the "Aristotelians".) Galileo's last definitive statement on their location is quoted above in his August 1612 letter to Welser: "... a host of observations come to teach us that comets are generated in the celestial regions". Grassi had them nearly as satellites of the earth, moving in a circle around the earth at only about twice the moon's distance. Galileo saw in this error a strong possibility that Grassi was misusing parallax by tracking weather phenomena sensitive to the relative motion of the observer and the "comet". His suggestion through his student Mario Guiducci in the latter's 1618 *Discourse on Comets* that vapor clouds near the earth may have the appearance of comets is intended to cast doubt on Grassi's argument. Beyond this, Galileo denied that he had any theory of comets and asserted that he would not expect to understand them if he could live a thousand years.

62. from "Il Saggiatore" in *Discoveries and Opinions of Galileo,* trans. by Drake, p. 253-254.

63. Ibid., p. 253

statement made in a radio interview by a 20th century astronomer about a cometary nucleus: *"... you could break a comet apart with your hands; they are hardly put together with anything at all."* [65] Does this sound like Grassi's solid body, or more like Galileo's toy planet?

The astronomer Jan Oort has postulated the existence of a cloud of icy material — something like slush — about a light-year distant from the sun, a cloud that is the source of comets, falling inward toward the sun. Beginning as an almost imperceptibly slow breaking away of a fragment from the main mass of the Oort cloud, a proto-comet continuously increases its speed, as predicted by Galileo's law of falling bodies, extended by Newton to cases of astronomical fall. Unless perturbed in a certain way by gravitational interaction with one of the large, permanent planets of the solar system, the iceball (having fallen through a light-year from the Oort cloud toward the Copernican center of motion) develops a vaporous tail as it approaches the sun and its surface material boils off.[66] It will round the sun above the threshold of escape velocity, and continue on a hyperbolic orbit to infinity, never to return. Occasionally, a comet is perturbed by large planets so as to be captured into an elliptical orbit. Like Halley's Comet, which returns periodically, it will play out its lifespan orbiting the sun until it fully vaporizes. The borderline between the two situations — the infinite open path of the hyperbola, and the closed eternal path of the ellipse — is the parabola, the curve to which Galileo will fit the path of freely falling bodies near the earth.[67]

Creativity and Posterity   Like a closed-path comet, Galileo has entered our lives repeatedly, as his students apply what they have learned: in dynamics, in fluid mechanics, in construction engineering, in acoustics, in aerospace engineering, in rocketry, in computer science, in operations research, in relativity theory and even in the experimental verification of the Copernican system. Many of these applications affect us daily. In Chapters 12 and 13, we will see examples of his recurrent influence. Individuals who are creative in pursuit of truth or goodness can remain with us through the abiding influence of their thoughts and actions.

---

64. "It is a great pity that Galileo never really tackled the rainbow problem, for he was even better equipped for it than was Kepler. He had an intense curiosity; he was quite unafraid of new ideas; he invented and used effectively many scientific instruments; and he believed strongly in the mathematical operation of nature." Carl B. Boyer, *The Rainbow*, Princeton U. Press, 1987, p. 195.

65. Spoken by David Levy, co-discoverer of Comet Shoemaker-Levy 9, the comet that collided with Jupiter in July 1994, in an interview broadcast 30 June 1994 on National Public Radio. Mr. Levy discovered 21 other comets and has written two books about comets.

66. I refer to the process of sublimation, changing directly from solid to gas, under intense solar radiation pressure. This is not boiling in the sense of vaporizing a liquid.

67. Like the hyperbola, the parabola is also an infinite, open path but is the least eccentric open path. For an ellipse the eccentricity e<1, for a hyperbola e>1 and for a parabola e=1.

Columbus of the Heavens?

What is the use of astronomy, the study of things seemingly remote from day to day experience? Galileo may have been first to recognize and to articulate the principle that events local to us in space and time can be understood from the study of remote events. Chemical engineering, for example, is in part an outgrowth of spectroscopic analysis, first applied to astronomy in 1842 when Becquerel photographed the solar spectrum using Fraunhofer's technique of directing a beam of sunlight through a narrow aperture before refracting it through a prism. In 1859, Bunsen and Kirchhoff developed the spectroscope to identify new chemical elements in various substances heated to incandescence, using the same technique by which chemical elements had first been found on the sun. In this way, helium was discovered spectroscopically on the sun by Janssen in 1868 *before it was discovered on the earth.* [68]

Returning to the year 1612 we can see a case in point: Galileo's astrometric investigation of sunspots. He first describes his inertial principle in the context of moving

Synoptic Vision of Two Domains

sunspots, far removed from any earthly observation. In his third letter to Welser, he throw us this rhetorical challenge:

> *"Who does not understand the periods and movements of the planets better than those of the waters of our various oceans? Was not the spherical shape of the moon discovered long before that of the earth, and much more easily?"* [69]

Galileo is a new kind of discoverer. Like Columbus, he ventures into new territory and examines it at close range. Unlike Columbus, an object of his inquiry is to stake a claim, not on some remote piece of real estate, but on the knowledge of the local and familiar gained through analysis of the remote and unfamiliar. His insight into nature comes from a synoptic vision of the two domains.

He has widened our window to the firmament. Others have used his telescope to make spectacular discoveries, like the Great Nebula in Orion, found by the Jesuit mathematician Johann Baptist Cysat while

Durability and Change in the Heavens

looking for the comet of 1618. Evolutes of Galileo's *occhiale* and his *celatone*, telescopes and binoculars used by amateur astronomers, allow you and me the opportunity of rediscovering visions he captured that first night from his Padovan villa. For aeons to come, humans can watch familiar patterns he watched on a

---

68. Cf. *Asimov's New Guide to Science*, Basic Books, NY, 1984, pp. 58-59.
69. from *Discoveries and Opinions of Galileo,* op. cit., trans. by Stillman Drake.

serene canopy stretching over turbulent seas on which earthly mortals strive to sail: the Medici moons; the Venusian *figurae*; the Saturnian cup handles; the buzzing beehive; the solar maculae; the Thirty-Six Sisters; the countless stellar colonies swarming in the white veil of our galaxy. And nearer to home shines our sibling planet with its vast plains, rugged mountains, and statistically smooth periphery. These things are all familiar and fixed. *Yet, Galileo taught us, there is change in the tenth sphere of Ptolemy: the supernova.*

The recognition of birth and death processes in the sphere of the fixed stars, discovered after Galileo but placed there by his and Tycho's parallax measurements of a supernova, implies that over a sufficiently long time scale, these familiar patterns do not hold. If you could have adjusted the time scale of your perception so that a year would pass as a second, you might have looked at the Leaning Tower over the first eighty "seconds" of this century and seen it falling. By an imaginative adjustment of time scale, those familiar celestial configurations can be seen as a consequence of earthbound perspective and the brevity of human life. Durability and change can be found in distinct interpretations of the same perception. In the same rivers we step, and we do not step.[70]

So that we can trace the events of the counter-revolution to the *Message from the Stars*, let's return now to the year 1610, before the announcement in *Macchiae Solari* of the phases of Venus, a disclosure critical to the undermining of the Ptolemaic world system. Galileo has challenged all of Rome to verify the announcements in his *Message from the Stars*. Those who see what he has seen through his telescope are hailing him as the New Columbus of the Heavens. Meanwhile, his enemies are preparing a counterattack. Unlike Galileo, and despite their pretenses, they operate in a non-scientific arena. Their authority derives not from the knowledge of nature, but from something quite different.

---

70. Paraphrase of Heraclitos' Fragment Twelve (see Ch. 1 "Empedocles") less paradoxically rendered by T. M. Robinson: "As they step into the same rivers, different and <still> different waters flow upon them." *Heraclitus: Fragments*, University of Toronto Press, 1987.

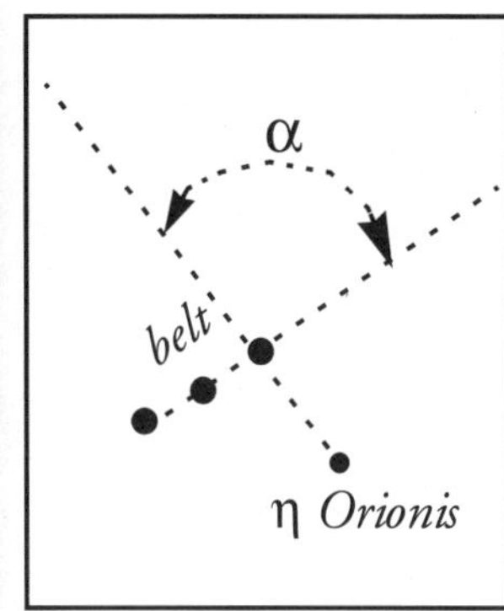

Orion at Dawn from Anza, California
Is the camera bewitched?
(photo by author)

Monument to Galileo's *Message from the Stars*
University of Padova, Italy
(photo by author)

Open Star Cluster M44: "Praesepe"
(photo by John Sanford)

Stellar Milky Way With Comet Halley (photo by Don Rado)
Galileo revealed the stellar composition of the Milky Way
and argued for the vaporous nature of comets.

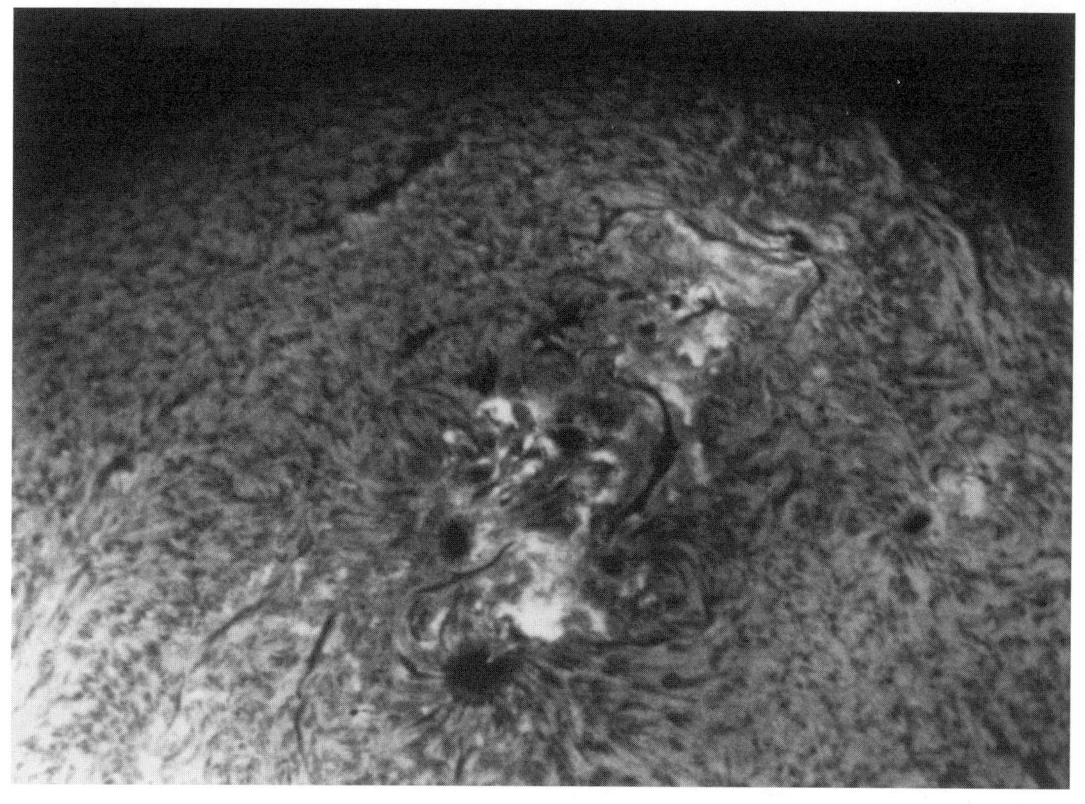

Sunspots (Solar Chromosphere) 1990 Aug 8 13:35 UT
Phenomena Galileo Discovered and Studied on the Sun and Correctly
Predicted of Other Stars; His Analysis of Their Motion Constituted the
First Observational Demonstration of the System of Copernicus.
(Courtesy, Mr. John Sanford; photo by Dr. Jean Dragesco)

The Pleiades: To Which Galileo Added More Than 29 "Sisters" to the Original Seven
Photographed through 400 mm lens @ F5.6, 15 min. @ ASA 400
(photo by Charles Morrill)

Jupiter and Companions
Discovered by Galileo

This is my attempt to approximate what he saw through his telescope on the discovery night of 13 January 1610 when for the first time he saw all four satellites. At that time, they were in the " mJmmm " configuration shown here.

(photo by author)

Chapter 8. Spiderwebs from the Sky

Galileo Champions a Condemned Doctrine (1610 - 1623)

"When a person finds no defense to be of any avail against his mistake and produces a frivolous excuse, people say that he is reaching for ropes from the sky. This author grasps not at ropes, but at spiderwebs from the sky, ..."

... Galileo Galilei, *Dialogo*, Third Day
(referring to the argument of an opponent)

Within four years of these watershed events, the counter-revolution to the *Message from the Stars* begins. Failing to recognize in Galileo its greatest benefactor, the Roman Catholic Church weaves complex webs, those same webs in which it will entangle itself, as it secretly prepares to move against him. In the meantime, Galileo takes two crucial steps that seal his fate. The first is to leave the safety of the neutral Venetian Republic. The second is to take up the banner of Copernicus, carrying it farther and faster than its creator ever did or even contemplated.

Galileo Returns Home, 1610

Like none other before it, the period 1609-1610 has marked a sharp discontinuity in the history of western civilization. Published within a year of each other,[1] but independent and seemingly unrelated, the astronomical discoveries of Kepler and of Galileo are the stuff of which Newton will, in less than a century, embody human comprehension of the universe in a system of laws. Even from the limited perspective of his contemporaries, Galileo's importance is already manifest in his telescope. Like Archimedes before him, he has applied new science to the defense of the Republic. As they stand at the Bell Tower and look out over the Adriatic in recognition of this result, the Venetians are preparing to offer him a raise contingent on his accepting a lifetime appointment. It is a condition to which he will prove unwilling to adhere. He sees the obligation to teach as a distraction from research and as a constraint to creativity. To Galileo, Florence is home. In the summer of 1610, he negotiates to be appointed First Philosopher and Mathematician to the Grand Duke of Toscana, using *Message from the Stars* as an application for a lifetime grant.[2] Cosimo II, son of Ferdinand de' Medici and the Grand Duchess Cristina di Lorena, had succeeded his father as Grand Duke only the previous year. A patron of the Arts and Sciences, Cosimo has brought lasting renown to Florence by:

---

1. Kepler's *Astronomia Nova*, announcing the first two laws of planetary motion, was published in the summer of 1609; Galileo's *Sidereus Nuncius*, on the 12th of March in 1610.
2. A lifetime grant, unlike the lifetime appointment offered him at Padova, carries no stipulations as to official duties. It might be compared to the IBM fellowship of today.

"... inviting back, protecting from persecution, and establishing in honour in his own country, the great Galileo, who had eighteen years before been compelled by jealous animosity to leave it. ... And in this capacity Galileo remained for twenty-three years, provided with a maintenance which left him free to prosecute his scientific studies, and shielded, under the personal protection of the Grand Duke of Tuscany, from the machinations of his enemies both at Florence and Rome; ... " [3]

Cosimo II will be succeeded in 1620 by his son Ferdinand II, whose deference to the papacy will weaken Galileo's protection. Having left the safe harbors of Venice to return to a territory more influenced by the Papal States, Galileo is gathering data for a book he has not yet written and that will trigger a torrent of trouble with the Vatican. The first problem he attacks at Florence is that of determining the orbits of the Medicean satellites of Jupiter,[4] a problem so difficult that even the mathematician Kepler, a founder of celestial mechanics, questions whether it is possible. But by April of 1611, Galileo has done it and is compiling tables of satellite positions. Meanwhile, he traverses the royal road between his villa and the ducal villa of Poggio Imperiale bearing this news to his hosts: he has seen Venus move through inferior conjunction, passing from waning to waxing crescent and growing in angular diameter to a full minute of arc, more than three times its size of the previous November.

Triumphal Visit to Rome, 1611 Seeking to exhibit his discoveries to a wider audience, Galileo leaves Florence for Rome on March 23. Night after night he stops and records observations of Jupiter's satellites: at San Casciano, at Siena, at San Quirico, at Acquapendente, at Viterbo and Monterosi. Each night, the crescent of Venus narrows and its disk shrinks as it moves toward the far side of the sun. He arrives on March 29 at Rome and is warmly received by Jesuits of the Collegio Romano. Galileo has brought his instrument of discovery. He discusses Copernican astronomy with Cardinal Bellarmino[5] and shows him the Medicean satellites and irregular features of the moon. Bellarmino is assured by Father Clavius[6] and others that they have verified Galileo's findings. But Clavius believes the lunar irregularities to be illusory. At a convocation of the Collegio Romano, the Jesuit mathematician Father Maelcote delivers an oration in praise of Galileo's *Message from the Stars*.

---

3. Col. G. F. Young, *The Medici*, pp. 651-652.
4. Cf. Chapter 5, "The Jupiter Satellite Ephemeris and Analog Computer".
5. Bellarmino, a nephew of Pope Marcellus II, had distinguished himself among Jesuits by his work *De controversiis* regarding heresies, which was nearly placed on the Index of Forbidden Books by Pope Sixtus V because it limited the Vatican's temporal powers, but was later approved by Pope Gregory XIII; see Thomas J. Campbell, S.J., *The Jesuits 1534-1921*, The Encyclopedia Press, NY, 1921, p. 110.
6. In addition to having helped bring about the Gregorian calendar, Cristoforo Clavio (Clavius) was also the leading mathematician of the Collegio Romano. See Ch. 7 "Lunar Features" for the account of the later bizarre naming of a prominent lunar crater for Clavius, who denied the existence of such objects.

On this visit, Galileo is elected to the Academy of the Lincei, founded eight years earlier by Prince Cesi to pursue knowledge about the observable natural world, in contrast to the "paper world" being promoted and fossilized by university philosophers. Cesi arranges a banquet on the 14th of April, at which Galileo is the focus of attention. Before dinner, Jupiter hangs low in the western sky near Aldebaran and the moon has not yet risen. Galileo has brought his *occhiale* and displays the magnified heavens for the guests. Giovanni Demisiani, Cardinal Gonzaga's mathematician, is present at the banquet and concatenates the Greek "τῆλε" ("far away") and "σκοπός" ("sentinel") to describe Galileo's instrument; after this, the new word, "telescope", much repeated by Cesi and by others, spreads beyond the banquet, and sticks.[7] Galileo has succeeded in bringing the attention of Rome to his discoveries, but not without controversy.

In June he returns to Florence and to multiple controversies brewing over his revelations: the telescope is being derided by Francesco Sizzi as a device for putting things into the sky that are not there, and the lunar mountains are denied by Ludovico delle Colombe who maintains that if the moon has surface irregularities, they must be encased in a smooth crystal. Galileo responds by offering a substantial prize to anyone who could make a telescope to create satellites around exactly one planet, and by postulating crystal mountains on top of Colombe's crystal casing. Then he is drawn into a controversy at Salviati's palace with some "Aristotelian" professors on the subject of floating bodies.

Floating Bodies

Why do floating bodies float? The "Aristotelians" attribute floating to the presence of air in bodies and to a flatness of shape that hinders descent through fluid. At Salviati's palace, a Pisan philosopher, Vincenzio Di Grazia, argues that a broad, flat object floats because it cannot overcome the water's cohesion. Galileo counters that the same object has no trouble rising when submerged and released. After his verbal duel with the "Aristotelians", Galileo publishes a treatise on hydrostatics [8] — not a stuffy one, but full of interesting and amusing experiments that anyone can perform with little equipment. *Discourse on Bodies on or in Water* holds surprises for the "Aristotelians" who argue deductively and do not test their conclusions. To explain floating he extends the Archimedean principle of the density difference between water and the bodies that float.[9] He proves, for example, that a cone of

---

7. See Willy Ley, *Watchers of the Skies*, Viking Press, NY, 1966; See also Drake, *Galileo at Work*, p. 196.
8. This was Galileo's *Discourse on Bodies on or in Water*, pub. 1612. Broader in scope than its title implies, it contains a new definition of moment: "Moment, among mechanics, means that force ... with which the mover moves and the moved resists, which force depends not simply on weight, but on speed of motion [and] on different inclinations of the spaces over which motion is made — for a heavy body descending makes greater impetus in a steeper space than in one less steep", trans. by Drake, *Cause, Experiment and Science,* op. cit., p. 28-29.

material twice as dense as water that would sink if placed base-down can float below the water's surface if placed vertex-down.[10] Still buoyed up by the *Message from the Stars*, his name rides a new wave of public acclaim as the *Discourse* sells out in 1612 and justifies a second printing.[11]

Galileo and the Jesuits

By 1613, and despite the Jesuit Collegio Romano's enthusiastic endorsement of the *Message from the Stars*, Galileo's philosophy and style of reasoning are already bringing him into potential conflict with this powerful order of Catholic priests and religious men. Since its founding by Ignatius Loyola in 1534, the Jesuit order has wielded enormous influence through its schools. From a 16 year training period during which Jesuits are called "scholastics", they emerge with a breadth and depth of knowledge that makes them leaders of the Counter-Reformation. They have become the premier schoolmasters of Europe. It is not just Galileo's non-Aristotelian theory of gravity, of hydrostatics, and of motion in general, that is bound to cause conflict with them. It is an even more distressing matter of style, one that threatens the qualitative approach to science that scholastics have made a tradition.

Education should decrease the likelihood of seeking to be bound by authority. Paradoxically, the Jesuits are superbly well educated in traditional subjects, yet surrender individual intellectual freedom to their superiors in the hierarchy, who in turn are bound by authority and tradition. Their intellectual life has been held together by the dynamic tension between education and submission. But note the commonality between the two poles of this paradox: love of tradition. Though the first six books of Euclid are included in the Jesuit curriculum even before the *Ratio studiorum* of Galileo's time, geometry is not for them the language of nature. Instead, the scholastic science of the Jesuits involves a kind of language game that uses logical rules, attributed to Aristotle, and maps qualities to phenomena observed or imputed to the physical world.[12] Now along comes this upstart Galileo, who insists on non-traditional measurements of real phenomena and on *quantitative* analysis based on those measurements. It is a more exacting approach, decidedly uncomfortable for vendors of vagueness and merchants of mysticism.

---

9. What about those needle-thin objects that are more dense than water and still float? Galileo tries to sink this "Aristotelian" objection by hypothesizing that such objects are buoyed up by a layer of air surrounding them. Though apparently unaware of surface tension, he is correct in treating this phenomenon as an exceptional situation that exceeds the pure conditions of Archimedean analysis.

10. Ibid., p. 141.

11. See Geymonat, *Galileo Galilei: A Biography and Inquiry into his Philosophy of Science*, op. cit.

12. Lattis gives this example of their qualitative science: "We might ... explain the blotchy appearance of the moon with this syllogism: the celestial and terrestrial regions are characterized respectively by clarity and obscurity; the moon is intermediate between the celestial and terestrial regions; therefore the moon shares in both clarity and obscurity." James M. Lattis, *Between Copernicus and Galileo*, op. cit., p. 33.

Soon after that second literary triumph *On Floating Bodies*, and while Galileo is debating publicly and forcefully with Father Scheiner about sunspots, comets, and scientific method, a social event sets the stage for a collision nearly two decades in the future. It is Thursday morning, the 12th of December in the year 1613, when his friend, confidant, and student Father Don Benedetto Castelli is having breakfast with Cosimo II, who inquires whether Castelli has a telescope. Yes, replies Castelli, and with his telescope he has verified the moons that Galileo recently discovered and named after this same Grand Duke. Meanwhile, Boscaglia, a philosophy professor at Pisa, is whispering into the ear of the Duchess [13] that, while Galileo may have found some new planets, his suggestion that they orbit the sun while orbiting Jupiter is beyond belief and contrary to Scripture. Armed with Boscaglia's arguments, after breakfast the G. D. Cristina begins to debate with Castelli about this matter. Castelli, though uncomfortable arguing with royalty, puts up a spirited defense using Galileo's arguments and makes a positive impression on his hosts.

The Royal Breakfast Meeting

What Boscaglia has told the Duchess includes an argument based on a passage from the Old Testament book of Joshua.[14] That same argument is evidently being whispered to Church officials around this time, but is not original either to the faculty at Pisa or to the Roman Catholic Church. One might well suppose that, if the Church could not come down on the right side of the issue, at least it would have originated its own argument. But most probably, it seems, the grounds on which Catholic priests will bring charges against Galileo before the Roman Catholic Inquisition originate not with the Church, but with that wily "heretic" (as they regarded him) Martin Luther.[15] For in the last days of Copernicus, while his ideas were spreading in advance of his book, Luther chided him for wanting to turn

The Lutheran Challenge

---

13. This is the Grand Duchess Cristina di Lorena, Cosimo's mother, whose title and name, for the reader's convenience, I will abbreviate as the "G. D. Cristina". G.D. is my abbreviation for "Grand Duchess".

14. "Then Joshua spoke to Yahweh, the same day that Yahweh delivered the Amorites to the Israelites. Joshua declaimed:

> 'Sun, stand still over Gibeon,
> and, moon, you also, over the Vale of Aijalon'.
> And the sun stood still, and the moon halted,
> till the people had vengeance on their enemies.

Is this not written in the Book of the Just? The sun stood still in the middle of the sky and delayed its setting for almost a whole day"; Joshua 10:12-13, *Jerusalem Bible*, Alexander Jones ed., Doubleday, NY, 1966, p. 287. Mr. Jones notes that the *Book of the Just* is an ancient collection of poems now lost. Here is a candidate for Ripley's *Believe It Or Not*: Galileo was accused of an anti-scriptural cosmology based on a passage — implicating God in vengeance — that a biblical scribe lifted from some old poetry book.

the whole of astronomy upside down by placing the sun at rest and the Earth in motion, in contradiction (thought Luther) to the scriptural passage in which God is said to have lengthened the day in answer to Joshua's prayer "Sun, stand thou still ..." in order to give favor to Joshua in battle. Why, asked Luther, would God stop a motionless sun?

On Saturday, 14 December, Castelli reports in a letter to Galileo the argument from Joshua that surfaced at the royal breakfast meeting. In a typical exercise of that forensic skill by which he first disarms an opponent, then uses the opponent's weapon against him, Galileo responds to Boscaglia's argument. The response is so ingenious and, though favorable to the Church, so crucial in understanding how Galileo was indicted by the Inquisition, that the reader deserves to hear it in its entirety: [16]

"... Let us then assume and concede to the opponent that the words of the sacred text should be taken precisely in their literal meaning, namely that in answer to Joshua's prayers, God made the sun stop and lengthened the day, so that as a result he achieved victory;...

I first ask the opponent whether he knows with how many motions the sun moves. If he knows, he must answer that it moves with two motions, namely with the annual motion from west to east and with the diurnal motion in the opposite direction, from east to west.

Then, secondly, I ask him whether these two motions, so different and almost contrary to each other, belong to the sun and are its own to an equal extent. The answer must be No, but that only one is specifically its own, namely the annual motion whereas the other is not but belongs to the highest heaven, I mean the Prime Mobile; the latter carries along with it the sun as well as the other planets and the stellar sphere, forcing them to make a revolution around the earth in twenty four hours, ...

Coming to the third question, I ask him with which of these two motions the sun produces night and day, that is, whether with its own motion or else with that of the Prime Mobile, ..."

Let's interrupt Galileo here to insure that the strategy of his argument is clear to us who live in another time. The Prime Mobile, or highest celestial sphere, is a mechanism of the old cosmology, not that of Copernicus. Here, Galileo begins by

---

15. Martin Luther, excommunicated in 1520 by the Roman Catholic Church (under Pope Leo X, Giovanni de' Medici) for his attacks on doctrine, is the earliest famous individual to express the objection to Copernicanism from Joshua, as far as I have determined, and though it is possible that Boscaglia's application of the passage is independent of Luther, the Church seems to have picked up on it without noticing or without being concerned about Luther's priority. Of Copernicus, Luther said: "... This fool wishes to reverse the entire science of astronomy; but sacred Scripture tells us that Joshua commanded the sun to stand still, and not the earth." Luther's statement is quoted in Ch. III Part II of A. D. White's *A History of the Warfare of Science with Theology in Christendom*. One wonders whether Erasmus, had he postdated Copernicus and Luther, would have directed his praise of folly at the former for having earned such a charming commendation from Luther, or at the latter for having called Copernicus a fool.

16. This quotation is from Finocchiaro's translation of Galileo's 21 Dec 1613 letter to Castelli.

assuming the Ptolemaic astronomy accepted by his opponents in order to derive a contradiction with the passage of Joshua, likewise accepted as literally true by his opponents. This is a standard method of arguing in geometry, but it seems to elude or baffle certain Church officials. Galileo continues:

" ... The answer must be that night and day are effects of the motion of the Prime Mobile, and that what depends on the sun's own motion is not night and day but the various seasons and the year itself.

Now, if the day derives not from the sun's motion, but from that of the Prime Mobile, who does not see that to lengthen the day one must stop the Prime Mobile and not the sun? Indeed, is there anyone who understands these first elements of astronomy and does not know that, if God had stopped the sun's motion, He would have cut and shortened the day instead of lengthening it? For, the sun's motion being contrary to the diurnal turning, the more the sun moves toward the east the more its progression toward the west is slowed down, whereas by its motion being diminished or annihiliated the sun would set that much sooner; this phenomenon is observed in the moon, whose diurnal revolutions are slower than those of the sun inasmuch as its own motion is faster than that of the sun. It follows that it is absolutely impossible to stop the sun and lengthen the day in the system of Ptolemy and Aristotle, and therefore either the motions must not be arranged as Ptolemy says or we must modify the meaning of the words of the Scripture; we would have to claim that, when it says that God stopped the sun, it meant to say that He stopped the Prime Mobile, and that it said the contrary of what it would have said if speaking to educated men in order to adapt itself to the capacity of those who are barely able to understand the rising and setting of the sun.

Add to this that it is not believable that God would stop only the sun, letting the other spheres proceed; for He would have unnecessarily altered and upset all the order, appearances, and arrangements of the other stars in relation to the sun, and would have greatly disturbed the whole system of nature. On the other hand, it is believable that He would stop the whole system of celestial spheres, which could then together return to their operations without any confusion or change after the period of intervening rest.

However, we have already agreed not to change the meaning of the words in the text; therefore it is necessary to resort to another arrangement of the parts of the world, and to see whether the literal meaning of the words flows directly and without obstacle from its point of view. This is in fact what we see happening.

For I have discovered and conclusively demonstrated that the solar globe turns on itself, completing an entire rotation in about one lunar month, in exactly the same direction as all the other heavenly revolutions; moreover, it is very probable and reasonable that, as the chief instrument and minister of nature and almost the heart of the world, the sun gives not only light (as it obviously does) but also motion to all the planets that revolve around it; hence, if in conformity with Copernicus's position the diurnal motion is attributed to the earth, anyone can see that it sufficed stopping the sun to stop the whole system, and thus to lengthen the period of the diurnal illumination without altering in any way the rest of the mutual relationships of the planets; and that is exactly how the words of the sacred text sound. Here then is the manner in which by stopping the sun one can lengthen the day on the earth, without introducing any confusion among the parts of the world and without altering the words of the Scripture."

Galileo is Denounced in Church

Galileo has lit a fuse that burns quietly for about a year. He writes on sunspots (*Macchiae Solari*) and engages in public debates with sundry "Aristotelians". Suddenly, in December of 1614, a young priest named Caccini, apparently having heard secondary whisperings of the slander that Boscaglia had whispered into the ear of the Duchess, denounces Galileo from the pulpit. He has not met the man he is denouncing and has not understood Galileo's argument. Galileo had provided a cover that protected the ecclesiastical interpretation of Joshua. Poised at the pulpit and drooling with ambition to rise in the institutional hierarchy of the Dominican order, Caccini seizes the opportunity and strikes, driving his fangs into that cover and injecting toxic venom into the lifeblood of intellectual discourse. Knowingly or unknowingly using the argument of the "heretic" Luther, he accuses Galileo of heresy. That the Earth turns on its axis to create night and day contradicts the sacred Scriptures, he claims. The logic is lacking, but the damage is done.[17]

Galileo Is Reported to the Inquisition

When the news of this event reaches Father Castelli at Pisa, he is in contact with a Dominican priest, Father Lorini, a professor of Church history at the University of Florence. Is Castelli thinking that Lorini can be won over by showing him Galileo's letter? Somehow, Lorini obtains a copy of Galileo's letter of the previous December, quoted above. Lorini's ambition is of a less mundane character than Caccini's; and while Galileo has reached the celestial realms directly through his telescope, Lorini hopes to do so by protecting the papacy from the propagation of perilous heliocentric doctrines. Filled with afflatus for defending the faith, and seemingly buoyed heavenward on ethereal vapors that flow from his pen, Lorini composes a submission to the Inquisitors, attaching a copy of Galileo's letter, which he has discussed with Caccini and other cronies. This complaint puts the Inquisition in motion against Galileo. It also shows that Lorini doesn't have a clue to Galileo's real meaning in this letter; his zeal for investigating potential heretics outstrips his ability to read plain Italian. He misrepresents the astronomical argument of Galileo against the Lutheran objection, saying of Galileo's letter to Castelli:

---

17. With charitable self-restraint, Galileo says of Caccini (letter to Picchena 20 Feb 1616) "I perceived not only his great ignorance, but that he has a mind void of charity, and full of poison" (Allan-Olney, *The Private Life of Galileo*, p. 83). A toxic void can sway the Congregation of the Index, as we shall see.

> "... it claims that when Joshua ordered the sun to stop one must understand
> that the order was given to the Prime Mobile and not to the sun itself." [18]

Of course the letter, his own inexact copy of which he attaches to his complaint, claims no such thing, but rather that it would be consistent with Ptolemy to consider the order as having been given to the Prime Mobile. Lorini ignores Galileo's demonstration that Copernicus provides a means for the Church to save the literal interpretation and to prevent the emergence from Scriptures of a God who creates and knows everything about the universe and then proceeds to babble blatant blunders in science. Knowing that Copernicus had been a respected clergyman whose astronomical calculations had been the basis of the "Gregorian" calendar and whose dedication of his book to Pope Paul III had been accepted, the Church missed an opportunity for positive change that would have followed from relinquishing its bureaucratic attempt to dominate science by force. Instead, the Inquisitorial machinery, blind and unthinking, rolls on, from that point gaining momentum against Galileo and the Copernican doctrine.

Galileo reacts to these denunciations by denying (in a letter to the G. D. Cristina who had been instrumental in starting all

To Cristina, 1615, the Boundary of Science and Faith

this trouble in the first place) that true propositions of religion can conflict with those of science. Just as he denies the duality of nature between the Earth and the heavens, so he denies duality of truth within the intersection of natural philosophy and theology. From the unity of nature, the unity of truth follows. Aiming to save the Church from interpreting Scriptures in a way that would prove untrue, he asserts that knowledge comes from the exercise of scientific method, not from authority. He does not say that one can attain one kind of knowledge about nature from Scriptures, and another directly from nature. Rather, he holds that the interpretation of Scriptures ought to be subject to what is revealed through nature:

"...in discussions of physical problems we ought to begin not from the authority of scriptural passages, but from sense experiences and necessary demonstrations; ... having arrived at any certainties in physics, we ought to utilize these as the most appropriate aids in the true exposition of the Bible, and in the investigation of those meanings which are necessarily contained therein, for these must be concordant with demonstrated truths. I should judge that the authority of the Bible was designed to persuade men of those articles and propositions which, surpassing all human reasoning, could not be made credible by science, or by any other means than through the very mouth of the Holy Spirit. ... But I do not feel obliged to believe that that same God who has endowed us with senses, reason,

---

18. Professor Finocchiaro's translation, from Lorini's complaint, 7 Feb. 1615.

and intellect has intended to forgo their use and by some other means to give us knowledge which we can attain by them." [19]

To deduce the absence of something is to deny the reality of its particulars in a single act of comprehension. Here Galileo has denied a convoluted natural universe in which reason and sense experience must be abandoned when contradicted by a separate claim to knowledge, whether it be called revelation, dogma, or faith. While science pursues the discovery of truth, the Bible aims at the preservation of belief according to commonly accepted opinions of the times. He writes as follows:

"Among a thousand ordinary men who might be questioned concerning these things, probably not a single one will be found to answer anything except it looks to him as if the sun moves and the earth stands still, and therefore he believes this to be certain. But one need not on that account take the common popular assent as an argument for the truth of what is stated; for if we should examine these very men concerning their reasons for what they believe, and on the other hand listen to the experiences and proofs which induce a few others to believe the contrary, we should find the latter to be persuaded by very sound arguments and the former by simple appearances and vain or ridiculous impressions.

It is sufficiently obvious that to attribute motion to the sun and rest to the earth was therefore necessary lest the shallow minds of the common people should become confused, obstinate, and contumacious in yielding assent to the principal articles that are absolutely matters of faith. ... Speaking of this, St. Jerome writes: ...'It is the custom for the biblical scribes to deliver their judgments in many things according to the commonly received opinion of their times.'" [20]

Accommodating "shallow minds of the common people" is the aim of biblical scribes, not of his own. Galileo's writing appeals to the reader's capacity to corroborate. Consider the treatise on floating bodies, written in the vernacular and full of simple experiments. It radiates awareness that claims about the physical world, unlike the pronouncements of religion, can be tested independently of any authority. On the heels of his triumph in astronomy, Galileo is offering a deal: separation of function.[21] By adhering to his advice, the Church can avoid pronouncements that would discredit its authority. But the warning is wasted on theocrats preoccupied with political pressures. The Counter-Reformation against Luther and Calvin is in full swing, and political storm clouds that precede the Thirty Years War (1618-1648) between Catholics and Protestants are brewing. In a vain attempt to flaunt its authority in biblical exegesis, the Church chooses to ignore Galileo's advice, a choice it will come to regret.

---

19. Stillman Drake's translation.

20. trans. by Stillman Drake in *Discoveries and Opinions of Galileo*, pp. 200-201.

21. Stripped of diplomacy, his argument might be stated: "Don't concern yourself with the Earth's motion, and I won't sell indulgences".

Remember that during Galileo's 1611 visit to Rome, Bellarmino inquired of the Collegio Romano [22] about Galileo's interpretations of what his telescope shows in the

Copernicanism Condemned

heavens. The faculty agreed with Galileo's findings — except that Father Clavius held the lunar surface irregularities to be only apparent.[23] Nevertheless, to stem the current of conversions to Galileo's epistemology flowing from his letters to Castelli and Cristina, the Inquisition hastens its procedure against him. In December, 1615, Galileo rushes to Rome to foil the plots against the system of Aristarchos/ Copernicus, on whose triumph he is now staking a reputation by vigorous public defenses. He is temperamentally indisposed to heed the advice of Sagredo: [24]

> "Philosophize comfortably in your bed, and let the stars alone. Let fools be fools; let the ignorant plume themselves on their ignorance. Why should you court martyrdom for the sake of winning them from their folly?"

Winning them from their folly was what mattered to Galileo.

Catholics Use Luther's Argument To Silence Galileo (1616)

At Rome he walks into an ambush. Favaro tells us that there ...

> he busied himself making new converts, and while expecting to be called by the Inquisition to defend others, and to enlighten the Holy Office on the scientific merits of the Copernican system, he found them instead acting against him as the pr......l accused, and as an accused so dangerous as to have to deny him even the right of self-defense.[25]

After a week's deliberation, a report of Inquisition Consultants dated 24 February 1616 declares the Copernican doctrine "foolish and absurd in philosophy". Copernicus is not mentioned by name, but the "foolish and absurd doctrine" is described by two tenets of which the second is undeniably that of Copernicus:[26]

(1) "The sun is the center of the world and completely devoid of local motion."
(2) "The earth is not the center of the world, nor motionless, but it moves as a whole and also with diurnal motion."

Tenet (1) is declared "formally heretical" and tenet (2) "at least erroneous in faith".

-----

22. a forerunner of the modern Gregorian University, which operates from the same Vatican site.
23. See Brodrick, *Robert Bellarmine*, Newman Press, 1961.
24. Allan-Olney, *The Private Life of Galileo*, op. cit. p. 97. Giovanfrancesco Sagredo, a student of Galileo, was an active scientific experimenter who avoided controversy and urged Galileo repeatedly against public disputes. Galileo will ignore his advice but immortalize his name in various dialogues.
25. Paraphrase of my translation; cf. Appendix A.
26. All quotations from Inquisition documents in this section are translations by Finocchiaro (see Bib.).

Decree of the Index (Edict of 1616)

On March 5, the Congregation of the Index suspends certain books and bans others that try "to show that the ... doctrine of the sun's rest at the center of the world and the earth's motion is consonant with the truth and does not contradict Holy Scripture." Again, the name of Copernicus is not attached to the doctrine, rather it is "the false Pythagorean doctrine"; Copernicus' book *On the Revolutions of Spheres* is suspended pending correction. A book by Father Paolo Antonio Foscarini, mathematician and founder of the Carmelite monastery at Montalto, portrays Copernicus as consistent with Scripture; it is prohibited and condemned. In order that this Copernican opinion "may not creep any further to the prejudice of Catholic[27] truth", all books which teach that the opinion is true and does not contradict Holy Scripture are likewise prohibited. No mention is made of Galileo. But slanderous rumors — that the Church has censured him — begin to circulate.

Bellarmino Warns Galileo (1616)

On March 11, he is warmly received by Pope Paul V, who deplores the slanders against him and promises that Galileo can feel safe as long as he (the Pope) lives.[28] Trusting in the Pope's statements, Galileo writes a reassuring letter to the Tuscan Secretary of State. Apparently though, in seeking truth and justice, it does not suffice even for the world's greatest scientist to "kiss the feet of His Holiness". Inquisition minutes will later reveal that only two weeks before making this promise to Galileo, this same Pope has given an order to Cardinal Bellarmino[29] requiring Galileo to abandon his Copernican views, and *if he refuses*, to imprison him. Paul V's order explains why, despite the lack of reference to Galileo's work in the Church edict of March 5th, Bellarmino personally warns him. Privately, Bellarmino tells Galileo that the Copernican system may be taken hypothetically and taught as a convenient mathematical model for predicting the motions of the planets, though not held or defended as physically true.[30] Publicly, summoned before Cardinal Bellarmino and before witnesses "of whom one would almost say that the official presence had been hidden from him",[31] Galileo is warned to abandon the opinion of Copernicus; he promises to obey.[32] Even more slanderous rumors ensue: not only was he censured but, they say, he recanted.

---

27. Note that truth is qualified as Catholic as if to suggest the existence of a different kind of truth.

28. Paul V died in 1621 and, after the short papacy of Gregory XV, was succeeded in 1623 by Urban VIII.

29. I follow the spelling in the Inquisition documents. Many writers and American Jesuits insist on "Bellarmine", on the curiously wimpy ground that Bellarmino's dispute with James I of England caused his name to be anglicized in the 17th century (see Sharratt, *Galileo: Decisive Innovator*, Blackwell, 1994, p. 38). Apparently, American Jesuits (and the Vatican, when using English) set aside Bellarmino's own spelling in favor of that arbitrarily adopted by a Protestant king.

30. Besides the fact that Galileo claimed this, it is clearly Bellarmino's position from his 12 April 1615 letter to Foscarini, translated by Finocchiaro in *The Galileo Affair*.

31. Here I quote my translation of Favaro's Preface; see Appendix A.

Galileo stays on in Rome and hears the new rumors. The truth is that he has been neither censured nor gagged. To stifle the rumors, he asks Bellarmino to put the warning in writing, and receives a certificate dated 26 May clearing him of slanders to the effect that he had recanted. The certificate states that he has not abjured any opinion nor received any penance and was merely notified of the censors' opinion that the "doctrine" of Copernicus "is contrary to Holy Scripture and therefore cannot be defended or held." But the censors' opinion is not official Church doctrine. Teaching the Copernican view as a hypothesis has not been condemned. The censors have caused some minor revisions to the book of Copernicus to be made as a condition of approving further publication. The effect of the revisions is to make it appear that Copernicus advanced his theory as a convenient mathematical model, not as a physical truth. This is the real sense in which the Church action condemns the theory — not by banning it from publication but by attempting to remove it from the physical world. Galileo has been warned not to teach the view of Copernicus as if it were physically true.

Bellarmino's Affidavit

Fueling the engine that plots the downfall of his influence on men of his own time, and drawing closer to condemnation, Galileo is making bolder forays into uncharted territory while he builds the foundation of modern physics. As in all progress, some of these ventures are not immediately fruitful; his hypothesis on terrestrial tides is a case in point. Though it will prove inadequate to the aspirations that its creator has for it, it will inspire new science and generate new controversy into ages beyond, including our own.[33]

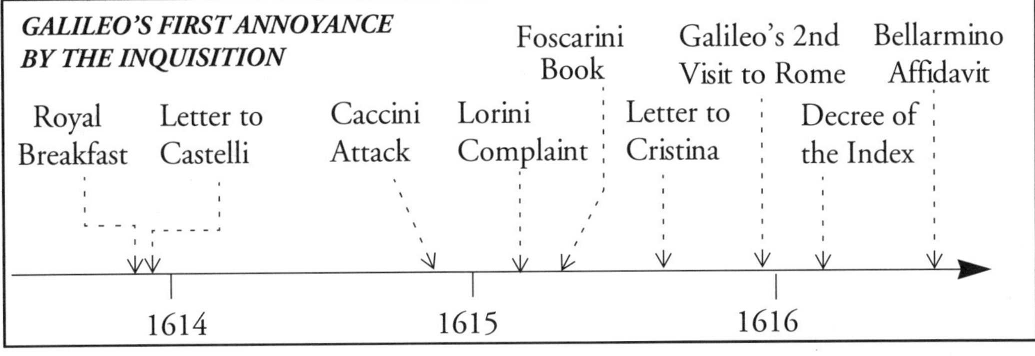

---

32. Inquisition Minutes, 25 February and 3 March 1616; see Finocchiaro, op. cit., pp. 147-148. The "Special Injunction" (26 February 1616) conflicts with these Minutes and with Galileo's behavior and is therefore disregarded here as a bogus document.

33. See Epilogue and Appendix B.

Galileo's Hypothesis of Ocean Tides (January 1616)

"The sea Yes teaches the canefield:
to advance in a creeping line,
to spread itself out
hole by hole up to the tideline.
The canefield Yes teaches the sea
the horizontal eloquence of its verse,
georgics of the news-stand, uninterrupted,
spoken aloud and parallel in silence. ..."

... from *O Canavial e O Mar*
by Joao Cabral de Melo Neto [34]

Let's travel back to Galileo's home town in the late 1570s and set our imaginations down on the bank of the Arno, a short walk from his birthplace. Here we see tide marks, and the young Galileo perched on the riverbank pondering them. At Padova, he will find reminders of these old questions as he commutes to Venice along the Brenta River by boat. In this curiosity, he will be joined by Kepler. Each man will work out his own hypothesis to explain the tides. Galileo seeks a mechanical explanation for the tides; that it might accidently turn the tide of public opinion in favor of Copernicus is a possible effect and not a cause of his hypothesis. In 1616 he writes out a discourse on the tides addressed to Cardinal Orsini, and later sends it to Archduke Leopold of Austria. Revised and recast in dialogue form, it will be included in his *Dialogo* of 1632.

Mediterranean Tides

A starting premise of Galileo's hypothesis is the dual motion of the Earth — and it is a good start. But his limited observations of Mediterranean tides do not suggest a direct effect of the moon's motion. An ancient superstition held that the moon dominates the waters; Galileo criticizes what he thinks is Kepler's assent to it. Each of Kepler's and Galileo's hypotheses implies a single daily ebb and flow; both explain intervening cycles of the tides by local secondary effects. Here is how Stillman Drake compares them:

"Whatever one may think of Galileo's tidal theory, it should be noted that, taken as a whole, it accounted coherently for the phenomena known to him about tides around the Mediterranean, his description of which in the *Dialogue* is fairly detailed and accurate. Although Kepler's theory appears closer to that which is now accepted, it was in many ways less so, principally by reason of its omission of various considerations included by Galileo in his list of further factors ..." [35]

---

34. trans. by Louis Simpson; *An Anthology of Twentieth Century Brazilian Poetry*, Wesleyan U. Press, 1972.

Galileo's mature theory is discussed in Appendix B. Oversimplified accounts of it ignore the moon's effect on Earthly tides, but Galileo does not. In the *Dialogo* he cites it as a secondary periodic effect: "*The second period is monthly, and seems to originate from the motion of the moon; ...*".[36] He asserts that this effect alters the magnitude of the ebb and flow of the Mediterranean at six-hour intervals, an alteration due to the moon's effect on Earth's motion rather than directly on its waters. He denies that the moon attracts the water and says of the lunar influence on water depth: "*... the moon travels over the whole Mediterranean every day, but the waters are raised only at its eastern extremity and for us here at Venice.*" [37]

Hydrodynamics and the Removal of False Premises

Looking back on his early attempt at a theory, we see in hindsight that he did not account for the synchroneity of global tides with the orbital motion of the moon. This theoretical lacuna suggests Galileo's resistance to the concept of universal gravitation.[38] Later, Newton accounted for the tides being synchronous with motions of the moon and sun, in whose combined influence on tides Kepler also believed. But historians who focus on Galileo's rejection of lunar attraction on Earth's waters have been missing the point. First, Galileo related tidal irregularities to factors such as depth of water and shape of the basin, and began in his mechanical explanation of the tides a major extension of Archimedean hydrostatics that has led to hydrodynamics.[39] Secondly, the work involved in any revolution of thought lies as much in the removal of false premises as in the substitution of correct ones. Galileo cleared the way for Newton[40] by showing that celestial events can be understood in terms of terrestrial ones — against the Aristotelian postulate of essential differences, which implies that the heavens and Earth obey different laws.

Unity of Nature and the Grecian Urn

If scientific truth is the correspondence between all observations of a natural world and the totality of theoretic laws that describe that operation, then there are as many truths as there are worlds within which a consistent set of laws operates. And if the beauty of nature inheres in economy of design, then Galileo's unification of the Aristotelian plurality of worlds forges the link between beauty

---

35. Stillman Drake, *Galileo at Work*, p. 44.

36. *Dialogo*, trans. by Stillman Drake, op. cit., p. 418.

37. Ibid., p. 420.

38. Ibid., p. 462.

39. (rdc)

40. Bertrand Russell wrote of Newton: "If he had encountered the sort of opposition with which Galileo had to contend, it is probable that he would never have published a line" (quoted in Resnick and Halliday, *Physics*, Wiley & Sons, NY, 1966, p. 384).

and truth. In his *Dialogo* (Day 2), Galileo expresses the equivalence this way:

> *"Yet we had better listen to Salviati's answers, which if true must be even more beautiful; infinitely more beautiful, and the others extremely ugly, if that metaphysical proposition is correct which says that the true and the beautiful are one and the same ..."*

Later, John Keats re-articulates the equivalence in his *Ode On A Grecian Urn*: [41]

> *"... Thou, silent form, dost tease us out of thought*
> *As doth eternity: Cold Pastoral!*
> *When old age shall this generation waste,*
> *Thou shalt remain, in the midst of other woe*
> *Than ours, a friend to man, to whom thou say'st,*
> *'Beauty is truth, truth beauty,' — that is all*
> *Ye know on earth, and all ye need to know."*

✦✦✦✦✦

Physical Basis of The Copernican Cosmology

Any mechanical explanation for the system of Copernicus must address the question how the Earth can orbit the sun. In the twists and turns of the dialogue growing out of his tidal theory, Galileo hints to us how he must have wrestled in his own mind with motion of bodies affected by a centripetal force like gravitation. Against the sun's center-seeking or "centripetal" attraction, what "centrifugal" repulsion can be made to balance it so delicately that the Earth moves indefinitely in a circle, never increasing nor decreasing its separation from the sun? The question is not whether such a motion is possible — has he not seen something like this in the motions of his Medicean satellites? The question is: What are the mechanics of circular motion?

Extruding Tendency of Circular Motion (Centrifugal Acceleration)

In his Earthly observations, he has seen little stones adhere to the rim of a wheel until it reaches a speed sufficient to cast them off, or as we say, extrude them. As if Galileo had seen the system of Copernicus as a case of circular motion in which the Earth replaces the stone and the adhesion that causes the stone to cling to the wheel is replaced by gravitation, he digresses in his dialogue on two rival planetary systems to consider

---

41. cf. *The College Anthology of British and American Verse*, Hieatt and Park ed., Allyn and Bacon, Boston, 1964, p. 373.

the case of a stone stuck to the rim of a wheel from which it may be extruded as the wheel speeds up. His character Sagredo remarks that if two wheels of different size are turning with the same linear speed [42] (let's call it "rim speed"), the smaller wheel would have greater tendency to extrude a little stone lightly clinging to its rim. The large circumference, he says rather poetically, cloys *with dainty bites the appetite (so to speak) that the stone has for leaving the circumference; ..."* whereas the small wheel disfavors the direction of the tangent and *"tries too greedily to retain the stone, ..."*. Assuming that the wheels turn with equal rim speed, he conjectures how much the rim speed of the larger wheel must increase so that its centrifugal acceleration would equal that of the small wheel:

"Since the casting off [in case of equal rim speeds] *diminishes with the enlargement of the wheel, it might be true that to have the large wheel extrude things as does the small one, its* [rim] *speed would have to be increased as much as its diameter, which would be the case when their entire revolutions were finished in equal times* [i.e., their angular speeds are equal]." [43]

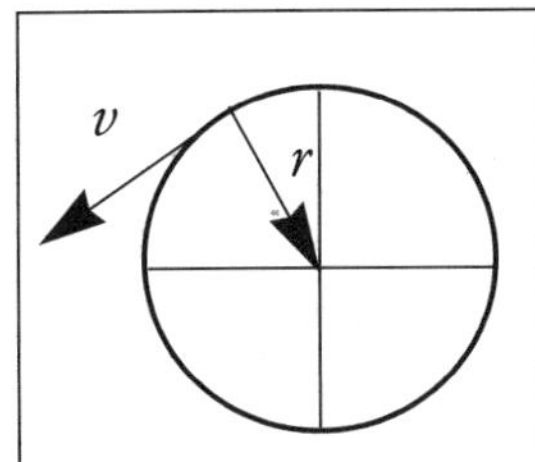

And it might not. Here he attempts to relate the magnitude v of the circularly moving object's velocity on the surface of the wheel, and the wheel's radius r to the acceleration a that triggers extrusion,[44] an admirable goal. In symbols, he suggests that a is proportional to angular speed v/r in radians, or to what we might call RPM's. After this tantalizing remark by Sagredo, Salviati changes the subject and does not go on to correct Sagredo with the observation that by the time the large wheel reaches the angular speed of the smaller it has greater extruding capacity, having already passed through the speed at which the centrifugal accelerations are equal.[45] In Appendix C, we equate the acceleration of extrusion to v^2/r. It is left to Huygens to convert Galileo's conjecture to theory not easily cast off.

His mechanical explanation of the tides was to be the key to a Galilean system of the world that failed to materialize. Though he had analysed

Universal Gravitation
(Centripetal Acceleration)

what would happen if God dropped the planets,[46] Galileo did not extend his analysis to a concept of universal gravitation. Yet he had extended the concept of

---

42. Literally, " velocità " = linear speed component, v, of the velocity vector at the rim.

43. *Dialogo*, (Drake trans.), p. 217. The bracketed [] expressions are mine, not Drake's.

44. Galileo refers to this acceleration as "casting off"; here I call it "centrifugal acceleration".

45. That is, a multiple of the original speed equal to the ratio of square root of the larger to the smaller radius. Christian Huygens' discovery that $a = v^2/r$ builds on the foundation Galileo laid here and lets me calculate this multiple. I thank jm for pointing me toward Huygens' work on the centrifugal conatus.

46. See Epilogue, "Myth Number Three: The Enormous Gulf Myth".

the interaction of terrestrial matter into his law of falling bodies and discovered a principle of inertia. Looking back at this situation from a 20th-century perspective, Galileo and Kepler each held a key to the doubly locked secret of universal gravitation, and each was oblivious to the other's key. By his remark "Galileo says quite clearly that there must exist some kind of interaction (tendency to mutual approach) of the matter constituting a star",[47] Einstein reminds us how close Galileo came to identifying a principle of centripetal acceleration that operates throughout the universe. Galileo did not generalize such a view, however. Rather, the pathway to universal gravitation[48] had to be walked later by Newton.

By 1623, Galileo has disposed of some major "Aristotelian" impediments to Earth's motion, including the astronomical objection from the doctrine of immutability of the heavens. In Chapter 9 we will see how he has found (in his own mind, though he couldn't yet publish) a means to eliminate the physical objection from the misconception about kinematic frames of reference. Having smoothed out some rough terrain in his theoretical approach, he begins to address philosophical obstacles to popular acceptance of Copernicus' world view.

Il Saggiatore (1623), Galileo's Philosophy of Science

Galileo's 1612 letters *Macchiae Solari* were about much more than spots on the sun. He embarked there on a philosophy of science that he completes in *Il Saggiatore (The Assayer)*. In contrast to the "Aristotelian" doctrine of essences, he held that essences are unknowable. Knowledge of nature, derived from measurable properties of observables, is the domain of science. With this epistemology, Galileo drove a wedge between natural philosophy (science) and "philosophy" as a set of propositions that are metaphysical in the sense defined by Ayer.[49]

In *Il Saggiatore* he introduces, behind the shield of the 1618 comet controversy, the next round in his philosophy of science: his reply to Grassi's *The Astronomical and Philosophical Balance*. Published by the Lincean Academy, a private research institution founded by Federico Cesi at Rome, *Il Saggiatore* unveils modern scientific methodology in elegant expository passages interrupted by digressions of comic relief in which Galileo buries Grassi beneath a barrage of polemic.

---

47. in Einstein's foreword, trans. by Sonja Bargmann, to Galileo's *Dialogo*, Drake trans., p. xiii.

48. Newton's universal gravitation law: every body in the universe attracts every other with a force proportional directly to the product of their masses and inversely to the square of the distance separating their centers of mass.

49. "... We may accordingly define a metaphysical sentence as a sentence which purports to express a genuine proposition, but does, in fact, express neither a tautology nor an empirical hypothesis." See A. J. Ayer, *Language, Truth, and Logic*, Dover, undated (I purchased my copy in 1962).

Galileo represents the Linceans in a debate with Jesuits of the Collegio Romano. The Jesuit side, argued by Grassi, represents philosophy produced and consumed in a closed loop by academia, while the Lincean side represents a flow of knowledge between individual scientists and from them to the educated layman. Let's consider here four concepts from that part of Galileo's philosophy of science enunciated within *Il Saggiatore*: (1) his dictum of bird droppings; (2) his linguistic proscription of the incomplete dichotomy; (3) his distinction of primary and accidental phenomena; (4) his view of mathematics as the language of nature.

Galileo's Dictum of Bird Droppings

Galileo symbolizes the individual thinker by the assayer who measures the purity of gold with a fine instrument. In contrast, he represents Grassi's "balance" as a crude attempt to determine scientific truth by weighing a consensus of opinions among the multitude of academics:

"I say that the testimony of many has little more value than that of few, since the number of people who reason well in complicated matters is much smaller than that of those who reason badly. If reasoning were like hauling I should agree that several reasoners would be worth more than one, just as several horses can haul more sacks of grain than one can. But reasoning is like racing and not like hauling, and a single Arabian steed can outrun a hundred plowhorses. So when Sarsi brings in this multitude of authors it appears to me that instead of strengthening his conclusion he merely ennobles our case by showing that we have outreasoned many men of great reputation." [50]

He compares their consensus of opinion to bird droppings:

"Perhaps Sarsi believes that all the host of good philosophers may be enclosed within four walls. I believe that they fly, and that they fly alone, like eagles, and not in flocks like starlings. It is true that because eagles are rare birds they are little seen and less heard, while birds that fly like starlings fill the sky with shrieks and cries, and wherever they settle befoul the earth beneath them." [51]

Not only is the majority often wrong, but as Galileo has already stated in his reply to Scheiner (*Macchiae Solari*):

> *"... in the sciences the authority of thousands of opinions is not worth as much as one tiny spark of reason in an individual man."*

As the history of science is the history of expansion in the scope of human endeavor

---

50. Cf. Galileo, "The Assayer", trans. by Drake, *Discoveries and Opinions of Galileo*, op. cit., p. 271.
51. Ibid., p. 239.

that is conducted scientifically, this dictum promises to assume more general utility as that expansion continues into the future. Through the study of the history of science, the fallacy of preponderant opinion as a determinant of truth can be eliminated. And for one who studies that history, the fallacy presses to be eliminated.

The Language of Natural Philosophy

Not only have a preponderance of philosophers fallen into falsehood by spreading about each other's mistakes in a collective pile of intellectual bird excrement from which they freely borrow, but much of what they have to say is neither false nor true. Using as an example Sarsi's claim that objects are magnified by a telescope depending on whether they are near or far, Galileo remarks:

"... Sarsi has created an incomplete dichotomy (as logicians call this error) when he divided visible objects into "far" and "near" without assigning limits and boundaries between these. He has made the same mistake as a person who should say, 'Everything in the world is either large or small.' This proposition is neither true nor false, and neither is the proposition 'objects are either near or far.' ... A courtier calls the trip from Rome to Naples very short, while a great lady grieves that her house is so far from the church."[52]

Recognizing classes of propositions that are neither true nor false, but meaningless by virtue of being indeterminate, Galileo anticipates the 20th-century Principle of Verification as expounded by the Vienna Circle.[53] Here stands revealed in simple language the philosophical underpinnings of Ayer's assertion: *"The traditional disputes of philosophers are as unwarranted as they have been unfruitful."* If science is concerned with truth about nature, then it cannot admit classes of propositions for which there is not, at least in principle, a test for their truth or falsehood. In a few strokes of his pen, Galileo eliminates much of what is considered philosophy by the theologians and scholastics of his day.

Primary and Accidental Phenomena

In the midst of polemic humor, Galileo announces a phenomenological principle in the philosophy of science. It will become an issue in his conflict with the Church.[54] To understand a natural phenomenon requires separating its sensible essential properties from accidental properties that depend on the observer. If I

---

52. Ibid., p. 249.

53. The principles of the Vienna Circle, and logical positivism in general, are currently "out of fashion" in academia. In light of Galileo's Dictum of Bird Droppings, lack of fashion has nothing to do with the correctness or utility of those principles.

54. Documents on file with the Holy Office indicate a misapplication of his separation of phenomena to the "Doctrine of the Eucharist", in an attempt to fit his premises (by distorting them) to the conclusion that he is a heretic on other grounds than will be used for his conviction. I refer here to Redondi's document G3 in *Galileo: Heretic*, Princeton U. Press, 1987.

walk around my writing desk, I may see it in a variety of hues as I move about. Another observer may, in the same incident light and from the same perspective, report a color different from that which I perceive, due to a difference in retinal structure: clearly, not a property of the object being observed. And so, when we speak of the color of an object, what inheres in the object itself that has this property? [55] Galileo answers: *What inheres in an object to excite our senses exists there in the form of number, shape or motion.* [56] Phenomena measurable independently of the observer include quantity, size, shape, position and motion of particles that comprise a body, and its contiguity with other bodies. All of these can, in principle, be agreed upon by observers. Other phenomena, such as color, taste and odor, depend on the observer. By identifying inherent properties of a body with those that are measurable and can be expressed mathematically, Galileo revises Aristotle's distinction between essential and accidental properties, rendering it serviceable to mathematical physicists. [57] In this way, he supplies a starting point for Eddington's principle of selective subjectivism. [58]

Among Galileo's revelations on the philosophy and methodology of science is the famous passage:

Mathematics, the Language of Science

"Philosophy is written in this grand book, the universe, which stands continually open to our gaze. But the book cannot be understood unless one first learns to comprehend the language and read the letters in which it is composed. It is written in the language of mathematics, and its characters are triangles, circles, and other geometric figures without which it is humanly impossible to understand a single word of it; without these, one wanders about in a dark labyrinth."

Those philosophers who discuss phenomena only qualitatively have been croaking like frogs devoid of a language that supports rational thought about nature.

Two Worlds of Democritos Revisited

Anticipating molecular physics, Galileo identifies motion as the cause of heat and treats heat-sensation as a property measurable by a temperature in fluids or in solids:

Heat as Molecular Motion

---

55. For a lucid and extended discussion of this problem, see Bertrand Russell, *The Problems of Philosophy*, Henry Holt and Company, New York, undated, Chapter 1.

56. He says, for example: "To excite in us tastes, odors, and sounds I believe that nothing is required in external bodies except shapes, numbers, and slow or rapid movements." *Discoveries and Opinions of Galileo*, Drake, op. cit., p. 276.

57. (rdc).

58. Sir Arthur Eddington, *The Philosophy of Physical Science*, U. of Michigan Press, 1958.

"Those materials which produce heat in us and make us feel warmth, which are known by the general name of 'fire,' would then be a multitude of minute particles having certain shapes and moving with certain velocities. Meeting with our bodies, they penetrate by means of their extreme subtlety, and their touch as felt by us when they pass through our substance is the sensation we call 'heat.'" [59]

Though he could not have directly observed molecules, the behavior he ascribes to his tiny corpuscles is consistent with that of molecules in the modern sense:

"The operation of fire by means of its particles is merely that in moving it penetrates all bodies, causing their speedy or slow dissolution in proportion to the number and velocity of the fire-corpuscles and the density or tenuity of the bodies. Many materials are such that in their decomposition the greater part of them passes over into additional tiny corpuscles, and this dissolution continues so long as these continue to meet with further matter capable of being so resolved." [60]

The Atomic Genesis of Light

He goes on to connect heat with light at a sub-corpuscular level, the level of the smallest possible quanta of matter:

" ... And perhaps when such attrition stops at or is confined to the smallest quanta, their motion is temporal and their action calorific only; but when their ultimate and highest resolution into truly indivisible atoms is arrived at, light is created." [61]

Later Galileo will postulate a previously unsuspected material that can give off light and heat like fire. He believes "its brilliance is very different from the brilliance of our burning substances." [62] His comment will anticipate a new energy source in the interiors of stars. [63]

The Return of Democritos

You may recall (from Chapter One) the two worlds of Democritos: the world of the very large and the world of the very small. Just as Galileo has opened up the macro-world with his *Message from the Stars*, in *Il Saggiatore* he paints the micro-world of atoms in high relief against the macro-world of astronomy. It is a portrait that will not be completed by him or in his lifetime. How the Church and its philosophers will manage to divert the otherwise unstoppable Galileo from his mission will unfold as we resume our story on the road that began in Florence and leads to Rome.

---

59. Stillman Drake's *Discoveries and Opinions of Galileo*, p. 277.
60. Ibid.
61. Ibid.
62. in response to Ingoli (see Ch. 9).
63. It will not be until the 20th century that the interiors of stars are recognized as a fourth state of matter: neither solid, nor liquid, nor gas, but gaseous in atomic structure, in a highly pressurized state called "'plasma", resembling a solid and radiating energy through continuous nuclear fusion.

Galileo's Drawing
Explaining Retrograde
Motion from a
Copernican Perspective

(from Favaro's
National Edition)

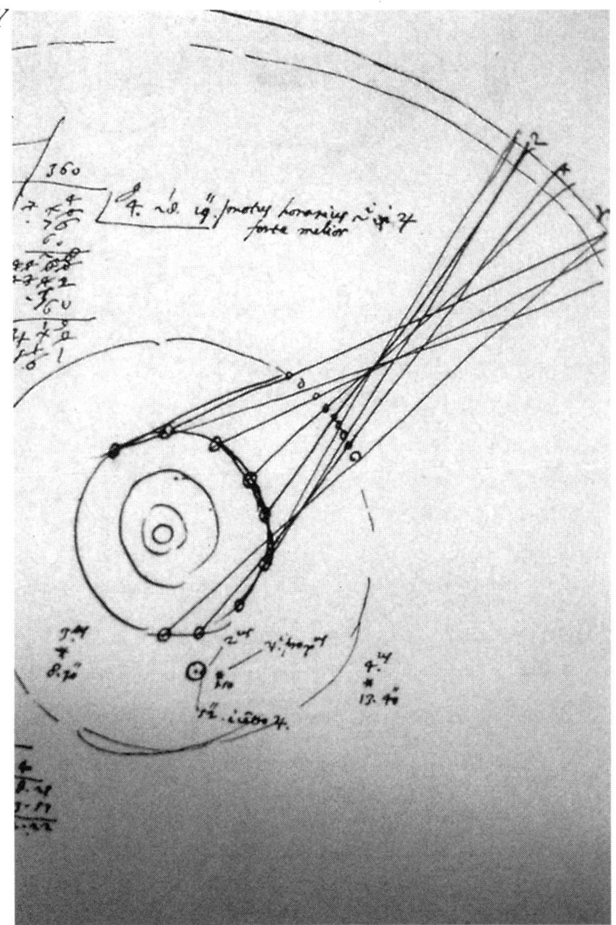

Tide Marks at the Bank of the Arno near Birthplace of Galileo,
Creator of a Theory of Effects of the Earth's Motion on the Tides.
(photo by author)

Galileo's Design of a
Pendulum Clock

This sketch is based on
an original diagram by
Vincenzio Viviani
derived from Vincenzio
Galilei's notes and
incorporating Galileo's
design of a pendulum-
driven clock. Galileo
described such a design
in his June 1637 letter to
Lorenzo Realio.

(Sketch by
    Vincent Di Canzio)

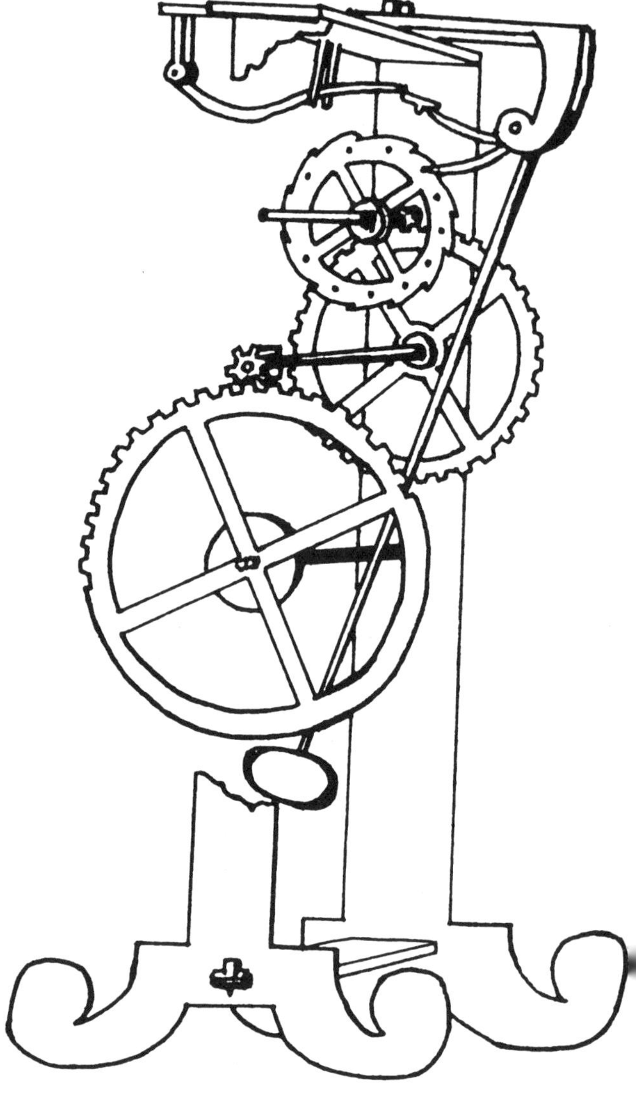

Magnificent
Anachronism:
A Geocentric
Calendar Clock at
Piazza San
Marco, Site of the
Overturn of the
Geocentric World
System

(photo by author)

Chapter 9. The Blue Balcony

Galileo Renews the Copernican Campaign (1624-1632)

"... Full Moon,
Let you add,
His lute of yellow strings;
And, our Night
Is square, nay,
Our Night
Is round, nay,
Our Night
Is a blue balcony —
And therewith close your Inquisition! "

... Alfred Kreymborg [1]

That rough and unroyal road along which Galileo has led himself, from a boyhood home and the vibrating strings of a lute to the balcony of the Padovan villa where, under the azure blue of waning twilight, he examined the geometry of the night sky, will lead inexorably to the Inquisition, a process that Church authorities do not know how to close. Are they deaf to his message and blind to his achievements? Does their posture mask a fear that nature might conform to Galileo's vision and thus befoul the "infallible" [2] pronouncements? Or has any threat to their authority become unthinkable in their own minds? [3] Whichever is the case, the Medieval Church is now entwined in a fictional world of its own creation. As if in a desperate attempt to realize Aristotle's speculation about the existence of an other-worldly realm with separate laws of nature, this consensus-driven institution has created the ghastly Inquisitorial examining rooms, dark dungeons of despair, and shrieking torture chambers: places where geometry is not only useless but dangerous, and where truth is an enemy to life. Not content with mimicking Dante's fearful fables in its doctrine, the Church has proved the reality of hell by creating it on Earth. For that same mathematician who measured the Inferno, the road from Padova back to Toscana now runs irreversibly on to Rome and to those infernal halls of terror.

---

1. from "Nun Snow: A Pantomime of Beads", *A Comprehensive Anthology of American Poetry*, ed. by Conrad Aiken, The Modern Library, New York, 1944.
2. I speak here not of "ex cathedra" infallibility (a post-Galilean Church doctrine) but of the certainty ascribed in Galileo's time to the consensus of the Holy Fathers.
3. To denote this hypothesis (of rdc), I will later refer to it as "Challenges Are Unthinkable".

Bellarmino's Enduring Legacy

At the time of Galileo, the concept of hell included man-made processes that began on Earth. To understand the still-latent philosophical revolution that Galileo made possible, we may ask: What is the contemporary Church's concept of hell? In a fusion of Matthew's gospel and Cardinal Bellarmino's macabre imagination, the Church graphically depicted it in 1961 with approval of censors: [4] *"It remaineth that we consider the justice which God will use in punishing sinners in the uttermost depths of hell ...* Into fire, *that is, into the furnace of burning and unquenchable fire (Matt. 13:42) which shall not consume one member alone but all the members together with horrible punishment.* Everlasting, *that is, into a fire which is blown by the breath of the Almighty and therefore needeth no fuel to make it always to burn, that as your fault shall still remain, so your punishment shall forever endure."* Evidently the best characterization of hell still extant within the Church reaches back to the time of Galileo, as Brantl attributes this description to Cardinal Bellarmino. If his quote is correct then, in this unpoetic but otherwise Dante-like passage, Bellarmino has ventured beyond the literal meaning of Matthew's gospel to make a pronouncement on science: the universe has an infinite energy supply, as God is a breathing Being who apparently has nothing better to do with his breath than to fan eternal fires that exact infinite punishment from creatures who are not even capable of being rehabilitated by this process.

Heaven Hotter Than Hell

To gain a sense of the strident clash of epistemology between the Church and science, it may be instructive to pose the question: how might Galileo respond to this description of hell? Though the question is unanswerable in this form, we do know how a 20th century scientist, following the path laid down by Galileo, has characterized hell based on his scientific analysis of Scriptural passages. In it, Mr. Wensel of the United States Bureau of Standards has invoked the Stefan-Boltzmann law, a product of science more than two centuries beyond Galileo:

"The temperature of heaven can be rather accurately computed from available data. Our authority is the Bible: Isaiah 30:26 reads, *Moreover,* [in heaven] *the light of the moon shall be as the light of the sun and the light of the sun shall be sevenfold, as the light of seven days.* Thus heaven receives from the Moon as much radiation as we do from the Sun and in addition

---

4. See *Catholicism*, ed. George Brantl, Washington Square Press, NY, 1961, *imprimatur* by Francis Cardinal Spellman. This appears to be a quote from Bellarmino containing a paraphrase of Matthew; although Brantl has not used quotation marks, he attributes the entire section containing this passage to Bellarmino, citing Bellarmino's *On the Ascent to God*, trans. T. G. Gent, 1616 (London, 1928). A contrasting description of hell is attributed to St. Augustine, who says: "When God punishes sinners, He does not inflict His evil on them, but leaves them to their own evil." Apparently, then, an Augustinian view of Bellarmino's hellfire would limit its application to arsonists and destructive pyromaniacs. Does this group include those who burn humans at the stake? (On Bellarmino's role in the execution of Giordano Bruno, see de Santillana, *Crime of Galileo*, p. 28 and Singer, *Bruno, His Life and Thought*, Henry Schuman, NY, 1950, p. 177).

seven times seven (forty-nine) times as much as the earth does from the sun, or 50 times in all. ... With these data we can compute the temperature of heaven: The radiation falling on heaven will heat it to the point where the heat lost by radiation is just equal to the heat received by radiation. That is, heaven loses fifty times as much heat as the earth by radiation. Using the Stefan-Boltzmann fourth-power law for radiation,

$$\left[\frac{H}{E}\right]^4 = 50$$

where E is the absolute temperature of the earth — 300°C [*sic*][5] (273+27). This gives H as 798° absolute (525°C).

The exact temperature of hell cannot be computed but it must be less than 444.6°C, the temperature at which brimstone or sulfur changes from a liquid to a gas. Revelations 21:8: *But the fearful, and unbelieving ... shall have their part in the lake which burneth with fire and brimstone.* A lake of molten brimstone means that its temperature must be below the boiling point, which is 444.6°C. (Above this point, it would be a vapor, not a lake.)

We have, then, temperature of heaven, 525°C (977°F). Temperature of hell, less than 445°C (833°F). Therefore, heaven is hotter than hell." [6]

However silly this conclusion may sound, it emerges from valid reasoning applied to fantastic premises. In this way, it differs from Cardinal Bellarmino's perpetual-malice machine, which enjoys only fantastic premises.

Six years after the 1960 publication of this analysis, Catholic censors approved a new translation of the Bible in which the passage quoted by Mr. Wensel from Rev. 21:8 referring to the site of punishment as "the burning lake of sulfur" was footnoted stating that the fire is symbolic.[7] This is a fine footnote to the story of Galileo's suppression by the Church; later we will see that at the time of his trial in 1633 Galileo was accused of violating Bellarmino's warning against treating the Copernican system not merely as a convenient hypothetical model for explaining planetary motions but as a physical reality. In an action parallel to its treatment of Galileo, the Church by this footnote dismissed Saint Bellarmino's representation of eternal hellfire as a physical reality, rendering it merely a convenient hypothetical model for discouraging what it calls "mortal sin". Apparently, qualifications for Sainthood do not include competence at Biblical exegesis.[8]

---

5. It is evident from the context that 300°K (300° absolute) is intended.
6. from *Applied Optics*, vol. 11, A14, 1972. This article, entered in the August issue by the Managing Editor (P. R. Wakeling), was attributed to "an unnamed environmental physicist of several decades back", and was forwarded to me via e-mail by Dona Morgan. The December issue contained a response by Kurt Nassau of Bell Laboratories stating that he had seen it in the *Register of Phi Lambda Upsilon*, Vol. 45, Fall 1960 as a reprint from *Saturday Review* by H. Allen Smith, who attributed it to "a Mr. Wensel of the Bureau of Standards some thirty years previously."
7. Cf. *The Jerusalem Bible*, Doubleday & Co., Inc., NY, 1966 with *imprimatur* by John Cardinal Heenan.

Can there be a pressure differential (of hell over heaven) raising the boiling point of sulfur above the temperature Wensel calculated for heaven? This question, raised by Nassau in an attempt to "save the appearances" of hell being hotter than heaven, is debatable.[9] For heaven to receive seven-squared times as much radiation from the sun as does Earth, it must be 1/7 as far from the sun: heaven's orbit would be well inside Mercury's, not beyond the tenth sphere and all the fixed stars. Perhaps for Biblical fundamentalists there is a Vulcan after all.[10] In the Copernican system, Galileo might argue, heaven would then be assigned a lower place in the universe, and by the Ptolemaic dictum *denser and heavier bodies occupy lower places,* the atmosphere of heaven should be at a higher pressure than hell, not a lower one. What human exposed for a second to a temperature of 525°C (or 798°K, more than double the absolute temperature of boiling water), and surviving the experience, would not describe it as "hotter than hell"? Certainly, this is hot enough to vaporize Dante's spheres of ice, un-suspending the traitors and giving them the run of hell.[11]

Wensel's analysis illustrates not the usefulness of analyzing Scriptural passages mathematically but the absurdity that can result from treating them literally. To demonstrate this absurdity, it applies quantitative methods to literary allegory, a procedure pioneered by Galileo in his measurement of Dante's *Inferno.* The story of Galileo's influence is the story of the gradual abolition, by application of quantitative methods and individual reasoning, of various forms of man-made hell, many of which have survived to our own time and depend on confusion about the position of an individual human in the universe. Let us now revisit that story.

---

8. Bellarmino's characterization was reprinted as late as 1965, only one year before the *Jerusalem Bible* downgraded hellfire to symbolism. I assert not a causal connection between Wensel's argument and that downgrading, but only the temporal priority of the former. In an apparent attempt to cool off heaven, the Jerusalem Bible translation of Isaiah 30:26 has also downgraded the heavenly incident sunlight from 49 times Earth to 7 times Earth (or 8 times Earth including incident sunlight reflected from the moon); again, using Stefan-Boltzmann, this new translation lowers heaven's temperature only to 232°C. By comparison, water boils at 100°C.

9. Kurt Nassau acknowledges that Wensel's argument leading to the temperature of heaven "appears to be impeccable". He points out, however, that Wensel assumes equal pressure between heaven and hell. He then calculates from the Clapeyron equation that, if the pressure in hell were assumed higher than 4.2 atmospheres, "the boiling point of sulfur rises above the temperature calculated for heaven"; *Applied Optics*, Dec. 1972, p. A14. In the 1966 *Jerusalem Bible*, Bellarmino's source, Matt. 13:42-43 is re-rendered as "The Son of Man will send his angels and they will gather out of his kingdom all things that provoke offences and all who do evil, and throw them into the blazing furnace, where there will be weeping and grinding of teeth". Now, grinding of teeth could pose a serious physical difficulty, at a temperature high enough to melt sulphur, for persons wearing dentures. I would also challenge Mr. Nassau to explain weeping against a tear-suppressing pressure above 4.2 atmospheres.

10. Vulcan, named for the Roman god of fire, was an intra-Mercurian planet hypothesized to account for the precession of Mercury's perihelion solely by gravitational perturbations until Einstein explained that precession as a consequence of General Relativity.

11. See Ch. 3, "Measuring the Inferno (1588)".

Evidence That Galileo Had Permission To Uphold Copernicanism

Our story resumes as the year 1623 draws to a close. Cloistered in his Florentine villa, Galileo has been pondering the underlying physics by which the motion of the Earth can be shown consistent with that of objects in free fall near its surface.[12] His discovery of the relativity of motion, which we will discuss shortly, and his analysis of sunspot rotation have advanced the underlying physics that favors motion of the Earth. Sensing that ideological victory is on the horizon, Galileo journeys again to Rome to see whether it might be permissible for him to introduce his new physics in the context of a hypothetical treatment of Earth's motion and its effect on the tides.

His new optimism follows a concourse of unpredictable events. That protector of the faith and bridler of Galileo, Bellarmino, has passed from the scene: he died in 1621. *Il Saggiatore* has met with a groundswell of acceptance among Rome's most vocal intellectuals. Duke Cesarini, a star student of the Jesuits and former protégé of Cardinal Bellarmino, has been won over by Galileo and others to the Linceans. A member of the Lincean Academy, Cardinal Maffeo Barberini, has been elected Pope, taking the name Urban VIII. Not only has he supported Galileo in scientific disputes, he has even written a Latin sonnet in praise of Galileo's discoveries.

Soon thereafter, the new Pope appoints as papal secretary a Lincean and friend of Galileo, Monsignor Ciampoli. With his newly acquired influence inside the Roman Curia, Ciampoli will secure the appointment of Father Castelli, Galileo's scientific protégé, to a prestigious university post. It is a propitious time for Galileo to test whether he might now openly resume his mission, suppressed since the Edict of 1616, of bringing the pronouncements of theologians into line with his discoveries. He seeks and obtains an audience with Urban VIII.

In a series of talks at the Vatican gardens in the spring of 1624, Galileo and the new Pope discuss the issue of Earth's motion face-to-face and alone. Although Urban defends the Aristotelian/Hipparchan view, Galileo returns to Florence encouraged to publish a comparative work on the rival systems of planetary motions. He may show how the Copernican system explains the observed motions of the planets more neatly or more accurately than the Hipparchan and Tychonic alternates. Is he also permitted to present the arguments for the physical reality of the Copernican system, provided that he stops short of declaring it true? Setting aside all the nonsense that has been written about Galileo's supposed defiance of the Church, this question can be viewed in light of the following facts:

---

12. Following his announcements of 1610-1615, and the fierce public debates in which he advanced evidence and scriptural arguments for Earth's motion, he remained reticent on this topic in the decade following the Decree of the Index. Public discussion of the doctrine now demands care lest one seem to "defend" it.

One. The March 1616 Decree of the Index does not mention Galileo or his works. It refers to the "doctrine" that the Earth moves and the sun is motionless as a false "Pythagorean"[13] doctrine and contrary to the Holy Scripture; however, this language expresses only the fallible opinion of a Congregation and does not represent a declaration by the Church.[14] The work of Copernicus is not banned; it is merely suspended pending correction. The required "corrections" are intended to make it appear that Copernicus did not present his hypothesis as physical truth.

Two. Inquisition minutes dated 25 February 1616 state that "His Holiness" ordered Bellarmino to warn Galileo to abandon the opinion of Copernicus and that, *if he should refuse,* he is to be ordered to "abstain completely from teaching or defending this doctrine and opinion or from discussing it; and further, if he should not acquiesce, he is to be imprisoned." Inquisition minutes dated 3 March 1616 state that Galileo acquiesced when warned of the order.[15] The Papal order and Galileo's non-incarceration show that he is prohibited only from holding the opinion and not from teaching it or from discussing it.

Three. A few months after that decree, Cardinal Bellarmino certifies that Galileo has neither been censured nor required to abjure. Bellarmino's affidavit goes beyond the Papal order, stating that the "doctrine" of Copernicus is "contrary to Holy Scripture and therefore cannot be defended or held." It does not forbid him to "teach [Copernican theory] ... in any way whatever", as Inquisitors will later claim that Galileo was told privately. Nor does it forbid him to present Earth's motion as a hypothesis. Bellarmino has no reason to omit from the certificate anything he had told Galileo privately; such action would be inconsistent with Galileo's purpose in requesting the certificate and with the Christian precept against false witness. As a derivative of the Decree of the Index, Bellarmino's warning does not represent the

---

13. Added to the "Pythagorean" theorem, we now have the "Pythagorean" doctrine of terrestrial motion. Dantzig apparently believes this attribution a diversion of blame from a churchman to a pagan philosopher; Galileo adopted the diversion as his own ploy, for in the preface to the *Dialogo* he refers to the "Pythagorean opinion that the earth moves". But whatever the Pythagoreans may have believed about Earth's motion did not include its motion around the sun. Galileo knew very well who originated the heliocentric hypothesis. At the end of the Second Day in the *Dialogo*, he clearly attributes this hypothesis to Aristarchos and Copernicus, not to the Pythagoreans.

14. Cf. "Galilei, Galileo" in the *Catholic Encyclopedia*: "The actions of the Holy Office and the Congregation of the Index in no way represented a commitment of the infallible teaching authority of the Church against the new astronomy. The decree of the Index received papal approval only *in forma communi* and therefore was only the fallible decision of a Roman Congregation."

15. In 1618 and again in 1624, Galileo himself confirms his acquiescence by referring to the Copernican opinion as one "held by me as true at that time [1613]" and "which I then [before the 1616 Decree] considered true" (see his letter to Archduke Leopold 23 May 1618, and his response to Ingoli translated by Finocchiaro in *The Galileo Affair*, op. cit., pp. 199 and 155, respectively). His use of the past tense shows that he wishes to be considered compliant with that Decree.

official position of the whole Church any more than that Decree does.

Four. Soon after Galileo's meeting with Urban VIII, Cardinal Zollern [16] reports to Galileo that Urban VIII had told him the Church has not condemned the opinion of Copernicus, nor is about to condemn it, as heretical, but only as rash, and that no one should fear that it might be proved necessarily true.

Five. By the next year, Galileo has available an expert theological opinion that he may write a book suggesting Earth's motion. In 1625, Father Giovanni di Guevara, an Inquisition Consultant, examines a complaint against Galileo's *Il Saggiatore* and declares "that the earth's suggested motion is not by itself blameworthy". [17]

Six. When the censoring of Copernicus obstructs the conversion of some German Protestant noblemen, Urban VIII remarks to Campanella about the Decree of the Index: "It was never our intention, and if it had been up to us, this decree would not have been made". These were his exact words, conveyed verbatim to Galileo by Prince Cesi.

Seven. Galileo does not blithely assume that he may write a book presenting evidence for the physical reality of the Copernican system. Shortly, we will see how he responds to an attack by the clergyman Francesco Ingoli on the Copernican system with arguments that establish its physical probability. The effect of his response is to test Urban's statement, and *the result of the test is no objection from Urban VIII*. On the contrary, Urban lets it be known that he appreciates Galileo's response.

Eight. If Galileo had any doubt about the foregoing, he could further validate the Church's concurrence with his publishing a discussion of the Copernican system by seeking the pre-publication approval of Church censors. This in fact he did. The result of the test was that he obtained all the required ecclesiastical approvals.

No wonder he did. The Pope is anxious for Galileo himself to quiet the intellects of educated Church members who see in Galileo's discoveries the possibility that the Church blundered in issuing the Decree of the Index. Encouraged by Urban VIII to demonstrate to the world that the Church did not issue that edict in ignorance of arguments in favor of Copernicus, Galileo decides to introduce his new physics in a dialogue. This is the book that, in the words of Drake, "historically did the most toward breaking down the religious and academic barriers against free scientific thought". [18] That is its long-term effect. Its short-term effect is a brutal backlash against Galileo and against freedom of philosophical expression.

---

16. Inexcusably, his name has been spelled "Zollern"," Hoenzollern", and "Hohenzollern" in the literature. This is a good example of how to frustrate looking up a work by surname in an index or reference catalog.
17. de Santillana, *The Crime of Galileo*, p. 212.
18. See his Translator's Preface to the *Dialogo*.

Galileo's Scientific Test (1624) of a Church Edict

In a scientific manner, Galileo has observed Urban's attitude toward his presenting the kinetic arguments for the system of Copernicus, has hypothesized that he may do so, and has designed a test to verify this hypothesis. He treats the statements of the Pope as if reading facts from nature. After all, has not this Pope assumed the role of interpreter for the divine Architect of nature? Just what is the test by which Galileo verifies that he may publish the *Dialogo*? It is his reply to Ingoli.

Galileo's Reply to Ingoli Galileo has quietly waited eight years to respond to an open letter to him in which Ingoli attempted to refute the Copernican system. In that time, Ingoli has risen from the post of an obscure cleric to that of secretary of the Congregation for the Propagation of the Faith. Galileo opens his letter with the reason he has not responded: he wanted to spare Ingoli the embarrassment of having all his arguments refuted. During his recent visit to Rome, however, having learned that his unresponsiveness is being interpreted as agreement with Ingoli's demonstrations, he must now rescue his own scientific reputation. He wants to present, he says, all the counter-arguments in support of terrestrial motion so that heretics will realize that he, being Catholic, is certain of the truth taught by the sacred authors, not for lack of scientific understanding, but rather because of reverence for the Scriptures and allegiance to the doctrine of the Church, a doctrine that he says is higher than physics and astronomy in dignity and authority. In contrast to human reason, the "higher sciences" teach us what we could never learn by recourse to reason and experience, he claims.

Epistemological Height This may be the most ambiguous claim in all of Galileo's writings. If the ambiguity were intentional, then it would be the rhetoric of *praeteritio* elevated to a meta-Ciceronian dimension, ingeniously and marvelously endowed with that quality of adapting itself to the reader's prejudices. How could any theologian blame him for wanting to show, by laying out convincing arguments for Earth's motion, and then passing over them, that his faith in the Church's pronouncements is so strong as to overcome the intellectual appeal of these arguments? Even Urban VIII was impressed when he read this. What piety, what Catholic zeal!

But this passage is capable of bearing other interpretations, assuming that Galileo intended no ambiguity. What can it mean to proclaim the doctrine of the Church higher than physics in dignity and authority, when he has already told Cristina that the truth taught by sacred authors is subject to any certainties in physics? Taking the relation of verifiability as more primitive than the concept of the "height" of an epistemology, the proposition "X is higher in dignity and authority than Y" evidently equates in Galilean language to the proposition "X is subject to verification by Y". Like the height of a physical object, the "height" of an epistemology is measured by its energy potential for downfall. Authority, being higher, is more distant from certainty than physics, and can only be grounded in physics, the underlying validating process. The "higher sciences" teach us things we cannot learn from sense experience or reasoning — things that are nonsensical and unreasonable until they are validated. Though not asserting that Galileo necessarily deduced this concept of epistemological height from his own writings, I consider it compatible with his view of the physical universe. His behavior is that of a man who believes Church doctrine on strictly religious matters remote from the observable, comprehensible universe. Just as in the physical universe, higher objects are those that are remote from the center of attraction, what he calls the "higher sciences" are remote from that universe. In deference to these "higher sciences", he has made a public declaration of abandoning impressive rational arguments to faith. What are those arguments?

He introduces them by scolding Ingoli for trying to overturn Copernicus with hastily considered objections. Unlike his flanking maneuvers against Sarsi, where he attacked at the boundary between science and philosophy, here Galileo deals with purely scientific issues and executes a frontal attack on Ingoli's arguments, rolling over them like a Sherman tank over chicken wire.

The Argument from Solar and Lunar Parallax

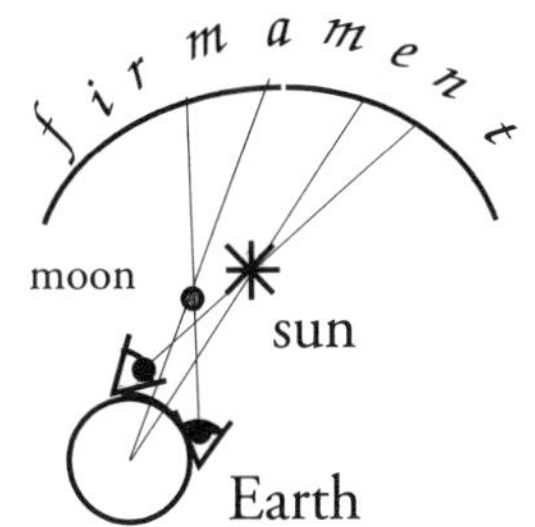

The first of Ingoli's arguments crushed by Galileo is that the sun cannot be at the center of the firmament because that would place it farther from the firmament than the moon, which shows us a lesser parallax. Galileo corrects the picture, defining the parallax of a celestial object as the angle formed at that object by lines passing through the observer and Earth's center: this definition gives the moon a larger parallax. He then shows that Ingoli "begs the question" by referring their apogee and perigee to the firmament rather than to Earth. Ingoli's misuse of parallax leads him to assume what he is trying to prove.

Stellar Magnitudes

Would not the brightness of a star vary as Earth is alternately far from and near to the star? But that is not what we see, Ingoli has argued, therefore Earth does not move. Here Galileo responds that the distances to the stars are greater than assumed by opponents of Copernicus, and that such variations would be undetectible (by the methods of the time). Then — like a falling cat — he manages to land on his feet after taking an astonishing tumble in the course of this argument, stating that Earth is 30 times bigger than Jupiter.[19]

The Jupiter Somersault

Galileo's misstatement points to the assumed value of Earth's distance from Jupiter, a derivative of its distance from the sun. He estimates Jupiter's angular diameter to be "barely" 40 seconds of arc, an accurate midrange value of a variable that ranges from 30.8 to 50 in modern measurements. His value for the radius of Earth's orbit can be deduced from his statements to be a factor of 20 too small, suggesting to me that he accepted the faulty measurement of Aristarchos, who had underestimated the Earth-sun distance by the same factor.[20] Ironically, Isaac Newton will make a similar mistake in a calculation based on Galileo's law of falling bodies.[21] Noteworthy of these great scientists is that they traverse the rough terrain of human fallibility to reach true conclusions, testing their conclusions and correcting their mistakes as they go — occasionally succeeding in spite of those that remain uncorrected.

Instantaneous Visibility of Half the Celestial Sphere

Among those of Ingoli's anti-Copernican arguments that are false but *prima facie* have some semblance of truth, Galileo says, is one that he took from Ptolemy. The argument is that if one of two fixed stars rises when the other sets, and vice versa, the part of the sky above Earth is equal to the part below, and this being true of all horizons, it shows that Earth is at the center of the assumed celestial sphere. Here, Galileo notes, a rotating Earth would exhibit this effect anywhere in the universe.

Much of Ingoli's misunderstanding flows from the notion that the physical universe cannot be infinite, or finite and large enough that his arguments from the symmetry of celestial events fall beneath thresholds of measurability. Though Galileo underestimates the distances to the fixed stars,[22] the universe as he views it is much larger than what Ingoli and most of his contemporaries envision — and Galileo reserves the possibility that the universe is infinite.[23]

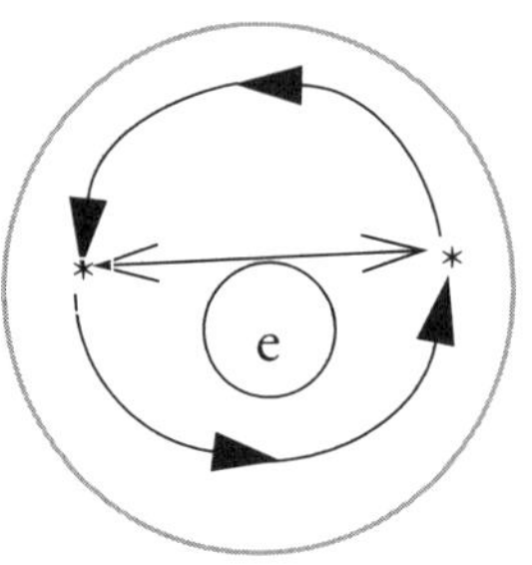

---

19. I have not spotted the academic vultures, who circle continually around the master's literary remains, picking on this point. Are they reading what he wrote?

20. See Berry, *A Short History of Astronomy*, op. cit., par. 32. Although Galileo tended to discover by thought experiment and to verify by actual experiment, this same caution did not necessarily extend to verifying the results of others.

21. Newton tried to show the effect of Galileo's law of falling bodies on the moon; his calculation did not agree with the moon's observed orbital period. This mistake was rectified 20 years later when Newton realized he had taken an inexact value of Earth's radius; See Hans Reichenbach, *From Copernicus to Einstein*, Philosophical Library, NY 1942.

Galileo's successor in developing the mathematics of infinity, Georg Cantor, will assert that one cannot understand God without a proper concept of infinity: "*... were it the case that I am right in asserting the truth or possibility of the Transfinitum, then (without doubt) there would be a sure danger of religious error in holding the opposite opinion ...*".[24] The ecclesiastical stumbling block against infinity seen in Ingoli's arguments suggests to me Aquinas' proofs of God's existence, taught with reverence in Jesuit schools in our own century, that deny infinite regression in nature.

After passing from astronomy to terrestrial physics, Galileo takes a short and spectacular digression: [25]

Nature Ignores a Vacuum of Intelligence

"*Here, before I go any further, I must tell you that in natural phenomena human authority is worthless. Like a lawyer, you seem to capitalize on it; but nature, Dear Sir, makes fun of constitutions and decrees of princes, emperors, and monarchs, and at their request it would not change one iota of its laws and statutes. Aristotle was a man, saw with his eyes, heard with his ears, and reasoned with his brain. I am a man and see with my eyes much more than he did; as regards reasoning, I believe he reasoned about more things than I, but whether he reasoned better than I about those things which we have both examined will be shown by our arguments and not by our authorities.*"

Will later generations heed Galileo's argument? In the *letter to Cristina*, he had already demonstrated the futility of attempting to legislate nature. In 1897, an Indiana legislator will introduce House Bill No. 246 to modify the value of π by statute! The Indiana House of Representatives will unanimously pass the Bill. [26] Whether trying to alter a constant of Euclidean geometry, to oppose laws of supply

---

22. Unlike his estimate of the Earth-moon distance, which was accurate to within about 6% of the modern accepted value, Galileo's estimate of the Earth-sun distance was, in close conformity to that of Copernicus and Ptolemy, 1,208 Earth radii, too small by a factor of about 19.4. (Having accurately estimated the average angular diameter of the sun at 30 minutes of arc, he thought the sun smaller than it is by the factor 19.4.) His estimate of the distance to a 6th magnitude star equal to the sun in absolute brightness was 2,160 Earth orbital radii; cf. *Dialogo*, Drake trans., pp. 359-360. By comparison, the nearest fixed stars to Earth, those of the Alpha Centauri system, are about 270,000 A. U., or 125 times 2,160 Earth-orbital radii. Now, Alpha Centauri, while it is nearly equal in absolute luminosity to our sun, is not a 6th but a -0.27 magnitude star; hence $(\sqrt[5]{100})^{6.27} \cong 322$ times as apparently bright as Galileo's hypothetical star. Assuming that brightness attenuates as the square of distance, it would have to be removed to a point $\sqrt{313} \cong 17.9$ times as far away to appear as a sixth magnitude star. Galileo thus underestimated the distance to the fixed stars by a factor of approximately $(19.4) \cdot (125)(100)^{(0.5)(0.2)(6.27)} \cong 43522$. His arguments against Ingoli based on the size of the universe are far more compelling than even he realized.
23. See *Dialogo*, Drake trans., p. 319: "... or whether it is infinite and unbounded."
24. Cantor to Jeiler, Pfingsten, 1888, as quoted by Joseph Warren Dauben in *Georg Cantor: His Mathematics and Philosophy of the Infinite*, Princeton U. Press, 1979, p. 232.
25. Finocchiaro, *The Galileo Affair*, p. 178.

and demand with price controls, or to enact some similar folly, legislators of times beyond Galileo will routinely pass edicts that attempt to defy nature.

Having cited nature's indifference to political decrees, Galileo declares it a fact of nature that the sun lies at the center of the planetary orbits, calling this fact as "clear and glaring as the sun itself". By this juxtaposition of topics he contrasts the political decree [27] in favor of Ptolemy with his reasoned argument for Copernicus.

Density Ordering the Universe

Ingoli has contrived an argument from the ordering of the universe by the density of its elements. Galileo reduces the argument to a syllogism that can be paraphrased thus: [28]

> DEFINITION: The "center" of anything is its lowest place.
> (1) Denser and heavier bodies occupy lower places;
> (2) The Earth is the densest celestial object;
> (3) Therefore the Earth occupies the center of the universe.

Heavenly bodies were considered incorruptible in Ptolemy's system. Turning Ptolemy's argument against him, Galileo notes that gold and diamonds are less corruptible than are lighter and less rigid bodies. Therefore, the sun is likely to consist of a heavy substance, and to occupy a lower position under the first premise of Ingoli's argument. Here we see the extent to which Galileo himself occasionally wanders in a labyrinth of sterile deduction in order to deal with the entrenched Aristotelian/Ptolemaic view on its own terms.

Licking the Tower

If the Earth were rotating daily, says Ingoli, the vertical fall of bodies would appear affected by its motion, unlike the fall of an object that grazes a tower on its way down. Galileo replies:

"When you and Aristotle infer that the rock falling from the top of the tower moves in a straight line perpendicular to the earth's surface, you do this, and can do this, only from seeing how during its fall it goes on licking (as it were) the surface of the tower, erected perpendicularly on the earth's surface; this makes one perceive the line traced by the rock as being also straight and perpendicular ... if with Copernicus I say the earth turns and consequently carries with it the tower and also us who are observing the rock, then we can say the rock moves with a motion composed of the general circular motion toward the east and the accidental

---

26. See Petr Beckmann, *A History of* π *(Pi)*, St. Martin's Press, NY, 1971. A "new value" of π can be inferred from the Bill: "... the ratio of the diameter and circumference [of a circle] is as five-fourths to four.". This would imply a value of 16/5 or 3.2 for π .

27. I refer here to the 1616 Decree of the Index, which had been solicited by Paul V.

28. Here I pass over an ambiguity that Galileo detects in Ingoli's idea of the center toward which heavy bodies are attracted, and consider the case that the center of the universe, not the center of Earth, is meant.

straight motion toward the whole to which it belongs, and the result of these is a motion inclined toward the east; in this case, the motion which is common to me, the rock, and the tower is for me imperceptible, as if it did not exist, and only the other remains observable, which the tower and I do not share, namely the motion of getting nearer the earth."

These and other astronomical and physical arguments in Galileo's letter to Ingoli appear to defend the system of Copernicus. The Pope does not object to this rebuttal of Ingoli and lets it be known that he appreciates Galileo's response.

Galileo's *Dialogue Concerning the Two Chief World Systems (Dialogo)*

Over three centuries before Galileo, John of Hollywood [29] had combined elements of Aristotelian and Hipparchan views into a single cosmology, while his younger contemporary St. Thomas Aquinas packaged that secular cosmology with Christian theology. The result was a system of the world that resembles a fantasy ride at Disneyland but was taken very seriously by generations of later philosophers. Like giant clockwork, a nested set of concentric crystalline spheres rotated around a central and motionless Earth. The original eight spheres of Aristotle carried the moon, Mercury, Venus, the sun, Mars, Jupiter, Saturn and the fixed stars. After Hipparchos' discovery of precession, astronomers added the ninth and tenth spheres to account for precessional changes in the apparent positions of the stars. Under the later system of Copernicus, these changes are explained by gradual drift in the orientation of Earth's axis, but in the Hipparchan/ Ptolemaic system, which has Earth at rest, the ninth and tenth spheres are required.

To this world machine, Church philosophers led by Aquinas added heaven, the home of God beyond the tenth sphere, and a fleet of angels to drive the rotating spheres.[30] It was the Hollywood model merged with the fantasy of Aquinas and backed by centuries of religious and philosophical fervor that Galileo had to penetrate with concepts like inertia and composite motions that might have sounded as fantastic to his contemporaries as angel-driven crystal spheres sound to us today. Galileo's *Dialogo*,[31] his final explicit rebuttal to opponents of Copernicus, was an outrageous novelty. Ironically, the change that it brought to the

---

29. Under the name John de Sacrobosco ("holy wood" and "sacro bosco" are synonymous), this English mathematician published *Treatise on the Spheres*, used by Galileo as the basis of his early (1586-1587) lecture notes at Pisa, which he organized into an unpublished manuscript. Galileo remarked in it that "though there had been great mathematicians and astronomers who had assigned motion to the earth, he would follow the customary opinion." Drake, *Galileo at Work*, p. 12.

30. See Alan G. R. Smith, *Science and Society in the Sixteenth and Seventeenth Centuries*, Harcourt Brace Jovanovich, Inc., London, 1972.

world view of the educated who came after Galileo poses a challenge for modern readers to appreciate how revolutionary it was. The dialogue, a lively debate between three gentlemen (Sagredo, Salviati and Simplicio) [32] who meet in a house on the Grand Canal in Venice, spans four days. In the debate, an irresistible intellect collides with a seemingly immovable world view.

The First Day examines the doctrine of essential differences between celestial and earthly realms put forth by Aristotle in *De Caelo*. Galileo portrays Earth as a celestial body, part of a single universe rather than (as Aristotle viewed the earth) the whole of one compartment in a dual universe. On the Second Day, the main attack begins at the prongs of a fork. Cycles of day and night can be explained by rotation either of the heavens or of Earth. Galileo's character Salviati then adds: [33]

"Now if precisely the same effect follows whether the earth is made to move and the rest of the universe stay still, or the earth alone remains fixed while the whole universe shares one motion, who is going to believe that nature (which by general agreement does not act by means of many things when it can do so by means of few) has chosen to make an immense number of extremely large bodies move with inconceivable velocities, to achieve what could have been done by a moderate movement of one single body around its own center?"

Dance of the Planets

The outer planets, Salviati observes, have longer periods than the inner ones, as in the miniature system of satellites orbiting Jupiter. The periods of the major planets progress up to 30 years for Saturn, the farthest planet known to Galileo. To assume, then, that the incomparably more distant fixed stars would rush around the Earth in 24 hours through immense orbits would not only grant them a preposterous speed but would greatly upset this observed relationship of longer periods for larger orbits. The stars near the pole would have to move in tiny circles at immense distances from the center of motion. The precession of the equinoxes would have to be shared and precisely choreographed in the dance of myriad distant bodies around Earth. The assumption of a rotating Earth resolves all of these difficulties.

Simplicio, who often represents the "Aristotelian" side of the debate, retorts that the difficulties connected with moving the stars in unison at great speed are no difficulties at all for an infinitely powerful God. Salviati prudently leaves this

---

31. The major title is *"Dialogo Sopra I Due Massimi Sistemi Del Mondo: Tolemaico, e Copernicano"* (orig. pub. 1632); in English translation by Drake *"Dialogue Concerning the Two Chief World Systems — Ptolemaic and Copernican"* (see Bibliography). An alternative source is the older Salusbury (1661) English translation of the *Dialogo* published by U. of Chicago Press, 1953.
32. Galileo's ideas are presented at different times by all three characters. Simplicio appears often a spokesman for the "Aristotelians" and Salviati for the Copernicans.
33. *Dialogo*, Drake trans., pp. 116-117.

argument to stand on its own merit, or lack thereof. He merely berates Simplicio for his cavalier treatment of infinity. Lest he offend his licensers, Galileo avoids having Salviati say the obvious. Simplicio has created a divinely driven perpetual-motion machine to do in a complex way what inertia could do more simply without external intervention, while denying the motion of Earth on the ground that it would require an external force to keep it perpetually in motion.

Cannonballs and Cotton Pendula

Aristotle assumed that a continuously applied force is required to sustain motion, like that exerted by an ox pulling a cart uphill. If once made to rotate or revolve, the Earth would wind to a stop; hence, the Aristotelians say, it does not move. Again, Galileo seems to be exercising self-restraint. He does not point out that any one angel from the fleet assigned by Church philosophers to move all the celestial spheres around Earth could just as easily (and more efficiently) move the earth while the other angels take an eternal coffee break.

Galileo not only enumerates and answers the standard anti-Copernican arguments, he even adds his own arguments *against* the earth's motion, strengthening the Ptolemaic side of the controversy. For example, if the earth were rotating west to east, it would carry a cannon toward a projectile fired horizontally from that cannon to the east, and away from one fired to the west, *ceteris paribus*.[34] But experience shows that shots fired horizontally have the same range to the east and west. The cannon, and therefore the Earth, is motionless.

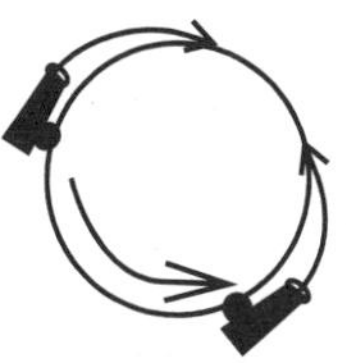

Showing that Copernicus could be right despite his opponents' arguments requires a different interpretation of sense experience. To this new interpretation, Galileo brings his discoveries that will be commonplace to us but probably sound like science fiction to readers of his day. Consider his concept of air as an inhibitor of motion. In the view of the Aristotelians, a cannonball projected with forward motion must be assisted by the air to retain that motion. In the case of a stone simply dropped from a mast, there is no forward projection, hence no "air assistance", and the stone should fall behind the mast.

Galileo's Copernican character counters with a question. Which would oscillate longer: a pendulum of lead or of cotton? If air were to conserve impressed force, the cotton pendulum should remain in motion longer. Once he has admitted the contrary, the Aristotelian begins to see the medium through which the stone falls as an interference rather than a mechanism to sustain motion. By introducing an

---

34. "Other things being equal": Galileo here implicitly assumes that the cannon is situated at the Earth's surface, but not at either pole, that wind is not a factor, and that the terrain in the direction of the shots approximates a smooth arc so as not to introduce obstacles as a factor in the range.

inertial motion, Galileo replaced "air assistance" with "air resistance", a term familiar to us today, but strikingly new to his contemporaries.

Galilean Relativity and the Earth's Motion

In these arguments for the motion of cannonballs and stones dropping from masts, Galileo identifies an overlooked relativity of motion which will spark Einstein's imagination three centuries later. In the case of the falling stone, Galileo's opponents had argued: if the Earth rotates west to east under an object that is simply dropped rather than thrown, why doesn't the object land west of where it is dropped?

To answer this objection to the Earth's motion, Galileo conceives a postulate that can be stated in modern (Einsteinian) terms:

> *The conformity of observed events to physical laws is unaffected if the observer is in uniform translatory (straight and steady) motion relative to the events being observed.*

In terms of his analogy, a fish swimming in a fishbowl borne by an oceangoing vessel in straight and steady motion cannot detect a difference in swimming toward or against the vessel's direction. And in general, there is no test by which an observer can detect his own motion by observing events in uniform translatory motion relative to himself.

After *Sidereus Nuncius*, Galileo writes in Italian. He adopts a vernacular style unlike that of the typical scientist addressing colleagues. He could have posed his relativity postulate in the language of academia. Instead, assuming the persona of Salviati, he takes our imaginations below the deck of a ship with a water clock, some fish (in an aquarium), some butterflies, and a human friend:

"With the ship standing still, observe carefully how the little animals fly with equal speed to all sides of the cabin. The fish swim indifferently in all directions; the drops [of the water clock] *fall into the vessel beneath; and, in throwing something to your friend, you need throw it no more strongly in one direction than another, the distances being equal; jumping with your feet together, you pass equal spaces in every direction. When you have observed all these things carefully, ... have the ship proceed with any speed you like, so long as the motion is uniform ... You will discover not the least change in all the effects named, nor could you tell from any of them whether the ship was moving or standing still. ... And if smoke is made by burning some incense, it will be seen going up in the form of a little cloud, remaining still and moving no more toward one side than the other."* [35]

---

35. *Dialogo*, op. cit., pp. 186-187.

If one stands on the deck of a ship in straight and steady motion relative to the water and observes a stone dropped from the mast, one will observe it to fall in the same manner as if the ship were not moving. Galileo's law of falling bodies is not altered for the members of this system: at the instant of release, ship and stone are moving uniformly relative to the observer. Under his inertial concept, nothing is required to keep the stone advancing with the ship so that it grazes the mast as it falls. Earth's motion is shared by all its parts: the ship, the mast, the falling stone. A force is required not to keep the falling stone advancing with the ship around Earth but to alter the motion that the stone after its release continues to share with the ship and the rest of Earth.[36] Similarly, when cannonballs fired east and west under identical conditions are viewed as continuing to participate in a motion shared with Earth's surface, it is easy to see that they would have the same range, and without the help of a medium.

Suppose that we replace the ship moving through water by Earth moving around the Copernican center of motion and around its own axis. Then, in a space sufficiently local to the observer, the trajectory of a falling body is not observably altered by Earth's motion; hence, the difficulty of Earthlings directly observing Earth's motion. Small wonder that Galileo is able to perceive a motion of the earth that earthbound "Aristotelians" cannot observe. Though bound to the earth physically, he is able to leave it in his imagination.

Which is easier to stop: a heavy chariot, or one half its weight but moving twice as fast? Galileo questions the role that weight and speed play when an object resists a braking action. He posits a quantity of motion, or momentum, as the product of these two variables:

Galilean Momentum

"Now fix it well in mind as a true and well-known principle that the resistance coming from the speed of motion compensates that which depends upon the weight of another moving body, and consequently that a body weighing one pound and moving with a speed of one hundred units resists restraint as much as another of one hundred pounds whose speed is but a single unit. And two equal movable bodies will equally resist being moved if they are to be made to move with equal speed. But if one is to move faster than the other, it will make the greater resistance according as the greater speed is to be conferred upon it."[37]

---

36. Here Galileo conceives of the inertial component as a circular motion around Earth. Under the modern concept, the inertial motion would carry the rock along the horizon plane and hence away from the center of Earth. Later in the Second Day, however, he discusses motion along a tangent line to a circle at the point of release of a whirling object.

37. *Dialogo*, Drake trans., p. 215. Drake notes "A hint of the concept of *force* as the product of mass and acceleration may also be seen here, if we are liberal in interpreting the ideas of 'resisting restraint' and of 'conferring velocity'."

Galilean momentum, the product of a weight and a speed, is a direct analog of the modern concept of momentum as the product of a mass and a velocity. In this discussion, he anticipates Newton's second law of motion. From here, the introduction of Galileo's physics into the Copernican controversy continues with a law of nature he deduced from experiments at Padova before 1608.

A Cannonball Falls From the Moon's Orbit

"... But you, Salviati, descending sometimes from the throne of His Peripatetic Majesty, have you ever toyed with the investigation of these ratios of acceleration in the motion of falling bodies?" With this rhetorical question, Galileo prepares to announce his new law.[38] The catalyst for the announcement is a calculation by Joannes Locher, a student of Scheiner, in which a cannonball requires more than six days to fall from the moon's orbit to the center (!) of Earth at a constant speed that it would have in that orbit: 12,600 German miles per hour.[39]

Salviati proceeds as if the drop commences from a stationary position in the moon's orbit opposite the moon. The cannonball satisfies Locher's assumption while falling directly to Earth along the radius of the moon's orbit.[40] He attacks Locher's argument at its weakest point: its neglect of acceleration. To rectify this neglect, he introduces Galileo's law of falling bodies: [41]

"... the spaces passed over by the body starting from rest have to each other the

---

38. ...after some intervening discussion of underlying principles: this comment appears in the Drake translation on p. 164; the law of falling bodies is enunciated 58 pages later.

39. Drake follows Strauss in assuming that Galileo's "German mile" is 1/5400 of Earth's equatorial circumference (*Dialogo*, note to p. 219), or 7.5 km, based on Eratosthenes. This assumption ignores the considerable difference between the lunar and sidereal day. For the sake of argument, Galileo here adopts Locher's Ptolemaic premise that the moon orbits Earth once every 24.878 hours, and his initial premise that the cannonball falls at the speed it would have in the moon's orbit (40500H km/24.878 hours = 91,163 km/hr. where H is the Earth-moon distance in Earth radii that Galileo here says is 56). These premises and his reckoning the speed at 12,600 German miles/hour jointly imply that one German mile = 7.235 km. (Drake has here taken even greater liberty with units: he translates "braccia" as "yard"; this would normally imply a British yard of 0.9144 meter. By contrast, Galileo evidently reckoned Earth's circumference at c. $66 \cdot 10^6$ braccia; equating this to 40500 km by Eratosthenes' estimate yields 1.63 braccia/meter, implying that one braccio = 0.614 meter. Twenty-one years after his *Dialogo* translation, Drake will admit "the Florentine braccio of Galileo's time was 58.4 cm." in the Glossary of his *Two New Sciences*, p. xxxix.)

40. Ibid., p. 221. Though Galileo had (ibid., p. 165 ff.) described fall of a body from a tower as occurring in an arc that intersects the center of Earth (corrected to a parabolic path in *Two New Sciences*), here he calculates the fall of the cannonball as if rectilinear from a point on the moon's orbit to Earth's center. Perhaps Platonic clinging to circles and inattention to Kepler blinded him to the possibility of a transfer orbit; i.e., an ellipse tangent to the assumed circle of the moon's orbit and connecting to the center of Earth. By analogy to his tower drop, he might have here described a circular transfer, in which case he would have had to account for a multiple of π times the path length of the straight radial drop.

41. *Dialogo*, Drake trans., p. 222.

ratios of the squares of the times in which such spaces were traversed. Or we may say that the spaces passed over are to each other as the squares of the times."

Using this squared-time rule for distance fallen, Salviati recalculates the elapsed time of the cannonball drop from the moon's orbit to the center of Earth to be half the square root of the distance fallen of 588,000,000 braccia, yielding 3 hours, 22 minutes and 4 seconds.[42] Here he implicitly assumes the acceleration [43] due to gravity to be 8 braccia/sec^2. Galileo's calculation treats astronomical fall like local acceleration of a falling body, a vast improvement over Locher's constant speed of fall, yet still lacking the notion that the acceleration itself is variable with distance.

Among other topics, the Second Day visits the weighty objection that elephants do not fly off the Earth. Here we see Galileo approach and miss uniting the physics

Galileo's Close Brush with New Discovery

of terrestrial and celestial motions. Had he calculated by his own rules, he might have shown conditions under which objects gravitating toward Earth's center would have orbited Earth or otherwise departed it along a conic-sectional path.

Simplicio argues that the Earth does not move because elephants do not fly to the heavens. This is a more serious objection than might be supposed on the casual basis that we do not directly sense Earth's motion.[44] The speed of Earth at the equator in the Copernican system is so large in relation to anything on Earth conceived heretofore by medieval minds that it has become a cause of concern that people and things might be cast off the Earth. Just when Salviati has nearly won him over, Simplicio realizes that as Earth turns in the system of Copernicus, an object situated on its surface would have a motion that resolves a tendency to

42. Salviati uses a constant acceleration of gravity (see note above) throughout the drop. Following Newton's law varying gravitational acceleration with the inverse square of distance, a correct calculation would start the fall with a greatly attenuated acceleration (by a factor of $59^2 = 3481$) due to the altitude of the drop (360,963 km between the center of Earth and moon as reckoned by Galileo; the actual mean distance is about 384,400 km), and would reckon continuous change of acceleration over the drop of the cannonball. This would yield about 4 days, 19 hr., 30 min. (cf. Castelnuovo, cited by Sosio, p. 272). Ironically, Locher came accidentally closer to this value, though Galileo's approach was a great advance over his.

43. The acceleration due to gravity at Earth's surface is about 9.8 meter/sec^2. For this calculation, Galileo has chosen a value of half that amount (8 braccia/sec^2 = 4.9 meter/sec^2), which might be construed to indicate an awareness that the acceleration due to gravity would be attenuated at astronomical distances. Yet his example of a 100 braccia fall in 5 seconds assumes this same value and is said to be an experimental result. An unresolved question here is whether Galileo was using distances and times of fall that he thought to be real, or just making up numbers for the sake of example. Another intriguing possibility is that Galileo had calculated the acceleration due to gravity from his inclined plane experiments and either used a braccio half as big as I calculate above, or miscalculated by a factor of exactly 2.

44. Any point on the equator of a sphere the size of Earth which rotates once in 24 hours moves in a circle at a rate about twice the normal cruising speed of a commercial airliner. Airplanes flying westbound routes under $\cos^{-1}(1/2) = 60°$ absolute latitude (for example, New York to Chicago or Istanbul to Madrid) cannot quite "keep up with the sun", an idiom that itself hints at Ptolemaic clinging.

move straight along the horizon plane and a downward tendency as the arc of Earth "falls away" from that horizon plane. Comparing them, he declares:

"... And I allowed myself, simple-mindedly enough, to be convinced that stones would not be extruded by the whirling of the earth! I take it back, then, and declare that if the earth did move, then stones, elephants, towers, and cities would necessarily fly toward the heavens; and since that does not happen, I say that the earth does not move."[45]

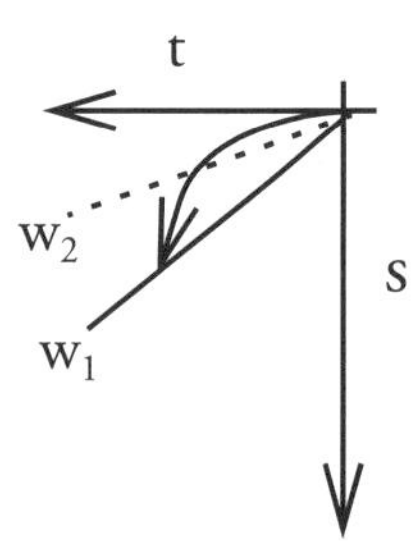

Salviati retorts that when particles are extruded from the surface of a circularly whirling object, they move along a tangent line at the point of separation. Here Galileo corrects his earlier description of "circular" inertia.[46] To show that motion along the tangent line to Earth at an arbitrary time cannot prevail over the downward motion due to gravity along the secant line to Earth's center, he introduces what may be the first multidimensional graph in history, showing the speed (s) of a moving body of weight (w) on the surface of Earth in relation to time (t).[47] Galileo poses the problem and provides tools for solving it, but does not solve it.

With modern methods combining principles of Galileo and Newton, we can calculate the extrusion velocity. No doubt Cardinal Bellarmino's hellfire-fanning deity could spare some of its infinite supply of breath to blow in the appropriate direction on some large paddles anchored in a vertical plane at Earth's equator. If Earth could thus be made to spin faster by a factor of about 17 so that the speed of rooted surface objects near the equator approaches $\sqrt{GM/r} \cong 7.9$ km/sec,[48] then such objects would approach a condition familiar to artificial-satellite passengers as weightlessness. This would make the new-day about 85 old-minutes long. As the Earth begins to spin still faster, spike-heeled "anti-extrusion" shoes and summer homes in the polar regions might become fashionable among humans. At $\sqrt{2}$ times this speed,[49] or about 11.2 km/sec, objects at the equator would reach parabolic speed, or "escape velocity", and the dodging of flying elephants might

---

45. *Dialogo*, op. cit., p. 196.

46. Cf. Drake's Essay 13 "The Case Against Circular Inertia" in his *Galileo Studies: Personality, Tradition, and Revolution*, U. of Michigan Press, 1970.

47. Although there are two coordinate axes, there is a third variable represented: the angle between the time axis and an oblique line representing the weight of the object in question. The prototype of this graph is Galileo's (*Dialogo* p. 199); the main difference in my representation of it here is the labelling of the axes.

48. I derive this expression from Sterne's (1.3-7), the energy equation of the one-body problem, by setting the semi-major axis equal to the radius of Earth (Theodore E. Sterne, *An Introduction to Celestial Mechanics*, Interscience Publishers, Inc., NY, 1960). Here, G is the gravitational constant, M the attracting mass (Earth), and r the distance of the object from the center of Earth.

49. Ibid.; this factor is required to cause the linear speed of the object to approach zero as the distance *r* from Earth approaches infinity, a condition for the parabolic orbit.

then become a popular concern. The "day" would be just under an hour (of 3600 SI seconds) [50] long. In this argument, Galileo has clearly recognized a rectilinear inertia that would tend to carry objects along a tangent line away from a physical point of tangency on Earth's surface.

As the Second Day draws to a close, Simplicio attempts to undermine this new inertial concept by denying the existence of a physical point of tangency: "*a straight line would go for tens and hundreds of yards touching even the surface of water, let alone the ground, before separating from it.*" Taking Simplicio's argument as a given, Salviati considers its effect on the probability that an object would be projected from a rotating surface, and responds: [51]

The Resilience of Truth

"*But don't you see that if I grant you this, it will be so much the worse for your case? ... Do you not see that in this way the projection would take place along the very surface of the earth, which is as much as to say that it would not be made at all? So you see that the power of truth is such that when you try to attack it, your very assaults reinforce and validate it.*"

However much Simplicio may have suffered as a foil to the brilliant Salviati, he has shown himself an intelligent and learned (though miseducated) spokesman for the prevailing philosophy of his day. In the Third Day, Simplicio's thinking is even characterized as Archimedean. His method of reaching conclusions, by inference from premises accepted on authority, is not rare among academia, even today. The hypothesis that Urban took umbrage at the *Dialogo* because it portrays the Aristotelian spokesman for his own views as a simpleton is too simplistic.[52]

It is fashionable among academic pigeonholers to present Galileo as a physicist and not a mathematician, as if the two were mutually exclusive. Passing now to the Third Day of his dialogue, we can catch the first of many glimpses of his interdisciplinary versatility. Galileo digresses temporarily from his topic of Earth's motion, taking a step back to view the role of mathematics in inferring from a sample of celestial observations. As we did before, [53] here we encounter a science

---

50. In order to establish a unit of time independent of Earth's rotation and its vagaries, as of 1972 January 1 the Bureau International de l'Heure in Paris has defined the SI second, the fundamental unit of international atomic time, as "the duration of 9 192 631 770 periods of the radiation corresponding to the transition between two hyperfine levels of the ground state of the cesium 133 atom". *The Astronomical Almanac*, pub. jointly by the U.S. Naval Observatory and the Royal Greenwich Observatory, 1995, p. L2.

51. *Dialogo*, op. cit., p. 203.

52. Galileo introduces Simplicio as "a Peripatetic philosopher who seemed to have no greater obstacle to the understanding of the truth than the fame he had acquired in Aristotelian interpretation"; see Finocchiaro, op. cit., p. 216. This description does not fit Urban VIII. Finocchiaro speculates that Simplicio represents the Padovan professor of Philosophy, Cesare Cremonini, a scholar and a far cry from Urban VIII.

that begins in an application to astronomy and branches out to diverse applications much closer to home. As examples, consider the connection time of your telephone calls, the service time you perceive on the internet, and the illusion you may have at work of having the service of a large multiple-user computing system all to yourself; all three depend on applications of statistical theory to the design of software and machines.

✦✦✦✦

> *"... upon the shore I lately viewed myself,*
> *When the sea stood still, unruffled by the winds."* [54]

Statistics and Pseudoscience

Summers by the seashore can do more for a student's appreciation of the mathematical basis of nature than any formal course in mathematics. Properly positioned by a pier, one can easily observe the crest of a wave as it passes a surface marking on one of the vertical wooden piles that support the pier. With a watch that has a second hand, it is easy to monitor the time between traversal of a wave and its successor at the same point on the pier. Over multiple samples separated in time, the mean wave amplitude and interarrival time might vary with wind conditions, but after recording a sample of thirty or so consecutive waves, the mean of that sample becomes a reasonable estimator of the height of the next wave.

Predictability does not come so easily when the arrival of an event depends on preceding like events, as when customers arrive for amusement rides. Here the apparent wait time creates a psychological bias against entering the queue that increases with its length. But when one takes a random sample from a distribution of independent events, like the crashing of waves against a pier, the growing sample tends toward assuming the characteristics of the population, as in the familiar "bell curve". On the Third Day, Salviati examines a random sample cited by Antonio Lorenzini, an anti-Copernican.

In 1606, Kepler complained in *De Stella Nova* that Galileo had not refuted Lorenzini when the latter attempted to portray the new star seen in 1604 as a sublunar phenomenon. Lorenzini's demonstration cites measurements of the nova's distance by 13 independent observers. Out of a multitude of observations, Lorenzini rejects as anomalistic those which would place the nova beyond the moon's orbit. Galileo now ends his silence, exposing Lorenzini's selective use of data to fit his presuppositions.

---

53. See Chapter 7, "Columbus of the Heavens?".
54. "*... nuper me in littore vidi,/Cum placidum ventis staret mare.*" Virgil, *Bucolics*, ii, 25 f., quoted by Simplicio in Galileo's *Dialogo*, Third Day, trans. by Stillman Drake, p. 328.

Had he dismissed Lorenzini on that point alone, he might not have anticipated a concept of dispersion that would later be formalized in statistical theory. Considering those observations that Lorenzini admitted, Salviati lists a dozen from which distances to the nova are calculated in the range of 1/48th to 32 Earth radii, with a mean of 9.8 radii, and (in modern terms) a variation coefficient exceeding 100%.[55] This lack of internal consistency in Lorenzini's sample is a clue to its bias. Galileo exclaims:

"... I see these results differing so much among themselves that some of them give me distances of the new star from the earth which are ... fifteen hundred times as great as others, so that I may well suspect that among those not calculated there might be some in favor of the opposite side."[56]

Salviati has noticed that Lorenzini rejected as erroneous observations from distances far beyond the moon sufficient in number that, in modern terms, they could have filled out the familiar bell-shaped Gaussian distribution if they were instead included. Might the sampling technique have better taken into account the dispersion of observational errors? Galileo answers:

"... astronomers, in observing with their instruments and seeking, for example, the degree of elevation of the star above the horizon, may deviate from the truth by excess as well as by defect; ... of these two kinds of error, which are opposite and into which the observers of the new star may equally have run, one kind when applied to the calculations will render the star higher than it should be, and the other lower. And since we are already agreed that all the observations are erroneous, what is the reason for this author wanting us to accept those which show the star to have been close as being more congruous with the truth than those others which show it to have been exceedingly remote?"[57]

Besides an unbiased sampling technique, Galileo suggests a better experiment: the

---

55. This means that the standard deviation of Lorenzini's sample exceeds its mean. Statisticians average all the deviations that individual points in a distribution have from their arithmetic mean by taking the quadratic mean (root mean square) of the deviations, thereby averaging their magnitudes without regard to sign. This average is known as the standard deviation of that distribution. To form the standard deviation of a sample taken from that distribution, the mean square is first formed by dividing the sum of squared deviations by one less than the sample size, before taking the square root. With a standard deviation of 10.5 Earth radii, the Lorenzini sample is so scattered that, by a principle due to Chebyshev (1821-1894), as many as 25% of additional observations can be expected to lie outside $2 \cdot (10.5/\sqrt{12}) \cong 6$ Earth radii from the mean of less than 10 Earth radii!

56. *Dialogo*, Drake trans., p. 284.

57. Ibid., pp. 291-292. Statistical terms had not been developed in Galileo's time; in modern terms, his comments point to the asymmetry of Lorenzini's sample as an indicator of bias. If his criterion for accepting observations formed an unbiased estimator, we would expect them to be more or less symmetrically placed around the mean of the distribution in the manner of a "bell-shaped" curve (also known as Gaussian or "normal").

nova's angular distance from the star χ Cassiopeiae, less than 1.5°, can be shown invariant over a distribution of observations. In this way, the negligible parallax and consequent remoteness of the nova can be determined with statistical confidence.

Having exposed Lorenzini's statistical bias, Galileo proceeds to the main argument, attacking his ostensible reason for rejecting the data he did not want — atmospheric refraction. Lorenzini has argued as if the images of stars separated by less than 2° and remote from the horizon would be refracted to a significantly differing extent, but this is not so. He has simply failed to quantify what turns out to be an effect so minor as to fall below the threshold of observability.

To justify his remark, quoted at the outset of the last Chapter, "*This author grasps not at ropes, but at spiderwebs from the sky, ...*", Salviati ends the assault with an explanation why he has charged Lorenzini with spiderweb-grasping:

> *"Truly, it was with too scant a store of ammunition that this author rose up against the assailers of the sky's inalterability, and it is with chains too fragile that he has attempted to pull the new star down from Cassiopeia in the highest heavens to these base and elemental regions."* [58]

Galileo has shown the nova represents a change in the heavens. But what about the Ptolemaic argument, immortalized in poetry by Virgil, that the changeless heavens revolve around the Earth?

> *"... A land that lies beyond the stars, beyond*
> *The year's path and the sun's, where, prop of heaven,*
> *Atlas upon his shoulder turns the pole*
> *Studded with burning constellations."*
>
> Virgil, *Aeneid* VI: 796 [59]

To answer this last question and to paint a contrasting picture to Virgil's is a major aim of the *Dialogo*. As we who, belonging to the first generation of humans that have been able to look back at Earth, read Galileo's verbal portrait of the new cosmology, we see the power of his science to predict what, centuries after him, will be the description of an eyewitness:

> **"WE CAN SEE FROM POLE TO POLE AND ACROSS OCEANS AND CONTINENTS, AND YOU CAN WATCH IT TURN, AND THERE'S NO STRING HOLDING IT UP, AND IT'S MOVING IN A BLACKNESS THAT'S ALMOST BEYOND CONCEPTION."** [60]

---

58. See *Dialogo*, op. cit., p. 318.
59. Trans. by James Rhoades in Brittanica *Great Books of the Western World*.

Now the dialogue takes a sudden and dramatic shift back to the topic of Earth's motion; in particular, the apparent annual rotation of the

Aggressive Resumption of the Copernican Defense

heavens *"generally attributed to the sun, but then, first by Aristarchus of Samos and later by Copernicus, removed from the sun and transferred to the earth."* By replacing the annual rotation of the heavens by the motion of Earth, Aristarchos and Copernicus put Earth into a circular motion that must be independent of the rotation they claimed it should have about its own axis. For the cycle of day and night is not explained by the same motion that advances the seasons even if the causes of both effects are transferred to Earth. The geocentrist philosophers objected that any independent circular motion would entail that Earth orbit about a central point removed from itself; hence, the "unthinkable" conclusion that Earth cannot occupy the center of the universe. Salviati now openly thinks the "unthinkable":

> *"I might very reasonably dispute whether there is in nature such a center, seeing that neither you nor anyone else has so far proved whether the universe is finite and has a shape, or whether it is infinite and unbounded."*[61]

To Simplicio's contention that Aristotle gives a hundred proofs that the universe is finite, bounded and spherical, Salviati gives this disarming reply:

The Universe: Finite or Infinite?

> *"Which are all later reduced to one, and that one to none at all. For if I deny him his assumption that the universe is movable all his proofs fall to the ground ... But in order not to multiply our disputes, I shall concede to you for the time being that the universe is finite, spherical, and has a center."*[62]

This concession allows the inquiry to proceed toward locating that assumed center. Here, Galileo displays what feat of the intellect is possible to one who disengages from dogma and re-examines his assumptions. Once Salviati has suspended

---

60. The voice of an unidentified NASA space traveller can be heard in an official NASA video expressing this reaction to looking back at Earth on the radio link from his Command Module in Space to Houston Mission Control. Galileo, never having seen Earth in space, had put forth his vision of it with a clarity and conviction that rivals this eyewitness account of a human traveller to the moon, 3½ centuries later. The vector from Virgil's Atlas to the inertial component (analogous to the "stringless" inertia of Earth's revolution noted by the space traveller) of Galileo's hypothetical stone falling from the top of a tower points to a future unified physics of field forces.

61. *Dialogo*, op. cit., p. 319. See also Drake's footnote, p. 486: "This was very dangerous ground for Galileo to tread; Giordano Bruno's conviction and execution had depended largely upon his having espoused the view that the universe was infinite." How "largely" may be arguable, but certainly Bruno's cosmology did not endear him to the Inquisition.

62. Ibid., p. 320.

Simplicio's Aristotelian premises for the sake of argument, Simplicio proceeds to deduce from observational data the positions of all the planetary orbits.

Aristotle has told us, notes Salviati, that the Earth is at the center of a bunch of planets orbiting it in variously sized circles. From any one planet, Earth should remain equidistant — but with the telescope we see big periodic changes in the apparent sizes of the planets. Moreover, *the cycle of changes is synchronous with their cycle of angular distance from the sun.* When the outer planets appear large, they are at or near opposition to the sun; when small, they are at or near conjunction.[63]

Simplicio Sketches the Copernican Solar System

On paper, Simplicio now marks arbitrary positions for the Earth and sun. Asked to locate Venus, he replies that it never reaches opposition, but remains within 40° of the sun while moving in cycles from one side of the sun to the other. While it goes through a cycle of phases, it contracts and approaches full phase nearing conjunction. *Its orbit cannot embrace the earth since it is never at opposition.* Nor can it be between Earth and sun when it appears nearly in full phase approaching conjunction. Nor can it be always beyond the sun, since it shows a horned crescent when it appears largest and is in retrograde motion. Simplicio draws its circle around the sun but not enclosing Earth.

The same is true for Mercury, and since Mercury remains even closer in separation angle from the sun, he draws its orbit inside that of Venus. Asked where to put Mars, he replies immediately: *"Mars, since it does come into opposition with the sun, must embrace the earth with its circle. ... But it always looks round; therefore its circle must include the sun as well as the earth."* [64] Likewise he places Jupiter and Saturn as outer planets, and in the order of their angular speed around the sun. Because the moon comes into conjunction and opposition and eclipses the sun, he places it in orbit around Earth. Thus it seems that Salviati has induced an Aristotelian to sketch the solar system of Copernicus. But has he? The sketch shows Earth with an orbit marked around the sun. Simplicio has not admitted that Earth actually moves in that orbit. Nor dare he.

The Inconclusive Conclusion

Apparently sensing the danger of having his Aristotelian character plainly say what his sketch shows, Galileo changes the subject while Simplicio remains unconvinced. But he *must* be unconvinced. In his preface, Galileo announced an intent to present

---

63. An outer planet is said to be in opposition when Earth is at an intermediate point on the line connecting that planet with the sun or, in other words, when it is opposite in the sky to the sun as viewed from Earth. It is in conjunction when the sun is at an intermediate point on the line connecting the planet with Earth; hence in the same direction as the sun and no longer visible because it is "behind" the sun as viewed from Earth.

64. *Dialogo*, Drake trans., p. 324-325.

the Copernican side as hypothetically superior, in order to show himself motivated by religion to accept that Earth is at rest without being ignorant of arguments to the contrary. To hedge against the possibility that Bellarmino's injunction might still be upheld despite Urban's ostensible disagreement with the intent of the Decree of 1616, Galileo must represent the entire dialogue as inconclusive. After a long discussion of his hypothesis of tides in the Fourth Day,[65] he does just that: he echoes Urban's admonition: *"it would be excessive boldness for anyone to limit and restrict the Divine power and wisdom to some particular fancy of his own."* [66] But here Galileo distances himself from these words, letting Urban's thought be expressed by Simplicio, who seemed to have lost the debate! Salviati concludes by saying that the innovation he introduced in the dialogue *"may very easily turn out to be a most foolish hallucination and a majestic paradox"* and that *"we cannot discover the work of His* [God's] *hands".*[67] Of all the passages in the *Dialogo*, this is surely the least credible.

The Pious Preface and the Publication

Early in 1632, Galileo secures approval of Church censors to publish, accepting all imposed variations to the text. He affixes a preface "written in accordance with the ideas of Father Riccardi, Master of the Sacred Palace, and signed by Galileo, in which the Copernican theory was virtually exhibited as a play of the imagination." [68] The *Dialogo* is printed in February at Florence. It draws the reader into the debate. We begin by laughing at Simplicio's pomp, and by the middle of the Third Day, we are cheering him on. We identify with Sagredo's burning curiosity about the secrets of nature, and we marvel at Salviati's intellect throughout. The *Dialogo* advances the best arguments on both sides of the debate about Earth's motion. It is readable by educated readers, it sparkles with wit and humor, it will generate sufficient market demand to sell out the original 1,000 printed copies, and the ensuing publication ban will prevent a reprint easily twice that size.[69] It will effect a revolution on the literature of its time:

---

65. See Ch. 8, "Galileo's Hypothesis of Ocean Tides (January 1616)" and Appendix B.

66. *Dialogo*, trans. by Stillman Drake, op. cit., p. 464.

67. Ibid., pp. 463-464.

68. A. D. White, op. cit.

69. "My publisher is disconsolate, the prohibition of the book has caused him a loss of more than 2000 scudi, for the sale of the first edition and a second twice as large was assured."; Galileo to Diodati, 15 Jan 1633 (J. J. Fahie, *Galileo: His Life and Work*, James Pott Co. NY, 1903, p. 284). Sales that would have come from a reprint will thus be unrealized due to Urban's edict squelching the *Dialogo*; the resulting financial damage to the publication of Galileo's book will be twice his annual salary (1000 Tuscan scudi per annum) in the service of the Grand Duke. At a moderate 2% average annual interest, that sum would today be worth about 2,600 times Galileo's annual salary, enough to fund for several centuries a private organization along the lines of the *Accademia del Cimento* (founded by Leopold de' Medici to pursue Galilean science) to maintain and research Galileo's equipment and discoveries and to extend them.

"In the period between Machiavelli and Manzoni, Galileo is the master of Italian prose as well as the creator of its classic style; to discover the roots of subsequent Italian prose one must seek them in Galileo's writings. ... His writings not only put an end to a basically erroneous method of scientific thought, but also rejected strained style in artistic language. Conventionality in content and expression gave way to the adventure of creative thought, and a personal confrontation between nature and man." [70]

"The *Dialogue* is and remains a masterpiece of Baroque style, which knows how to move effortlessly through tight passages of reasoning, unroll with a rustle of silk, sparkle with malice and restrained good humor, maintain its cadence through vast reaches of syntactic intricacy, and rise without break to the solemnity of prophetic invective. It was not only ruthless analysis, it was the magic of the Italian language handled by a master, which broke the monopoly of stuffy Latin learning, which took the people into camp and revealed to them the new unimagined power of mathematical physics." [71]

Galileo's Tribute to the Invention of Writing

Three hundred copies of Bernegger's 1635 Latin translation of the *Dialogo* will be sent to Paris.[72] In 1661, Thomas Salusbury's English translation will appear in London.[73] The timelessness and universality of this book confirms Galileo's own comment about the invention of writing:

"But surpassing all stupendous inventions, what sublimity of mind was his who dreamed of finding means to communicate his deepest thoughts to any other person, though distant by mighty intervals of place and time! Of talking with those who are in India; of speaking to those who are not yet born and will not be born for a thousand or ten thousand years; and with what facility, by the different arrangements of twenty characters upon a page!" [74]

By those same arrangements of characters that can leap across continents, oceans, and epochs, it is possible to invite trouble with one's own neighbors.

---

70. from Olschki's essay, *Galileis literarische Bildung*, translated by Thomas Green, S.J. and Maria Charlesworth, as edited by McMullin in *Galileo, Man of Science*, op. cit.

71. from Giorgio de Santillana's essay "Galileo in the Present", published in *Homage to Galileo*, Morton Kaplon ed., M.I.T. Press, 1965.

72. James MacLachlan, *Children of Prometheus*, Wall & Emerson, Toronto, 1988, p. 114.

73. Most copies were destroyed in the 1666 great fire of London, but the Salusbury translation was reprinted in the 20th century as: *Dialogue On the Great World Systems*, U. of Chicago Press, 1953.

74. Galileo Galilei, *Dialogo*, op. cit., end of The First Day, trans. by Stillman Drake, p. 105.

Galileo's masterpiece of expository writing does not save him from his contemporaries: to wit, Father Scheiner, who believes that he has discovered spots on the sun and is ill-disposed to see "his" discovery used as support for Copernicus by a rival claimant who has publicly overturned his arguments about the nature of sunspots. On 19 June 1632, Castelli writes to Galileo about an incident in a Roman bookshop. Scheiner, he says, "on hearing a priest praising the *Dialogue* as the greatest book ever printed, turned livid and shook so badly that the bookseller who told Castelli about the event had been astonished. Scheiner had said that he would give ten gold ducats for a copy in order to compose an answer to it immediately. (When he did, it was so violent that his fellow Jesuits prevented its publication until after his death in 1650.)" [75]

In August, Tommaso Campanella, a Dominican who has been deeply influenced by the empiricist philosophy of Telesio and the ideas of Giovanbaptista Della Porta, and who has courageously written and published a defense of Galileo, reports to Galileo that "a congregation of irate theologians ignorant of mathematics had been convened to prohibit the *Dialogue*".[76] By September, these stirrings and rumblings are generating formal complaints that reach the Pope.

Papal Reaction to the *Dialogo*

In roguish and roughshod reaction to the rumblings, Urban orders a halt to printing and distribution of the *Dialogo*. He institutes a recall of the original manuscript to Rome. He appoints a Preparatory Commission of persons unfriendly to Galileo to investigate its content and manner of publication and to prepare an indictment. We have seen the similarity of content between Galileo's Reply to Ingoli in 1624 and the *Dialogo* of 1632, and are left in wonderment at the disparity of Urban VIII's reactions to these two documents. What events occurred between the two dates that might explain this disparity?

Has Urban VIII suddenly "discovered" that Galileo had withheld a material fact from him in their talks? At the time of the letter to Ingoli, Urban was probably unaware of an unsigned memorandum, dated 1616, which forbade Galileo to teach the Copernican theory "in any way whatever, either orally or in writing".[77] Fr. Scheiner may have been responsible for bringing this document to his attention; the same could be said of Fr. Grassi, especially since Galileo has crushingly counter-attacked him in *Il Saggiatore*.[78]

An Unsigned Memorandum

---

75. Drake, *Galileo at Work*, op. cit., pp. 337.
76. Ibid., p. 338.

No matter which wounded antagonist has called his attention to it, the Pope can easily see it as treachery not of Galileo but of his enemies: it is manifestly incongruous with the other events of 1616. It would be easy for Urban to check the Inquisition files and see that his predecessor Paul V ordered Galileo to be silenced on the Copernican issue *if he refused to obey an injunction to abandon the opinion of Copernicus*, but that he "acquiesced in this injunction and promised to obey".[79] Has Urban checked the files? Whether he has or not, he still knows (or thinks he knows[80]) that the memorandum would be a surprise to Galileo. Since Galileo has not signed it to show that he received it, it seems that he did not receive it; otherwise, there is no good explanation why he would openly defy a written order not to teach Copernicanism "in any way whatever, either orally or in writing". Such behavior would not only be uncharacteristically reckless but, to the Pope and the Inquisitors, unthinkable.[81] Except possibly as a desperate administrative prop by which prosecutors might create fear and extract a false confession, the memorandum is worthless to the Church, and Urban knows it.

Otherwise, what are we to conclude? That Urban believed Galileo thought himself bound to speechlessness about Copernicus? Hardly. Underlying the scheme that he and Galileo had cooked up in 1624 to compare Ptolemy and Copernicus was Urban's recognition of a distinct danger that Galileo's arguments might utterly destroy the system of Ptolemy. Consider those instructions tacked onto the text of the mandated Preface to the *Dialogo* and which the Master of the Sacred Palace,

---

77. Ibid.; this is the wording from the Special Injunction (26 February 1616), referred to here as the "unsigned memorandum". Drake has speculated first, that these words may have been read to Galileo by an overzealous cleric who ignored the Pope's instructions to prohibit Galileo's teaching the Copernican opinion only if Galileo had refused to comply with Bellarmino's injunction to abandon this opinion; secondly, that Bellarmino instructed Galileo privately to disregard the warning not to "teach, ... in any way whatever" (cf. the Drake-de Santillana controversy, Appendix A in Geymonat, op. cit.). But in this case, as modern dating methods indicate that the unsigned memorandum was prepared at the same time as Bellarmino's warning (see de Santillana's *The Crime of Galileo*), Bellarmino apparently failed in his responsibility to keep the erroneous injunction out of the Inquisition files.

78. "Gabriel Naude, then librarian to Cardinal Barberini ... wrote to Pierre Gassendi in April that 'Galileo has been cited by the Roman Curia through manipulations of Father Scheiner and other Jesuits who want him destroyed'." ; ibid., p. 343.

79. See Inquisition Minutes (25 February and 3 March 1616) in Finocchiaro, *The Galileo Affair*, op. cit., pp. 147-148. Here I respectfully disagree with Professor Drake's statement about the memorandum: "The pope had no reason to disbelieve it, ..." (*Galileo*, op. cit., p. 76). He had ample reason to disbelieve it, but may have *wanted* to believe it.

80. In a debate with de Santillana (Geymonat, *Galileo Galilei*, op. cit., Appendices A and B), Drake has advanced the interesting hypothesis that Galileo had received the injunction but was told by Bellarmino to ignore it. However, I consider de Santillana's rebuttal against Drake's hypothesis to be compelling. The probability that Galileo ever saw or heard of the injunction prior to the 1633 trial, I believe, is negligible.

81. Here I assume rdc's "Challenges Are Unthinkable" hypothesis of Inquisitorial behavior named in footnote 3 to this Chapter.

acting under Urban's ("Our Master's") direction, sent on 19 July 1631 to the Florentine Inquisitor:

> "At the end [of the *Dialogo*] one must have a peroration of the work in accordance with this Preface. Mr. Galilei must add the reasons pertaining to divine omnipotence which Our Master gave him; these must quiet the intellect, *even if there is no way out of the Pythagorean arguments.*" [82]

The "Pythagorean" arguments are those in favor of the motion of Earth. Here I find convincing evidence that Urban is expecting Galileo's arguments (in the book that he, Urban, wanted Galileo to write) in favor of Earth's motion to prove dominant, and is settling for "quieting the intellect" of the faithful after Galileo has won the debate in favor of Copernicus. In modern political parlance, Urban was preparing for "damage control".

For these reasons, neither the content of the unsigned memorandum nor Galileo's effectiveness in arguing for Copernicus is likely to have upset Urban's good will toward Galileo. Can the trouble be Urban's discovery that in 1616 Galileo was admonished to abandon the opinion of Copernicus and agreed to do so? Urban can easily interpret the *Dialogo* as evidence that Galileo still held the opinion he had promised to abandon. Of that promise, Galileo proffered no hint to Urban in their talks of 1624, or to Church censors. Might Urban feel tricked, perhaps even embarrassed, that Galileo has steered him into approving a scheme that requires Galileo to demonstrate the merits of a position he has promised to abandon?

Hardly. Urban has personally washed his hands of the Decree of the Index, and is now pretending, after the fashion of Pontius Pilate, to have nothing to do with that Decree while at the same time delivering the troublemaker to his enemies. Why would Urban not feel free to overturn his predecessor's (Paul V's) action against Galileo? Presumably, it is his allegiance to the principle that an ecclesiastical interpretation of Scriptures (however ill-advised) must take priority over a scientifically superior hypothesis with which it is discordant. This is the very mission he has given Galileo in approving the *Dialogo* — to stand up as a representative of science and endorse this principle. Urban can now see Galileo's silence about his agreement to abandon the opinion of Copernicus as a politically artful way of allowing himself to cling with impunity to the position he had taken in his *Letter to Cristina* and subtly sustained in his placing the principle Urban had wanted him to espouse in the mouth of Simplicio. But why should a politician, upon perceiving political cunning in someone else's behavior, not be proud of him? A double standard, you may say. Perhaps. But if we look for a still-better explanation of Urban's alienation from Galileo, we can find it.

---

82. Emphasis mine. See Finocchiaro, *The Galileo Affair*, op. cit., p. 354 note 57.

Urban's Conflict with the Medici
There was another series of events between the Letter to Ingoli and the *Dialogo*. The Medici, Galileo's employers, held something vulnerable to the avarice of Urban VIII: [83]

"... Pope Gregory XV, who had succeeded Paul V in 1621, died, and was succeeded by Urban VIII (1623-1644), whose main endeavor was to enrich in every way his family, the Barberini."

One way of enriching his family would be to seize land for the Papal States and appoint a family member to rule over it. Urban VIII presents a claim to the Duchy of Urbino, a future Medici inheritance. He claims that upon the death of its duke, it would become a "vacant fief" belonging to the Papal States. Galileo's employer, Grand Duke Ferdinand II, counter-claims: [84]

" First, he claimed it as being the lawful property of his betrothed wife Vittoria, she being the only child of the Duke's only son ... Secondly, if Vittoria's claim was set aside, then Ferdinand claimed Urbino in his own right as inherited from Catherine de' Medici ..."

With legalities stacked up against his lust for Urbino, the Pope simply seizes it: [85]

" ... Francesco della Rovere II, Duke of Urbino, at length died at the age of eighty-two. The Papal troops at once took possession of Urbino, almost before the breath was out of his body; ...and Ferdinand ... had to acquiesce in seeing his and his future wife's inheritance robbed from them. The matter created much bad blood between the Barberini and Medici families; ... The general result of the whole affair was that Pope Urban VIII nourished an undying hatred against the Medici throughout his pontificate, thwarting them on all occasions, making every priest and monk in Tuscany an enemy of the Government, and creating incessant difficulties in the administration of a country in which priestly influence was supreme; ...

In 1632 he [Galileo] published his *Dialogues*. By this time, however, the affair of Urbino had occurred, the Pope was incensed with the Grand Duke of Tuscany, and the weakness of the latter had been fully displayed."

There is a psychological phenomenon according to which one who has done violence to another hates the victim who serves as a reminder of the misdeed.[86]

---

83. Col. G. F. Young's *The Medici*, p. 664.
84. Ibid., pp. 664-665.
85. Ibid., pp. 670-671. The Papal States (shaded area in the author's sketch) after Urban VIII's seizure of Urbino nearly surrounded Toscana. Sketch is a composite derived from (1) "Estates of the Church" Map by R. Wilkinson pub. 1809 by Cornhill, London and (2) William R. Shepherd, *Historical Atlas*, 8th ed. Map 90a.
86. "Whom they have injured they also hate"; Seneca, "Moral Essays, On Anger, 2, 33" trans. by W. H. D. Rouse in *Bartlett's Familiar Quotations*, Little, Brown & Co., Boston, 1951, p. 1106. (rdc)

Cisalpine Territory, including Toscana and the Papal States, after Urban's Seizure of Urbino

(see footnote 85)

In 1632, Urban has reason to see publication of the *Dialogo* as a project of Toscana, not just of Galileo: Ferdinand himself has been involved in securing Church permission to print the book in Florence. Playing Galileo's vulnerability against Ferdinand's weakness, the Inquisition supports the acquisition — Urban's, that is.

Preparatory Commission Report

Let's return now to the firestorm that follows Urban's referral of the Galileo case to the Preparatory Commission. Early in September that Commission, stacked by Urban with Galileo's enemies, delivers its grounds for indictment. In the original manuscript, Galileo "overstepped his instructions" by asserting Earth's motion absolutely, and then submitted the manuscript to censors for review (and conversion to the hypothetical) without informing the censors of the Special Injunction of 1616 which — so they allege — forbade him to teach the opinion of Copernicus in any way whatever. The Commission also advances the "simpleton" myth, claiming that at the end of the book, Galileo put the Pope's argument "in the mouth of a fool".[87] These are the ostensible reasons for bringing Galileo to trial.

---

87. Finocchiaro, op. cit., p. 221. Would Galileo have escaped indictment had he let Salviati articulate the Pope's position? In choosing Simplicio to deliver that argument, did he consciously choose to risk his standing in the Church in order to protect the integrity of his work? I leave it to the reader to decide.

Doctrine of the Eucharist and Galilean Atomism

Was there another? What about the claim that this Commission indicted Galileo for having promulgated a scientific principle that allegedly renders the Doctrine of the Eucharist an absurdity? [88] Father Grassi's "Examination XLVIII", supposedly a basis for the hypothesis that Galileo's real crime was his atomism, here distorts Galileo's position and tries to draw a conclusion he never drew. "In the host, it is commonly affirmed, the sensible species (heat, taste, and so on) persist. Galileo, on the contrary, says that heat and taste, outside of him who perceives them, and hence also in the host, are simple names; that is, they are nothing. One must therefore infer, from what Galileo says, that heat and taste do not subsist in the host. The soul experiences horror at the very thought." And the mind recoils at such illogic from a Jesuit logician. Galileo said, in the *Assayer*, that heat and taste imply shapes, numbers, or motions of particles, which would remain in a body if "ears, tongues, and noses were removed". Here he denies both tenets charged by Grassi against him: that they are nothing, and that they do not subsist in the Host. [89] Shapes, quantities, and moving particles are not nothing, and since they remain in the object being sensed, they subsist there. Has Grassi consumed too much wine with his bread?

The Unspoken "Crime" of Galileo

If there were an unspoken crime, it is more plausibly that Galileo published a philosophical argument that was anathema to Urban VIII, who controlled the Commission. This argument can be summed up by viewing Galileo's position as if, beginning with the Letter to Cristina and culminating in the *Dialogo*'s exposition of scientific method, he had said: "Here is the scientific method by means of which you can approach certainty as nearly as you like. Now that I've revealed the method, it is dangerous for the Church to pronounce it a sin to believe what can be proved by it, as the tension between authority and truth will ultimately be resolved in the destruction of faith in the authority". Contrast that attitude with Urban's when he screamed at Niccolini, [90] who was pleading clemency for Galileo, that "One Must Not Impose Necessity On God!" OMNINOG was Urban's "pet" concept of divine omnipotence.

---

88. This conjecture is based on Redondi's Document G3 and Examination XLVIII, *Galileo Heretic*, op. cit., pp. 333 ff., and discussed as a myth in Chapter 13. Roman Catholics are required to believe that after transubstantiating into the body of Christ, the Eucharist (also called the Host) retains the appearance of bread. The contention was that by arguing in *Il Saggiatore* that accidents cannot exist apart from substance, Galileo had shown that transubstantiation is impossible. If Redondi thinks that Galileo's trial on the Copernican issue was a smoke screen to save Galileo from conviction of the more serious Eucharist heresy, he could think so only by failing to notice that the "Eucharist heresy" exists in the mind of the Document G3 author and not in Galileo's writings.

89. See Galileo's "The Assayer" in Drake's *Discoveries and Opinions of Galileo*, p. 276-277.

90. Francesco Niccolini, Ambassador from Toscana to the Vatican.

His angry outburst may well have been a "knee jerk" reaction to what he saw as a collision between OMNINOG and Galileo's attempt to show that God could not have designed the universe any way other than envisioned by Copernicus. If the Church is divinely guided in interpreting Scriptural passages, then is not limiting it in matters of exegesis also limiting God? The *Letter to Cristina* could be passed off as the whim of a cranky scientist, but the *Dialogo* brought forth powerful evidence to expose the dangers of dogmatizing science. Could Galileo's unspoken "crime" have been to advance an epistemology that, by converging on certainty in the real world, limits the Church from claiming otherwise? One easily wonders if Pope Urban might have felt more at home with 20th century developments that have put a murky nebulosity around mathematics and physics by running amok with interpretations too nimbly drawn from well founded conclusions such as the Gödel Incompleteness Theorem or the Heisenberg Uncertainty Principle. If transferred to the 20th century, might Urban invoke the Incompleteness Theorem to argue that if Galileo's reasoning is logically consistent, then it must be incomplete, perhaps lacking Urban's Theorem of Geostatic Geocentrism? Might he invoke the Heisenberg Principle to say that Galileo's telescopic observations are perturbing the phenomena he observes?

Interdiction and Summons to Trial

Ferdinand's instilled awe of the Papacy now mixes with his helplessness in the Urbino affair to embolden the Pope. In a quick succession of interdictive maneuvers to cut down the *Dialogo*, Urban places extraordinary demands upon the officials of Toscana. On September 25, he directs them to pre-empt shipments of the *Dialogo* within and from territory under their control so that Galileo will be powerless to distribute his book under the existing ecclesiastical sanctions. Over Niccolini's protest — that it makes no sense to open an inquisitorial procedure against a book that has already been approved — Urban refers the case to the Inquisition, secretly notifying both Niccolini and Ferdinand and binding them to secrecy.

On October 2, before witnesses who can be questioned if Galileo proves uncooperative, the Florentine Inquisitor summons him to Rome. On October 13, Galileo appeals to Cardinal Francesco Barberini requesting that, due to illness, he be allowed to respond by letter rather than make the trip to Rome. On November 13, Galileo receives the Pope's denial of this request. A week later, the Florentine Inquisitor grants Galileo a month's reprieve due to illness, but the Inquisitor's humane decision displeases the Pope. On December 17, three physicians examine Galileo and certify that he cannot be moved without jeopardizing his life.[91] On December 30, Urban threatens Galileo with examination by Vatican physicians to be sent *at Galileo's own expense*, with orders that if he were found able to travel (a foregone conclusion) he must be "bound in chains and transported bound to

Rome". The Grand Duke, terrorized by the Pope, advises Galileo against further resistance or delay. The famous philosopher could as easily escape to Venice. Never one to retreat from conflict, he chooses to confront his Church.

Even for young and healthy persons, travel is perilous to life. The Bubonic plague is ravaging the countryside. A member of Galileo's household, a glass blower who helped him make instruments, has already succumbed to the dread disease. It is now 20 January 1633. In the harshest of winter weather, having prepared a Last Will and Testament, he departs for Rome and for 24 days is exposed to the rigors of travel and the risk of fatal disease. Now frail, sick, old, tormented in body and mind, worn by worry, work and woe, the great Galileo, who alone stood on a blue balcony and thrust open a window into formerly unfathomed depths of the firmament, is being drawn into the depths of a process from which no man emerges unscathed.[92]

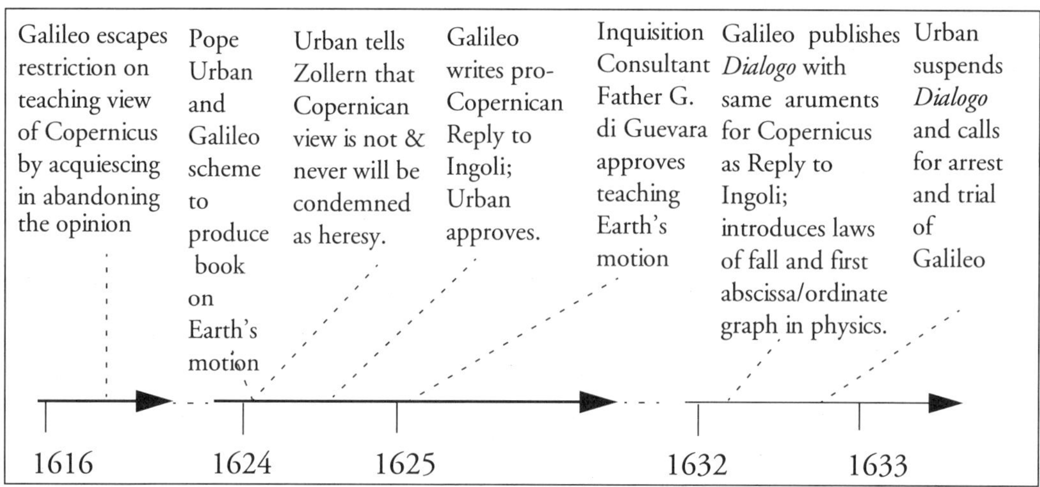

91. Favaro, *Galileo e L'Inquisizione*, p. 75. The report, signed by the physicians De Rossi, Ronconi, and Cervieri, cites multiple debilities, including frequent dizziness, insomnia, irregular pulse (skipping 3 or 4 beats at a time), severe hernia, and rupture of the peritoneum. It concludes with the alarming judgment "... *che per ogni piccola causa esterna potrebbero apportarli pericolo evidente della vita*" (that any small cause of worry from the world outside would pose a clear danger of fatality).

92. The institutional Church has created a conviction and sentencing machine whose process is irreversible and not even able to stop itself for its own protection. Like some modern institutions, it lacks a feedback loop and is capable of generating divergent oscillations that are in the long run self-destructive. But certain Inquisitors, *acting as individuals*, will understand that Galileo is right, and will refuse to act against him despite the danger to themselves.

Chapter 10. The Unexcluded Middle

The Trial of Galileo (1633)

"Sure he that made us with such large discourse,
Looking before and after, gave us not
that capability and god-like reason
to fust in us unused. ..."

... William Shakespeare (*Hamlet*, Act IV, Scene IV)

In the spring of 1633, the Roman Catholic Inquisition conducts Galileo's "trial",[1] a classic study in how a bureaucratic process allows Aristotelian reason to fust in itself unused, and to convict itself in the end.[2] Owing to his reputation and silent supporters in high ecclesiastical places, the defendant has been treated more courteously than the typical victim of these proceedings, having been allowed to await them in comfortable lodging at the Villa Medici, the embassy of Toscana at Rome. But his scientific fame, a tactical advantage, is also a strategic liability: the faster the academic world is populated with his pupils placed in powerful positions, the less the Church can tolerate open defiance from him. His request to see the Pope before the trial is denied.[3] His first interrogation opens in a ghostly, candle-lit committee room at *il palazzo del Sant' Uffizio* in Rome. The interrogators' questions are spoken in Latin, in the third person (Galileo is addressed as "he"). Each time he responds, the chill of a prolonged silence punctuates the fading echo of his voice, causing indescribable anxiety.

"How did he come to Rome?" "Who called him?" "Can he guess the cause of his having been called?" [4] The opening questions probe for an inadvertent confession.

First Interrogation (April 13)

Galileo seems to be anticipating them. In response to the ambiguity *"Explicitly what is in the book you imagine was the reason for the order that you come to the City?"* [5] he answers matter-of-factly as if the question means only *"What is in the*

---

1. No modern notions of jurisprudence should be associated with this word; a trial by Inquisition has a generally predetermined outcome, and does not "try" anything, except perhaps the endurance of the victim to abuse. The "defendant" is not presented with a charge, but must guess what it is. (dc)
2. "Plato's ideal of elite rule has been converted into fact by the Catholic Church. The Roman Church, under the Tridentine organization as it emerged from the Counter-Reformation, is a perfect bureaucracy." Ludwig von Mises, *Bureaucracy*, Arlington House, NY, 1969, p. 101.
3. *"... Galileo invano chiese udienza al Papa ... ";* Gino Loria, *Galileo Galilei*, p. 85.
4. Except as may be otherwise indicated, all quotations in this section are from the Stillman Drake translation in *Galileo at Work*, p. 344 ff.
5. Cf. *"That he explain the character of the book on account of which he thinks he was ordered to come to Rome."* Finocchiaro's translation (op. cit., p. 257) carries the dual meaning of the Latin original.

book?" and not *"What in the book is the reason ...?"*

Soon the interrogation begins to focus on the events of 1616, and the tribunal exposes a presumption fatal to its pre-planned strategy. On the miscalculation that Galileo will not be able to contravene the "evidence", the prosecution has rooted its case in the "special order" conveyed by the unsigned memorandum. To prove a violation of the Decree of the Index would require a more difficult argument than to prove a violation of this "special" order; indeed, the mere existence of the *Dialogo* is proof that Galileo taught the opinion of Copernicus "in any way whatever" — it suffices to show that he received this unsigned order.[6] A response from Galileo acknowledging it will be as good as a signature, but if the existence of the memorandum were revealed to him, he can deny any knowledge of that order. And so, without mentioning it, they ask:

"What were you told of the said decision [The Decree of the Index] *by his Eminence Bellarmine,[7] and did anyone else speak to you about it, and what?"*

This may be Galileo's first hint that something beyond Bellarmino's injunction is supposed to have been delivered to him. Seemingly oblivious to the ambiguity,[8] he answers only the first part of the question. Bellarmino informed him, he says, that the opinion of Copernicus could be held hypothetically *"as Copernicus held it"*, and knew that he (Galileo) held it hypothetically.[9] After the 1616 Decree, Copernicus was not banned, and Galileo was not reprimanded for hypothetically holding his opinion then, so why should he be now? He appears oblivious to another party or

---

6. The document containing this phrase is not signed by Galileo to show that he ever received it. It is not signed by anyone! Among other authors, Giorgio de Santillana has called it a forgery, asserting beyond doubt that it was "planted by the Inquisitors in their secret file in case it might come in handy" (see *Homage to Galileo*, Morton Kaplon, Editor, M.I.T. Press, 1965). He misses the point, however, when he goes on to say that he "had made a good case for pinning down the guilt on a small group of minor officials who had plotted and acted on their own." Regardless of the identity and motives of whoever planted the forgery, the trial judges admitted into evidence a document totally lacking in authentication, and they were certainly not a group of minor officials, nor were they acting on their own, but in the name of the Inquisition and of the Roman Church.

7. Drake has chosen to alter the spelling of Bellarmino's name, following the practice of American Jesuits who call him "Bellarmine". See Ch. 8 "Bellarmino Warns Galileo (1616)" footnote 28.

8. An ambiguity is contained in the original: *"Ut dicat, quid sibi notificaverit dictus Emin.[mus] Bellarminus de dicta determinatione, et an aliquid aliud sibi circa id dixerit, et quid".* The ambiguity centers around whether the subject of *"dixerit"* is *"Bellarminus"* or *"aliquid"*, which can mean "anyone" or "anything". Finocchiaro comes down on the other side of the ambiguity with his translation: *"What the Most Eminent Bellarmine told him about the said decision, whether he said anything else about the matter, and if so, what".* This translation contains a redundancy of the form "What else did he say about it besides what he said about it, and if so, what?" In Drake's (more intelligible) form, the question appears to be craftily designed to solicit information about another party having given Galileo an order, without appearing to assert that any other party exists.

9. Galileo cites Bellarmino's letter of 12 April 1615 to the Carmelite Father Foscarini, in which Bellarmino writes "It appears to me that your Reverence and Signor Galileo did wisely to content yourselves to speak hypothetically and not absolutely, ...".

another order. It is a formidable response.

Galileo has not taken the bait on the last question. Growing impatient, the interrogators become more specific as to the time of the event: *"What decision was made and then notified to you in the month of February, 1616?"* And now comes an answer for which the prosecution is totally unprepared:

"In the month of February, 1616, Cardinal Bellarmine told me that since the opinion of Copernicus absolutely taken contradicted the Holy Scriptures it could not be held or defended, but that it might be taken hypothetically and made use of. In conformity with this, I have an affidavit of the same Cardinal Bellarmine, ... of which affidavit I present a copy, and here it is." And he adds: *"The original of this affidavit I have with me in Rome, and it is entirely written in the hand of Cardinal Bellarmine."*

The existence of the affidavit is news to the tribunal. It confirms that Galileo has not abjured or received any "penance" (reprimand) but was notified only of the Decree of the Index. And while Bellarmino's affidavit states that the opinion of Copernicus "cannot be defended or held", it does not contain the phrase "teach, ... in any way whatever." By the rule of best evidence, Galileo has won this round. His document, of which he later presents the original, is signed by Bellarmino, who died in 1621; the prosecution's is unsigned. On further interrogation, he is pressed to speak of another order being given him; he does not remember another order, but Bellarmino told him *"a certain particular which I should like to speak to the ear of his Holiness before that of anyone else; ..."*. What was that particular? We may never know. Urban will never again allow Galileo in his presence.

He denies any recollection of the phrases "to teach" and "in any way whatever" within any precept intimated to him: they are not contained in the affidavit which he relied upon and kept as a reminder. He does not remember any precept being given him besides Bellarmino's. He adds:

> *"... I claim not to have contravened in any way the precept, that is, not to have held or defended the said opinion of the motion of the earth and stability of the sun on any account."*

This first interrogation triggers a frantic search for the original of the unsigned memorandum; the search is as unproductive as it is futile. As Paul V had ordered Bellarmino to forbid Galileo to teach the Copernican doctrine only if he refused to abandon belief in it, the memorandum, whether or not it is a forgery, is at best a record of some minor official's exceeding his authority and ducking the responsibility that a signature would have entailed. It fails to conform to the well documented conditions for communicating the Decree of 1616 to Galileo, and is useful only as a tool of intimidation. But the copy of the memorandum has sufficed as a decoy to divert the trial from consideration of that decree — until Galileo handed them a trickier replacement decoy, the Bellarmino precept.

The bogus memorandum, and Inquisition procedure which handicaps a defendant by limiting his responses only to questions posed by prosecutors, have steered the trial away from consideration of the Decree of 1616. Had the prosecutors attacked Galileo's obedience to that Decree rather than to an order which he had not seen, a stronger defense would have been possible. The Decree labelled as heretical only the tenet of the sun's immobility, for which he had not argued and against which he had brought evidence of solar rotation. The tenet of Earth's motion was "at least erroneous in faith" but not "heretical", and Galileo had presented arguments for and against it. Further, the Decree was approved *in forma communi,*[10] by a Congregation and not by the Church. Urban had remarked to Cardinal Zollern that the Church would not condemn an opinion that could not be proved wrong — that of Copernicus. His remark had been documented in a letter to Prince Cesi [11] to which Galileo had access but no opportunity to produce in his defense.

The trial hits an impasse as the Inquisitors, caught between the Pope's determination to prove the accused guilty of something and Galileo's unexpectedly strong defense, do not know how to deal with this new and treacherous situation:

> "At the first audience ... Galileo presented an impeccable line of defense. He had with him the statement, released to him in 1616 by Cardinal Bellarmino, which proved that during the famous meeting at the Vatican, Bellarmino had informed Galileo of Copernicus's sentence, and of his imminent inclusion in the Index, but which did not contain any further injunction as regards the exposition of Copernican ideas. Compared to that statement signed by Bellarmino, the list of charges in the hands of the judge ... had all the weight of confetti. The planned trial line was blocked; instead of collaborating, the defendant was making difficulties." [12]

"Impeccable" is too strong a qualifier for this line of defense. Galileo's testimony has bound him to a derivative of the Decree of 1616, Bellarmino's precept against defending the reality of the Copernican system. His claim of obedience depends on a tightly restrictive interpretation of the verb "to defend" and sounds disingenuous. Despite the mandated nonconclusion of the *Dialogo*, its Copernican characters cast such doubt on opposing arguments that the system of Copernicus, declared contrary to Scripture by the Congregation of the Index, appears physically probable. The tribunal now sees a clear course to pursue the issue of obedience to a precept of which there is no question that Galileo had received it. Like rats scuttling through a crack in the baseboard, the Inquisitors will rush to penetrate this chink in Galileo's armor.

---

10. See Ch. 9, note to item #1 of "Evidence That Galileo Had Permission To Uphold Copernicanism".
11. dated 8 January 1624; de Santillana, *The Crime of Galileo*, p. 163.
12. Redondi, *Galileo: Heretic*, op. cit., p. 259.

Their penetration, however, is not without risk. How can they declare that Galileo has erected a valid defense of Copernicus without seeming to acknowledge the victory **Inchofer's Dangerous "But"** of the Copernican system? If they cannot do that, then how can they convict him based on this new piece of evidence, Bellarmino's certificate? Nowhere is the difficulty of the Church's position and its ineptness in dealing with it better seen than in the report of an Inquisition Consultant, Melchior Inchofer. Inchofer is no Pope or Church politician but a scholar, perhaps the most learned of these technical Consultants; therefore, if anyone should get the case analysis right, it is he. For the Church, the more important question is not whether Copernicus is scientifically correct but whether the pronouncements of scientists can continue to be considered compatible with the Church's Ptolemaic interpretation of the Scriptures. Otherwise, Galileo's opening comment to Ingoli — that he believes the Ptolemaic interpretation of Scriptures while proving Copernicus — would not have comforted the Church so as to forestall any objection to his physical proofs.

The Decree settled the matter forever: by the Church's definition of possibility,[13] it is "impossible" after 1616 to show that Scriptures agree with Copernicus. Now along comes Inchofer on the 17th of April in 1633 trying to argue that Galileo is guilty of defending Copernicus. He refers to a lengthy pre-*Dialogo* treatise "which Galileo presented to the Grand Duke of Florence in support of his cause and in which he *not only proved* Copernicus' opinion *but established it by solving scriptural difficulties* as well as he could." [14] Thus speaks the same Inchofer who pronounced the opinion of Earth's motion the most abominable heresy, and worse than atheism. As Shakespeare might have said, he protesteth too much.[15]

His "but established it" is a very big "but". Defending Copernicus with physics is not nearly so dangerous as defending him with Scripture, as the outright banning of Foscarini's book should have made clear. Inchofer adds the uninformative qualifier "as well as he could" to cushion the impact of his big "but" which he has thrust into the matter of Galileo's alleged culpability while exposing his own. By claiming Galileo has proved Copernicus and solved Scriptural difficulties sufficiently to establish this opinion, he has set up against that "proved opinion" any Scriptures for which Galileo may not have completely solved the difficulties.

---

13. Note, for example, the wording of the Bellarmino affidavit, "... the doctrine attributed to Copernicus ... cannot be defended or held", as if to defend or hold it were impossible.

14. Finocchiaro's translation; italics mine.

15. quoted by A. D. White, op. cit., p. 139: "The opinion of the earth's motion is of all heresies the most abominable, the most pernicious, the most scandalous; the immovability of the earth is thrice sacred; argument against the immortality of the soul, the existence of God, and the incarnation, should be tolerated sooner than an argument to prove that the earth moves." (1631)

And as if he weren't in enough trouble already! Adding to his radical theological opinions, already including the cavalier declaration that Earth's motion is the most abominable of heresies, Inchofer's "work on a supposed letter by the Virgin Mary to the inhabitants of Messina was placed on the Index" [16] of Forbidden Books only the prior year; he had tried in vain to promote to the status of historical truth a legend that Mary had written them a letter "in which all the principal Catholic dogmas were listed".[17] It must be one of the great puzzles of history that this censure has not deterred the Inquisitors from employing an author of a banned work to render an "expert" opinion on Galileo's case. Nor has it alerted them to the political blunder that his opinion contains: the (correct) assertion that Galileo has proved the opinion of Copernicus and solved Scriptural difficulties that it poses. Surely this is reckless truth telling.

The Inquisitorial Predicament

The personal danger to the Inquisitors becomes clearer when Professor Drake tells us

"The Inquisition, however, could not let Galileo off scot-free; to bring a false charge of heresy was as serious a crime as heresy itself, and this charge had been brought by the highest authorities. Accordingly something more had to be done ..." [18]

Following a frantic and fruitless search of their files for a signed copy of the memorandum, the question *how to convict Galileo* has no immediate answer. Substituting for the charge of merely teaching the Copernican system the charge that his book defends its reality is a perilous and prickly problem for the Inquisitors. One thing is clear: either Galileo will be convicted or, for bringing a false charge of heresy, the highest authorities. The Pope gets to decide: not an especially tough choice for him. Stymied by the unexpected strength of Galileo's defense, the Inquisitors wait two agonizing weeks for a conclusion to secret meetings at Castel Gandolfo between the Pope and his nephew, Cardinal Francesco Barberini, one of the Inquisitors conducting the trial against Galileo. In these meetings, Barberini will steer a calculated course of diplomacy based on his apparent unspoken wish to protect Galileo as much as he can.

Apart from the difficulty of showing that Galileo proved a cosmology which they cannot accept, there are other thorny issues for the two prelates to ponder. Urban VIII has weakened the underlying Decree by telling Cardinal Zollern that the Church had not damned the opinion of Copernicus nor was it to damn it as

---

16. Finocchiaro, *The Galileo Affair*, p. 357 note 10. The Congregation of the Index is a department of the Church which censors books and can place them on an Index of Prohibited Books.
17. Redondi, *Galileo: Heretic*, op. cit., p. 254.
18. Emphasis mine; see Drake's *Galileo at Work*, p. 349.

heretical, but only as reckless.[19] The Church as a whole has not declared the opinion of Copernicus to be heresy. And it is not within the scope of the tribunal to take a defined position on the underlying doctrine. Even if it were, the trial judges, having read Galileo's book, can appreciate the risk of appearing to declare erroneous a theory which will probably be proved correct. Either they must commit the Church to labelling probable truth as theological error, or dismiss the charges and damage their authority. Awareness of this dilemma probably affects Cardinal Barberini's next move, which he intends to be favorable to Galileo as well as to the Church. At this point the safest course for the prosecution may be to obviate any proof of wrongdoing in order to insulate the process from potential backfire. And where a crime cannot be proved, he suggests to Urban, perhaps a confession can be obtained. Apparently, Urban smiles on this proposal.[20]

The Extrajudicial Conference

The strategy for extracting a confession from the uncooperative philosopher-scientist takes the form of an irregular extrajudicial procedure. On the afternoon of April 27, the presiding prosecutor, Maculano, has a private meeting with Galileo. Secrecy notwithstanding, what Maculano said to him we can guess from three evidences: (1) from the outcome of their tête-à-tête; (2) from Maculano's statement to Cardinal Barberini that *"after exchanging innumerable arguments and answers ... I made him* [Galileo] *grasp his error"*; and (3) from the usual procedure of the Inquisition for dealing with prisoners who must be convicted for political reasons, but for which it is impossible to prove the case against them. One of these methods is to deceive the prisoner as to the strength of the case against him, browbeat him with threats, and then offer leniency in return for a confession. It can hardly be doubted that the "error" which Maculano causes Galileo to see is that by not confessing to some wrongdoing, he will risk the Inquisition's methods of extracting confessions from impenitent prisoners. Bristling with petulant arrogance, Maculano fires off a missive to Cardinal Barberini:

"... in his deposition Galileo denied what can clearly be seen in the book he wrote, so that if he were to continue in his negative stance it would become necessary to use greater rigor in the administration of justice and less regard for all the ramifications of this business." [21]

In case of doubt about what is meant by the word "rigor", the Inquisition Manual

---

19. Redondi, *Galileo: Heretic*, op. cit., p. 144 footnote 10.
20. Superficially sinister, Cardinal Barberini's suggestion seems an adroit maneuver on his part and relatively favorable to Galileo. By this means, he can keep control over the outcome of the case. His behavior throughout the case indicates that he hopes to bring Galileo through it with the least damage to the hapless philosopher consistent with appeasing Urban's lust for some kind of conviction.
21. "Commissary General to Cardinal Barberini" (28 April 1633) in Finocchiaro, op. cit., p. 276.

"On the Manner of Interrogating Culprits by Torture" can be consulted:[22]

"THE CULPRIT HAVING DENIED THE CRIMES WITH WHICH HE HAS BEEN CHARGED, AND THE LATTER NOT HAVING BEEN FULLY PROVED, IN ORDER TO LEARN THE TRUTH IT IS NECESSARY TO PROCEED AGAINST HIM BY MEANS OF **A RIGOROUS EXAMINATION**; IN FACT, THE FUNCTION OF TORTURE IS TO MAKE UP FOR THE SHORTCOMINGS OF WITNESSES, WHEN THEY CANNOT ADDUCE A CONCLUSIVE PROOF AGAINST THE CULPRIT."

The "Voluntary" Confession

Galileo can remember certain precedents. There is the recent case of the impenitent Archbishop de Dominis, who had written about geometric optics and the tides and, having been charged of certain theological "crimes", died in Inquisition custody. There is the case of Tommaso Campanella, who has written a defense of Galileo himself, and, for political and theological "crimes", was twelve times tortured — once for 40 hours. And there is the case of Bruno, the farsighted firebrand who denied center and boundaries to the universe and, having intransigently declined to recant either his scientific hypothesis or his theological heresies, was roasted alive at the stake. The intimidation of Galileo's own trial to this point weighs ponderously on his spirit. Favaro tells us: *"Even by the way the letters of the signature are formed, the growing anxiety of the spirit of the dignified old man breaks through."* [23]

Undoubtedly terrified by thinly veiled threats of torture, imprisonment or condemnation, Galileo agrees to confess that his book gives the appearance of defending Copernicus. On April 30, he gives a second deposition in which he confesses that his arguments for Copernicus, in particular those based on sunspots and the tides, are more convincing than they should have been. He attributes this not to heretical belief in these arguments, but to ambition for showing himself cleverer than the average man *"by finding ingenious and apparent considerations of probability even in favor of false propositions"*. His distress at the success of his enemies in swaying Urban, the arduous and perilous journey to Rome, the terror of his first interrogation, the sickness and weakness of his body, fear as to his fate and that of his book: all contribute to his pathetic offer at the end of the second deposition to add one or two Days to the *Dialogo* in which he would refute his own arguments for the "false and condemned opinion". Luckily, it was not accepted.

What leniency is promised for this confession may forever remain unknown. But the tribunal's predicament, from which Galileo's confession rescues it, and the leniency that the presiding judge expects to obtain for him are suggested by Maculano's revealing statement: *"The Tribunal will maintain its reputation; the*

---

22. from Masini, *Sacro arsenale overo Prattica dell'officio della Santa Inquizione*, Pavoni, Genoa, 1621, as translated and quoted by Finocchiaro, op. cit., p. 363 footnote 84. Emphasis mine.
23. author's translation; cf. Appendix: Favaro's Preface to *Galileo e l'Inquisizione*.

culprit can be treated with benignity; ..." [24] Though Galileo has confessed to an error, and the trial might be expected to end now, Urban is not satisfied. One requirement for the charge "formal heresy" is "whether or not the culprit, having confessed the incriminating facts, admitted having an evil intention".[25] This Galileo has not done; he has confessed only to having inadvertently defended Copernicus. On the 10th of May, there is another hearing at which Galileo presents the original copy of Bellarmino's affidavit. He continues to stand his ground, resisting any admission of heresy or intentional wrongdoing, and placing himself at the mercy of the judges. The grounds for "formal heresy" are lacking; the Inquisitors rest their case in a "Final Report to the Pope".

The Final Report to the Pope

The persistence of the defense has pushed the prosecution into a pathetic posture. Falling back on its flimsiest evidence — the unsigned memorandum — the Inquisition suppresses mention of its own Minutes of 1616. Those Minutes tell us [26] how Galileo was permitted to teach the opinion of Copernicus because he agreed not to hold that opinion. An undated, unsigned "Final Report to the Pope", crawling with speculations presented as truths, omits this fact. Among other unsubstantiated claims, it states that the "tenor" of Bellarmino's warning to Galileo in 1616 was that he should not hold, teach or defend the opinion of Copernicus in any way. Here the tribunal clings to the language of the discredited memorandum. The Pope is persuaded that the Inquisitors must force the prisoner to confess error where his own evidence is weakest; viz., his denial of having defended Copernicus. Urban directs that he be interrogated again, this time under verbal threat of torture to "determine his intention"; i.e., presumably to extract a confession that he intentionally defended Copernicus.

For a prisoner to be submitted to the threat of torture, proof of heresy is not required, but only an accusation or any evidence however inconclusive. This is known as a "semi-proof", as if any missing part of a proof should be considered negligible. The "logic" of this procedure is stated in the Articles of Torquemada (for regulating the proceedings of the Inquisition). Article 15 states: [27]

"If a semi-proof exist against a person who denies the charge brought against him, he is to be put to the torture; if he confesses during the torture, and afterwards confirms his confession, he is to be punished as convicted; if he retracts, he is to be tortured again, or condemned to an extraordinary punishment."

---

24. Finocchiaro, ibid., p. 277. Note that Maculano signs his letters "Vincenzo da Firenzuola".
25. This is Finocchiaro's translation or paraphrase of the explanation of this charge by Masini; see *The Galileo Affair*, p. 15.
26. Cf. Ch. 9., "Evidence That Galileo Had Permission To Uphold Copernicanism".
27. William Sime, *History of the Inquisition*, Presbyterian Board of Publication, Philadelphia, 1834.

Fourth Deposition: Galileo Stands His Ground

There is no defense for the "defendant". A semi-proof is treated as a proof, in repudiation of Aristotelian logic, to say nothing of the obvious contempt, inherent in any torture mechanism, for any moral consideration. Under verbal threat of torture (carried out in the name of the gentle and merciful Jesus), Galileo is interrogated again. Standing resolutely before his accusers, he indicates willingness to die rather than admit to deliberate intent. As we have seen, Article 15 would require torture because he denied the heresy of which he was accused.[28] Cardinal Francesco Barberini, the Pope's nephew, presumably would have had to prevent such an unseemly outcome. Who else besides the Pope has the secure position to break the rules? Cardinal Barberini is also one of the three dissenters who will abstain from signing the Sentence.

Maculano's thoughts of treating the prisoner "with benignity" give way to Urban's frustration with Galileo's impenitence. The Tribunal is instructed to deliver a sentence of life in prison and to have Galileo publically renounce the Aristarchan/Copernican world view on which he had staked his scientific reputation. The renunciation is to be followed by public reading of his Sentence to professors of mathematics everywhere the long arm of the Vatican reaches.

Verdict of the Tribunal

The Sentence of 22 June 1633 is a case study in literary witchcraft. Mixed in its poison stew of "half-truths" generously seasoned with unprovable assertions, there is a verdict with explanation, and a sentence of punishment which we will discuss later. The verdict resounds the party line that in 1616 Galileo had been told not to teach Copernican astronomy in any way whatever. It is worded so as to invite the false interpretation that he had confessed to having received this injunction.[29] It states falsely that the Decree of the Index "prohibited books treating of such a [heliocentric, heliostatic, geokinetic] doctrine"; it actually prohibited Foscarini's book and "all other books that teach the same", namely, "that the above-mentioned doctrine ... is consonant with the truth and does not contradict Holy Scripture." [30]

---

28. Professor Drake tells us "as recently as 1942, a Dominican authority on canon law, Orio Giacchi, wrote that the only legal error in the proceedings was *failure* to put Galileo to torture"; *Galileo at Work*, p. 500 footnote 22.

29. The Sentence reads: "You confessed that about ten or twelve years ago, after having been given the injunction mentioned above, you began writing the said book, ..."; trans. by Finocchiaro (p. 289) who punctuated his translation just like the original. Except for the placement of commas, the phrase "after having been given the injunction" appears to belong to a "that" clause whose function is to state what he confessed. Galileo did not in fact confess to having received the injunction. This style of writing emulates that of the oracles at Delphi, whose counsel was notoriously ambivalent; e.g., asked to predict one's fate in war, the oracle might reply "You will come back not dead". (Supplied with punctuation, this can mean "You will come back, not dead " or "You will come back not, dead ".)

Given that the signers of the verdict cannot correctly represent an official Decree of a Congregation, what is the likelihood that they can correctly interpret the *Dialogo* on which they are passing judgment?

Of the Bellarmino affidavit, which supports Galileo's contention that he had relied on a document not containing the phrases "to teach" and "in any way whatever", it says only that it aggravates his case because it shows that he knew that the opinion he had presented as probable in the *Dialogo* had been declared contrary to Scripture. But Galileo's Second Deposition in which he confessed inadvertent error is accurately paraphrased, no doubt because it favors the prosecution.

So far, nothing surprising or of particular legal importance has been said. It is all controvertible and confirmed by confetti.[31] But now comes the peroration of the verdict:

The Unexcluded Middle

> "Because we did not think you had said the whole truth about your intention, we deemed it necessary to proceed against you by a rigorous examination. Here you answered in a Catholic manner,[32] though without prejudice to the above-mentioned matters confessed by you and deduced against you about your intention."

What can these deductions mean? What they cannot mean is that he "holds or has held, ... that the sun is the center of the world and the earth is not the center of the world but moves also with diurnal motion." That is the question, the only question, put to him repeatedly under threat of torture in his Fourth Deposition. And that is the question which the tribunal accepts that he answered "in a Catholic manner". The question put to Galileo in the Fourth Deposition is:

"Having been told that from the book itself and the reasons advanced for the affirmative side, namely that the earth moves and the sun is motionless, **he is presumed,** as it was stated, that he holds Copernicus's opinion, or at least that he held it at the time, therefore he was told that unless he decided to proffer the truth, one would have recourse to the remedies of the law and to appropriate steps against him." [33]

To which Galileo answered:

"I do not hold this opinion of Copernicus, and I have not held it after being ordered by injunction to abandon it. For the rest, here I am in your hands; do as you please."

---

30. (rdc).

31. That the Sentence against Galileo violates Aristotle's Law of the Excluded Middle is a proposition originally advanced by the author and later proved by rdc. This section, excepting its footnotes, is my summary and paraphrase of rdc's proof. For a discussion of the Law of the Excluded Middle, see the next section "Falsity of Galileo's Sentence".

32. "rispondesti cattolicamente"; cf. Favaro, p. 145.

33. See Finocchiaro, op. cit., p. 287. Emphasis mine.

The presumption was overturned by an answer accepted under threat of torture. By 17th century ecclesiastical epistemology, this is a valid method of discovery. Thus there can be no "deduction" of his holding or having held Copernicus' opinion that can stand against his "Catholic" answer. These "deductions" may refer to his "intention" regarding anything else (e.g., the violation of the unsigned special injunction or the failure to disclose it to the censors prior to publication), but they cannot refer to his intention regarding holding the opinion of Copernicus. Galileo denied holding the opinion of Copernicus and the tribunal accepted his denial.

Then there were three Consultant Reports of 17 April 1633, by Oreggi, Pasqualigo, and Inchofer. Oreggi concludes that in the *Dialogo*, the heliocentric opinion is held and defended by one or more dialogue participants, but does not claim that the author Galileo holds it. Pasqualigo concludes that Galileo "is strongly suspected of holding such an opinion". Only one consultant, forbidden-book author Inchofer, concludes against Galileo that he actually holds the condemned opinion. In all three reports, there is no "deduction" about Galileo's intention on any matter but that of whether he holds, teaches or defends the opinion of Copernicus.

After all this, the Inquisitors first accept Galileo's denial of holding the opinion of Copernicus by saying that he answered "in a Catholic manner", and then "say, pronounce, sentence, and declare that you, the above-mentioned Galileo, because of the things deduced in the trial and confessed by you as above, have rendered yourself according to this Holy Office vehemently suspected of heresy, namely of having held and believed a doctrine which is false and contrary to the divine and Holy Scripture: ..." [34] They may say "without prejudice" to their "deductions", but their acceptance of Galileo's answer has excluded precisely those deductions that they went ahead and used to convict him.

Falsity of the Verdict Against Galileo

These learned Roman Catholic Inquisitors who defend by brute force the system of Aristotle and Ptolemy against Galileo's Copernican proofs have surely learned Aristotle's Law of the Excluded Middle (*Tertium Non Datur*).[35] An assertion and its denial are not jointly true; e.g.,

> It is true that (i) Fido is a rabid dog or Fido is a non-rabid dog, and
> it is not true that (ii) Fido is both a rabid and a non-rabid dog.

If logic is requisite to diagnosis, then any veterinarian who makes the assertion (ii) about Fido may be presumed lacking in diagnostic skills. Yet that assertion is

---

34. Cf. Finocchiaro, *The Galileo Affair*, op. cit., p. 291. Emphasis mine.

35. "X is true" or "X is false": if X is a meaningful proposition then the disjunction is strict, and there is no third possibility. Today this tautologous biconditional is common to set theory and logic.

identical in form to the judgment that the Church Inquisitors made about Galileo. To see this, we may first note the following convention of language: that (C) to assert a suspicion that X is true is to assert a nonzero probability that X is true. Restating the pronouncement of the tribunal and applying the language convention (C), we have:

> (1) "We accept Galileo's denial of holding the opinion of Copernicus."

> (2) "We pronounce Galileo vehemently suspected of having held that opinion."

> (3) Combining (1) and (2), "He did not hold the opinion and we suspect him of holding it".

> (4) By the convention C, (3) is equivalent to: "We assert both a zero probability that he held the opinion and a nonzero probability that he held the same opinion."

Assertion (4) is an excluded middle in violation of Aristotelian logic. The law of the excluded middle disallows "both A and not-A"; a strict disjunction additionally disallows "neither A nor not-A". It is a strict disjunction that either Galileo holds and believes the opinion of Copernicus or he does not. If he does, then there is no reason for them to say that he answered "in a Catholic manner", and they have falsified the Sentence. If he does not, then their Sentence is false for claiming that he holds the condemned opinion. In either case, the false sentence implies a false charge of heresy. The Pope and the tribunal, excepting those who abstained from signing the sentence, are *prima facie* guilty of having brought a false charge as serious as heresy itself.[36] Acting as a collectively mindless institution, the Church in 1633 logically fails to convict Galileo even on its own terms and by its own rules, and instead convicts itself.

There is a risk to the Inquisitors at this point. Will Galileo "recant"?[37] His influence (earned by his accomplishments and especially their military value), the Medicean Court's

To "Recant" or Not To "Recant"

loyal support of him, the respect for him by a few well-placed churchmen, including the shrewd and manipulative Cardinal Barberini, and Galileo's reputation — all these factors weigh in the public relations risk of physical abuse or execution. But how does the risk to Galileo weigh against the cost of a

---

36. Cf. Stillman Drake, *Galileo at Work*, op. cit., p. 349.

37. I prefer "recant" to the historical term "abjure", against which I have an unrational and indefensible bias. (Besides, Grandma said "recant".) But is a sincere, voluntary disavowal of one's former belief fundamentally the same act as an involuntary disavowal performed under threat of torture? It would be an abuse of language to use the same word for both acts. Lacking a separate word for the involuntary act, I here resort to the use of quote marks; i.e., Galileo did not recant, he "recanted", or pseudo-recanted.

"recantation" travesty that anyone can plainly see is forced and not genuine? [38] Galileo will answer the question by submitting to this travesty.[39]

The Sentence of Punishment

The punishment pronounced upon Galileo includes:

the requirement to "abjure, curse, and detest" his heresy "with a sincere heart and unfeigned faith";

prohibition of his *Dialogo* "by public edict";

an indefinite term of imprisonment in the Holy Office;

the requirement "to recite the seven penitential Psalms once a week for the next three years."

Seven (out of ten) Cardinals on the Tribunal sign the verdict and sentence. Three Cardinals abstain from signing and deserve to be remembered:

> *Cardinal Francesco Barberini*
> *Cardinal Gaspare Borgia*
> *Cardinal Laudivio Zacchia*

Under Barberini's adroit maneuvering, the tribunal quietly commutes Galileo's imprisonment to house arrest. Galileo has thus escaped torture and the dungeon. His daughter, a Carmelite nun, offers to recite the psalms in his behalf, and this offer is accepted both by Galileo and by the religious authorities.

Galileo's Pseudo-Recantation

On 22 June 1633, Galileo is forced to swear that he has "been judged vehemently suspected of heresy, namely of having held and believed that the sun is the center of the world and motionless and the earth is not the center and moves". He shall never again assert anything that might cast upon him the suspicion that he holds this opinion, and he shall denounce to the Holy Office anyone whom he comes to know and suspects of a heresy. He abjures, curses, and detests his errors "with a sincere heart and unfeigned faith". Unlike the soldier who, in an act of random violence, slashed the face of Galileo's teacher Tartaglia, ineffectively muting him, the Church has, in a deliberate and calculated act of coercion, temporarily and most

---

38. Did Galileo consider not "recanting"? This consideration was brought to his attention by Friar Micanzio's passionate and eloquent letter to him during the trial, quoted by de Santillana, op. cit., as follows: "But what manner of men are these, to whom any good effect, and well-founded in nature, should appear contrary and odious ... **if this were now to prevent you from further work**, I shall send to the hundred thousand devils these hypocrites without nature and without God" (emphasis mine). Micanzio was an aide and later successor to Paolo Sarpi as theologian to Venice, and wrote an anonymous biography of Sarpi.
39. See the Epilogue "Myth Number Six: The Dishonor Myth" for my proof that Galileo never recanted.

ineffectively muted his student, Galileo.[40]

Prometheus: *"The time is past for words; earth quakes*
Sensibly: hark! pent thunder rakes
The depths, with bellowing din
Of echoes rolling ever nigher:
Lightnings shake out their locks of fire;
The dust cones dance and spin;
The skipping winds, as if possessed
By faction — north, south, east and west,
Puff at each other; sea
And sky are shook together: Lo!
The swing and fury of the blow
Wherewith Zeus smiteth me
Sweepeth apace, and, visibly,
To strike my heart with fear. See, see,
Earth, awful Mother! Air,
That shedd'st from the revolving sky
On all the light they see thee by,
What bitter wrongs I bear!

... Aeschylos [41]

Legend has it that after kneeling and pronouncing the Earth motionless in the center of the universe, Galileo rose and quietly uttered the words *"Eppur si muove"*.[42] It would not have been uncharacteristic. Unlike the character in an undeservedly popular play that misrepresents him as genuinely repentant, the real Galileo showed no betrayal of his mind. I had already formed this opinion when de Santillana brought the following statement of Galileo to my attention:

Nevertheless, the Earth Moves

"I do not hope for any relief, and that is because I have committed no crime. I might hope for and obtain pardon, if I had erred; for it is to faults that the prince can bring indulgence, whereas against one wrongfully sentenced while he was innocent, it is expedient, in order to put up a show of strict legality, to uphold rigor ..."[43]

40. Galileo was indirectly Tartaglia's student through his teacher Ricci, who had studied under Tartaglia, and directly through Tartaglia's textbook in geometry.
41. from *Prometheus Bound*, tr. by G. M. Cookson (*Great Books of the Western World*, op. cit.) In the time of Democritos, the Athenian playwright Aeschylos produced this tragic conflict of injustice and force against a just and intelligent man. Even Galileo's contemporaries had noted this situational parallel.
42. "Nevertheless, it does move". The quotation derives from a portrait circa 1640 by Murillo of Galileo in prison (cf. Gjertsen's *Classics of Science*, op. cit.); while he would not have audibly so spoken at the "recantation", it is plausible that he did so elsewhere.
43. Quoted in de Santillana's paper *Galileo in the Present*, included in *Homage to Galileo*, Morton Kaplon, ed., M.I.T. Press, 1965.

Dialogue Between Two Principal Spokesmen: Science and the Church

Considering only the proceedings that convicted Galileo, one cannot judge the scientific and epistemological conflict that formed the backdrop of the trial. The "defendant" was not allowed to defend himself. And with good reason: who could match wits with him? In 1633, who could have held his ground in a battle of logic with Galileo on the scientific or philosophical issues for which he had been indicted? Avoiding debate with him, the Inquisitors bound him to answer directed questions. For them, the only issue was whether he had disobeyed an order.

In the real conflict between the Church and Galileo, discussion of which the Church suppressed, two epistemologies emerge in sharp contrast. One establishes truth by common agreement among persons who view themselves as collectively endowed with divine inspiration. Common agreement requires only a predominance of opinion, a consensus of six out of ten will suffice; the dissenters (three of ten in the case of Galileo's trial verdict) risk being branded heretics. Certainty about reality is attained through faith. The role of the priesthood is to interpret reality consistently with faith, and not with scientific findings unless they can be physically proved. There are two worlds, one spiritual, the other physical. Knowledge of the first, to which scientific theories about the second must conform, comes from the Church.

The other epistemology, that of science, validates claims by reference to nature, not to consensus or authority. Its process is a quest for certainty to the extent that it can be approached by a finite number of observations.[44] The new science of Galileo advances monotheism by affirming the singularity of nature against Aristotle's dual worlds and multiple gods.[45] This modification of theology by a scientist is seen by "theologians" of Galileo's day as usurping their priestly role.

Simulating that Galileo had been allowed to debate in a truth-seeking forum and to support openly without threat of bodily harm a belief in the reality of Earth's motion, we can (imaginatively, of course) resurrect the Church's principal spokesman, the then-deceased Cardinal Bellarmino. You, the reader, are the judge of this debate. Let's listen as their conversation thrusts into high relief the points of contention between Galileo and the Church:[46]

---

44. Galilean scientific method is discussed in Ch. 13 "New Industries from Two Galilean Ideas".
45. See *Dialogo*, First Day, where Simplicio refers to his master's polytheism. Catholics, who at every Mass recite the Nicene Creed "We believe in One God, ..." came down on the side of the polytheists in persecuting Galileo, the monotheist.

Bellarmino: Your *Dialogo* has overstepped the bounds of prudence, Galileo. It would have been prudent to speak "hypothetically and not positively, as I have always believed Copernicus did. For to say that assuming the earth moves and the sun stands still saves all the appearances better than eccentrics and epicycles is to speak well. This has no danger in it, and it suffices for mathematicians."

Galileo: *Reverend Cardinal, this is not what Copernicus assumes. "Copernicus assumes eccentrics and epicycles; not these, but other absurdities, were his reason for rejecting the Ptolemaic system."*

Bellarmino: "But to wish to affirm that the sun is really fixed in the center of the heavens and merely turns upon itself without traveling from east to west, and that the earth is situated in the third sphere and revolves very swiftly around the sun, is a very dangerous thing, not only by irritating all the theologians and scholastic philosophers, but also by injuring our holy faith and making the sacred Scripture false."

Galileo: *"As to philosophers," Reverend Cardinal, "if they are true philosophers (that is, lovers of truth), they should not be irritated; but, finding out that they have been mistaken, they must thank whoever shows them the truth. And if their opinion is able to stand up, they will have cause to be proud and not angry. Nor should theologians be irritated; for finding such an opinion false, they might freely prohibit it, or discovering it to be true they should be glad that others have opened the road to the discovery of the true sense of the Bible, and have kept them from rushing into a grave predicament by condemning a true proposition."*

"As to rendering the Bible false, that is not and never will be the intention of Catholic astronomers such as I am; rather, our opinion is that the Scriptures accord perfectly with demonstrated physical truth. But let those theologians who are not astronomers guard against rendering the Scriptures false by trying to interpret against it propositions which may be true and might be proved so."

Bellarmino: Any attempt to interpret all the passages of the Bible that refer to the sun, Earth, moon and stars as support for Copernicus would be fraught with difficulty. As you know, "the Council [of Trent] would prohibit expounding the Bible contrary to the common agreement

---

46. This debate is constructed by juxtaposing actual quotations or paraphrase largely from Bellarmino's letter to Foscarini and Galileo's notes on that letter, and by adding connective paraphrase (outside quotation marks) to maintain conversational flow. Except where otherwise noted, all quotations and paraphrase in this dialogue are from Stillman Drake's *Discoveries and Opinions of Galileo*, pp. 162-170.

of the holy Fathers." If you "would read not only all their works but the commentaries of modern writers on Genesis, Psalms, Ecclesiastes, and Joshua, you would find that all agree in expounding literally that the sun is in the heavens and travels swiftly around the earth, while the earth is far from the heavens and remains motionless in the center of the world."

Galileo: *"It may be that we will have difficulties in expounding the Scriptures, and so on; but this is through our ignorance, and not because there really are, or can be, insuperable difficulties in bringing them into accordance with demonstrated truth." As to Earth's motion, if you care to examine a true copy of my letter to Father Castelli, you will find that I have demonstrated, not by consensus, but by observation and reasoning, that the passage of Joshua "shows clearly the falsity and impossibility of the Aristotelian and Ptolemaic world system, and on the other hand agrees very well with the Copernican one."*[47]

Bellarmino: "...If there were a true demonstration that ... the sun did not go around the earth but the earth went around the sun, then it would be necessary to use careful consideration in explaining the Scriptures that seemed contrary, and we should rather have to say that we do not understand them than to say that something is false which had been proven. But I do not think there is any such demonstration, since none has been shown to me."

Galileo: *"Not to believe that a proof of the earth's motion exists until one has been shown is very prudent .." But to say that we do not understand the Scriptures that may seem to impute motion to the sun, it is sufficient that falsehoods can be derived from the system of Aristotle/Ptolemy, while none can be derived from the system of Copernicus even though it may not be fully demonstrated. If the advocates on either side "are only ninety percent right, then they are defeated; but when nearly everything the philosophers and astronomers say on the other [Aristotelian/Ptolemaic] side is proved to be quite false, and all of it inconsequential, then this [Copernican] side should not be deprecated or called paradoxical simply because it cannot be completely proved." To assume the system of Copernicus does not contradict any observations of the planets, but to assume that of Ptolemy does; for example, Venus could not, as it does, go through a whole cycle of phases in the*

---

47. from Galileo's letter to Castelli; see Finocchiaro, *The Galileo Affair*, p. 53.

system of Ptolemy. Indeed, we see planets circling Jupiter that could not exist in any strictly geocentric world. These appearances are saved by the system of Copernicus.

Bellarmino: "To demonstrate that the appearances are saved by assuming the sun at the center and the earth in the heavens is not the same thing as to demonstrate that in fact the sun is in the center and the earth in the heavens. I believe that the first demonstration may exist, but I have very grave doubts about the second; and in case of doubt one may not abandon the Holy Scriptures as expounded by the holy Fathers."

Galileo: *"It is true that to prove that the appearances may be saved with the motion of the earth... is not the same as to prove this theory true in nature; but it is equally true, or even more so, that the commonly accepted system cannot give reasons for those appearances. That system is undoubtedly false, just as ... this one may be true. And no greater truth may or should be sought in a theory than that it corresponds with all the particular appearances. No one asks that in case of doubt the teachings of the Fathers be abandoned, but only that the attempt be made to gain certainty in the matter questioned. ... Solomon and Moses and all the other holy writers knew the constitution of the universe perfectly well, as they also knew that God did not have hands or feet or wrath or prevarication or regret."*

Bellarmino: If you are telling me "that Solomon spoke according to the appearances, and that it seems to us that the sun goes round when the earth turns, as it seems to one aboard ship that the beach moves away, I shall answer thus. Anyone who departs from the beach, though to him it appears that the beach moves away, yet knows that this is an error and corrects it, seeing clearly that the ship moves and not the beach; but as to the sun and earth, no sage has needed to correct the error, since he clearly experiences that the earth stands still and that his eye is not deceived when it judges the sun to move, just as he is likewise not deceived when it judges that the moon and the stars move."

Galileo: *"The mistake about the apparent motion of the beach and the stability of the ship is known to us after we have frequently stood on the beach and observed the motion of the boat, as well as in the boat to observe the beach. And if we could stand thus now on the earth and again on the sun or some other star, we might gain positive and sensory knowledge as to which moved. Yet looking only from these two bodies, it would always appear that the one we were on stood still, just as to a man who saw only the boat and the water, the water would always seem to run and the boat to stand still. ... It would be better to compare two ships, of which the one we are on will absolutely seem to stand still whenever we can make no other comparison than between the two ships. ... "*

Bellarmino: That is an interesting argument, Galileo, as long as you treat it suppositionally; but keep in mind "the declaration made by the Holy Father and published by the Sacred Congregation of the Index, whose content is that the doctrine attributed to Copernicus ... is contrary to Holy Scripture and therefore cannot be defended or held."

Galileo: *Here your higher sciences teach me what I could never learn from experience and reasoning! In conformity with your precept, my Most Reverend Cardinal Inquisitor-General "against theoretical depravity in all Christendom", I have abandoned "completely the false opinion that the sun is the center of the world and does not move and that the earth is not the center of the world and moves."* [48] *But as nature does not care that I have abandoned the opinion that Earth moves, it still moves ("eppur si muove").*

The assertion that (despite dogma to the contrary) Earth "still moves" around an external center implies a sustained acceleration at its center of mass in reaction to a centripetal force. Galileo's real *eppur si muove* transcends the mechanical issue of Earth's motion to address the ebb and flow of civilization in reaction to contrary forces: individual achievement versus the subordination of achievers to collectivist institutions by the false principle "might makes right". I refer to his handwritten note in the margin of his own copy of the *Dialogo*:

"In the matter of introducing novelties. And who can doubt that it will lead to the worst disorders when minds created free by God are compelled to submit slavishly to an outside will? When we are told to deny our senses and subject them to the whim of others? When people devoid of whatsoever competence are made judges over experts and are granted authority to treat them as they please? These are the novelties which are apt to bring about the ruin of commonwealths and the subversion of the state." [49]

---

48. Quotation from Galileo's abjuration; see Finocchiaro, op. cit., p. 292.
49. frontispiece of Galileo Galilei, *Dialogue on the Great World Systems*, trans. by Thomas Salusbury (1661), revised and annotated by Giorgio de Santillana, U. of Chicago Press.

Monument to Giordano Bruno at Campo di Fiori, Rome

The pigeons are real. So was the execution that took place here.

(photo by author)

Convent of Santa Maria Sopra Minerva, Site of Galileo's Pseudo-Recantation
(photo by author)

Il Gioiello, Galilei Villa at Arcetri
(photo by author)

Arcetri Astrophysical Observatory Viewed from Galileo's Villa
(photo by author)

Chapter 11. Two Baked Pears and a Rare Winter Rose
Galileo at Arcetri (1633 - 1642)

"A noiseless, patient spider,
I marked, where, on a little promontory, it stood isolated;
Marked how, to explore the vacant, vast surrounding,
It launched forth filament, filament, filament, out of itself;
Ever unreeling them - ever tirelessly speeding them.

And you, O my Soul, where you stand,
Surrounded, surrounded, in measureless oceans of space.
Ceaselessly musing, venturing, throwing, — seeking the spheres, to connect them;
Till the bridge you will need, be formed — till the ductile anchor hold;
Till the gossamer thread you fling, catch somewhere, O my Soul."

... Walt Whitman [1]

On a picturesque hilltop at Arcetri overlooking the city of Florence, Galileo will live out his remaining years under house arrest. Like Whitman's spider, he has a web of his own to weave and, in the wake of his Inquisition trial, must weave it patiently and ever so noiselessly. In the *Dialogo* he has connected the celestial spheres to the question whether they have a center; that is, whether they exist. Now, as in precise geometric argument he develops the laws of fall, he builds the bridge between his astronomy and his physics. Standing on that bridge, Newton will cause Kepler's ductile anchor that keeps Earth bound to its parent star, and surrounded in measureless oceans of space, to take hold in the minds of men. Galileo's most important work of the seventeenth century, integrating the research of a lifetime, lies ahead. Unlike the web woven by the Inquisition, this work is no spiderweb from the sky, but a solid latticework of interrelated natural principles grounded on a mathematical foundation.

The Revival of Galileo and the Suppression of His School

Immediately after the trial, Galileo is permitted to stay first at the Villa Medici in Rome, and then at the palace of the Archbishop of Siena, Ascanio Piccolomini. Here he remains a prisoner of the Inquisition. But under the Archbishop's respectful and compassionate care, and with letters of encouragement arriving from Galileo's beloved daughter Sister Maria Celeste, the resilient philosopher regains his spirit and composure in a few months. When his enemies notice that the Archbishop is treating him as an honored guest rather than as a prisoner, allowing him to receive visitors with whom he

---

1. *A Comprehensive Anthology of American Poetry*, Conrad Aiken ed., Random House, 1944.

engages in scientific discussion, complaints are lodged with the Vatican. Feeling obliged to nip in the bud any Galilean school of science, the Holy Office issues orders permitting him to retire to his villa at Arcetri with restrictions on visitors. This happens soon enough to hamper the propagation of his science, but too late to prevent him from regaining the momentum of his life's work. By helping to revive him, the courageous Archbishop Piccolomini has well served Galileo's posterity.

Public Humiliation

Is it enough that this magnificent benefactor of the whole civilized world has suffered an arduous trial without defense, threats of torture, "recantation", public humiliation, and imprisonment in his house? Not to Urban and his Church:

" He [Galileo] was forced to bear contemptible attacks on himself and on his works in silence; to see the men who had befriended him severely punished; Father Castelli banished; Ricciardi, the Master of the Sacred Palace, and Ciampoli, the papal secretary, thrown out of their positions by Pope Urban, and the Inquisitor at Florence reprimanded for having given permission to print Galileo's work. He lived to see the truths he had established carefully weeded out from all the Church colleges and universities in Europe; and, when in a scientific work he happened to be spoken of as 'renowned,' the Inquisition ordered the substitution of the word 'notorious.' " [2]

Science Is Smuggled to a Secular State

Confined to his villa, he busies himself with a task begun at the residence of the Archbishop: writing yet another book, his *tour de force*. Later, he is allowed two students and a few visitors. Still a prisoner of the Inquisition, he succeeds in exporting a critically important manuscript from his villa, not once but twice;[3] it is likely that the ruling Medici family is running interference for him. Though unable to publish under the control or influence of the Papal States, he arranges for Elzevir to publish it in Holland in 1638. The *Discourses and Mathematical Demonstrations Concerning Two New Sciences* [4] (*Discourses*) summarizes much of his physics. The two new sciences are the theory of mechanical stresses and the science of motion. The first lays a necessary foundation for the modern technology of bridge and building construction. The second contains a full development and proof of his laws of motion. While avoiding the forbidden topic of Earth's motion,

---

2. See A. D. White, op. cit., p. 143. White spells "Ricciardi" the name that is elsewhere spelled "Riccardi" — Niccolò Riccardi, Master of the Sacred Palace and licenser of the *Dialogo*, nicknamed "Father Monster" by Phillip III of Spain "for his enormous body, incredible eloquence, and phenomenal memory"; Stillman Drake, *Galileo at Work*, p. 464. Too bad none of these assets swayed Urban to support his licensing.
3. Galileo gave one copy to Prince Mattia de' Medici, the other to Louis Elzevir who made a visit to Arcetri.
4. The major title is *Discorsi e Dimostrazioni Matematiche Intorno a Due Nuove Scienze, Attenenti alla Mecanica & i Movimenti Locali,* orig. pub. in 1638 by Elzevir, republished in untranslated form in 1966 (see Bibliography). Galileo dedicated it to the Duke of Noailles, ambassador to the King of France; the Duke visited him while Galileo was a prisoner of the Inquisition.

he provides a theoretical base for the work of successors who will, without major risk of life and limb, prove that the Earth is a planet. In this way, he who appeared to have been vanquished in a battle with the Inquisition wins the war.

Galileo's Final Masterpiece: The *Discourses on Two New Sciences*

When one flips too casually through the pages of the *Discourses*, it is tempting to mistake it for some mere collection of theorems, perhaps left over from previously unpublished work.[5] It is far more than that. In Chapter 4 we saw how Galileo's kinematics and dynamics, beginning with the unpublished *De Motu,* had wound a tortuous path as he groped his way out of the Dark Ages. The *Discourses* are the culmination of that effort, a work that is modern in its methods. The conceptual gulf separating the *De Motu* and the *Discourses* suggests that this individual scientist, Galileo, modernized the whole discovery process that operates under the name of science. Comparing these two documents, McMullin notes:

> "Since the *De Motu* is the sort of thing that could conceivably have been written by one of the Paris or Oxford physicists who flourished two centuries before Galileo's time, and since the *Discorsi* contains the germ of Newtonian mechanics both in its results and in its methods, this mode of approach suggests that the transition from 'medieval' to 'modern' science was accomplished almost single-handedly by Galileo in the almost fifty years that separated his two writings on mechanics. "[6]

Then too, we can be led astray by the simple title, assigned by the publisher over Galileo's protest,[7] and by the superficially evident organization of the work into two new sciences. In reality, *Two New Sciences* embodies a rich diversity of topics both theoretical and practical, ranging over much of Galileo's entire life work and summarizing his mature thought on topics including:

Strengths of Materials ... Stresses on Beams ... Fracture of a Cantilever ... Non-Scalability of Machines ... Harmony and Musical Intervals ...The Continuum ...The Proper Design of a Sack ... The Screw as a Machine ... The Siphon ... Pendulum Oscillations ... Acoustic Resonance ... Properties of Infinity ... Inclined Planes ... Adhesion of Plates ... Cohesion of Solids ... Weighing of the Air ... Air Resistance ... Principle of Inertia ... Law of Falling

5.　In his review of the Crew and de Salvio translation, Galileo's successor in astronomy, Edwin Hubble, provides insight into why I had this impression: "... so numerous and delightful", he says, "are the digressions, ranging over the broad field of physics, that the whole seems rather a conversation, naturally developed, than a carefully worked-out treatise"; *The Astrophysical Journal*, vol. XLII, Oct. 1915, p. 283. Thanks to Laura Eklund for bringing Hubble's review to my attention.

6.　Ernan McMullin et al., *Galileo: Man of Science*, Basic Books, Inc., NY, 1967.

7.　The protest is mentioned in Favaro's Introduction to Galileo's *Dialogues Concerning Two New Sciences*, translated by Henry Crew and Alfonso de Salvio, Northwestern University Press, 1968, p. xii, and in Drake's *Galileo at Work*, pp. 386-387.

Bodies ... Principle of Virtual Work ... Role of the Medium in Free Fall ... Momentum ... Accelerated Motion ... Trajectories ... Range of Projectiles

Galileo's Mature Methodology of Science

The *Discourses* reveals a synergy of inductive, deductive, hypothetical, and empirical methodologies that is the hallmark of Galilean science. He refers an event like the fracture of a stressed beam to a theoretical framework within which he selects and interprets further observations. Filtered through existing theory, observations beget interpretations, some of which may fail to conform to further tests. As he alters them and combines them with conforming interpretations, weaving them together to construct a complete model of experience, the theory oscillates convergently until it explains the expanded ensemble of results. The results are announced in the *Discourses*.

The Paradox of Galileo

Galileo's first theoretical innovation in pure mathematics recognizes the unity of being not in permanence but in continuity of change. Boldly bridging the gap between Zeno's paradoxical infinity and continuous natural phenomena, he tells us that an infinite set of numbers is countable if (and only if) it can be put into one-to-one correspondence with the set of all non-negative integers — matching the two sets element by element. Any countable set, he notes, can be put into one-to-one correspondence with a proper subset, a new set that can be formed by removing elements from the original set. This discovery will come to be known by 20th century mathematicians as the *Paradox of Galileo*.[8] A cube, for example, is a special and rare integer, yet the cubes 0, 1, 8, 27, 64, ... can be mapped one-to-one with the integers 0, 1, 2, 3, 4, ... These two infinities are therefore of the same magnitude; that is, there are the same number of cubes as there are integers! Anyone can specify an operation that transforms any integer into its associated cube, and vice versa. Similarly, there are the same number of odd integers as there are integers, not half as many, as one might at first imagine.

Transfinite Mathematics: Galileo, Bolzano, Cantor

Galileo does not attempt to explain away the paradox; he accepts it and draws out its consequences. Having shown how an infinite part of an infinite set can be made to correspond element by element with the whole set, he eludes contradiction by avoiding the use of the relations "equal to", "greater than", and "less than" to compare infinite sets. In this way, he provides a starting point and a challenge for the 19th-century mathematician Georg Cantor, who will

---

8. See Birkhoff and MacLane, *A Survey of Modern Algebra*, The MacMillan Company, NY, 1963, p. 359, Theorem 2 (Paradox of Galileo).

define an arithmetic to enable ordering and comparison of these different infinite sets of numbers.[9] In his 1883 *On Linear Aggregates*, in which he treats infinities not as limits of convergent sequences but as actual quantities, Cantor will establish modern set theory. How daring a break this will be with mathematical tradition can be gleaned from an 1831 letter of Karl Gauss, perhaps the greatest of post-Renaissance mathematicians, containing a statement so dogmatic as to remind us that the dangers of authoritative pronouncements are not limited to theology: [10]

> "... I must protest vehemently against your use of the infinite as something consummated, as this is never permitted in mathematics. The infinite is but a figure of speech; an abridged form of the statement that limits exist which certain ratios may approach as closely as we desire, while other magnitudes may be permitted to grow beyond all bounds ... "

After the silencing and persecution of Galileo, transfinite mathematics languished in Europe. No one seemed to know what to make of Galileo's paradox, and anyone having an idea on the subject may have feared to discuss it; small wonder that the lead finally was taken by a priest with unpopular views. Not until 1820 did someone — a Czechoslovakian mathematician and priest, Bernhard Bolzano — address the contradiction that seems to result from ordering Galileo's infinities. By Bolzano's definition, two sets (finite or not) have the same power if they can be placed into one-to-one correspondence, while a set A is of greater power than a set B if the process of matching them element by element leaves unmatched elements in set A. Bolzano's power concept is the bridge between Galileo's pioneering correspondence principle and Cantor's mathematics of infinite sets. Cantor will identify the power of a set with its cardinality, the number of elements it contains, and to develop an arithmetic of infinite sets ordered by cardinality. From Bolzano to Cantor, in a time span of 63 years, was developed the fundamental arithmetic of infinite numbers, after being stalled by the Inquisition's treatment of Galileo.

Today, the mathematics of infinity remains a nursery for concepts that spill over into other disciplines, like "possibility" in mathematics and its application in the behavioral sciences, economics among them. As correct economics does not rest on faulty mathematics, users of infinite quantities might consult mathematicians as to what is and is not possible in dealing with them.[11]

---

9. Cf. Di Canzio, Albert G., *The Paradox of Self-Containing and Boundless Set Complexes*, Georgetown University Mathematics Society Yearbook, 1966. I now confess that, although aware of Galileo's contribution to the theory discussed, I absent-mindedly defaulted mention of Galileo in introducing Cantor's starting point. Here I correct this most embarrassing omission, with apology to the memory of Galileo. It is also noteworthy that Newton asserted that "the principle that all infinites are equal is a precarious one"; Newton to Bentley, January 17, 1692/3 (see Thayer, op. cit.).

10. Tobias Dantzig, *Number: the Language of Science*, MacMillan, NY, 1954.

The Dust-Covered Trail to the Calculus

Armed with his concept of one-to-one correspondence, Galileo ventured into the realm of infinitely large quantities to decipher that part of the book of nature which concerns continuous events — like the fall of a windblown acorn. His foray into pure mathematics did not end with the infinitely large. Arguing against extrusion from Earth, he considered an infinitesimal instant that it takes our planet to rotate through an angle bounded by an arbitrarily small quantity, and an infinitesimally small distance through which an object must fall in this time to remain Earthbound.[12] His physical analog of the mean-speed theorem,[13] which we will soon encounter as we revisit the topic of falling bodies, exemplifies a pre-calculus proof rivalling in rigor any post-Cauchy proof. In these ways, Galileo uncovered the trail of an Archimedean journey that leads to the theory of indivisibles and to elements of the calculus.

Galileo's younger disciple, the Jesuate[14] Bonaventura Cavalieri, also inspired by Kepler, pioneered the theory of indivisibles. Cavalieri, who had determined the focal length of elliptical lenses and also introduced logarithms in Italy, was lauded by Galileo for his work in geometric optics. As Galileo's papers on the theory of indivisibles have been lost, the scattered references in his dialogues are all that we have of his thoughts on the subject. It's enough to drive a man to drink.

Beer Sliding at the Lunar Pub

Imagine that you are enjoying a beer at the counter of a London pub. On a wet counter a slight nudge on the glass may cause it to slide a bit, but the slide will quickly decay due to surface friction. If the counter were varnished, it might slide still farther before stopping; if it were then coated with a lubricating substance, you might in this

---

11. The author has observed that the California Franchise Tax Board, not to be outdone in finding ways to separate people from their earnings, has attempted to use a class of infinity that taxes even the imaginations of mathematicians. If a Corporation pays $800 (minimum) tax on $1000 of profit, the ratio 80/100 is called an 80% tax rate. If it breaks even for a year, it will still pay $800 tax on zero profit, which makes the tax rate infinite. This is a form of infinity that mathematicians know about and have developed a limit concept to deal with. But if you lose money that year, you still pay $800 tax on that "profit". To what class of infinity does this rate belong? If we excuse their naivete in esoteric matters, what is their excuse for calculating capital gains by subtracting dollar amounts of different years as if the dollar were a unit unaffected by government-induced inflation? Even third-graders don't subtract apples from oranges.

12. argued by Salviati on the Third Day of *Two New Sciences* (cf. Boyer, op. cit., Ch. 16).

13. The theorem is proved in Galileo's *Two New Sciences*, Drake trans., op. cit., p. 165. In note 23, Drake makes this important observation: "Medieval writers assumed an ideal mean-speed to measure every uniformly accelerated motion directly. Galileo's proof matched elements in two infinite aggregates for each instant and all instants, conceiving that in uniform motion there is not one single speed but infinitely many, all equal, and corresponding to the infinitely many speeds, all different, in accelerated motion."

14. a religious order founded in the 14th century by Blessed John Colombini of Siena originally for the care of those stricken by the Black Death, not to be confused with the Jesuit order; cf. Howard Eves, *Great Moments in Mathematics Before 1650*, Mathematical Association of America, 1983.

manner pass your beer to the fellow at the other end of the counter, being careful of your aim, of course. Now imagine that, to top off the evening's entertainment, you commute to the moon, enter a lunar pub, order a beer, and give the glass a gentle nudge parallel to the counter's edge. Imagine further that this is a very special counter. It is built along the moon's equator and runs the whole circumference of the moon; further, it has a perfectly smooth, frictionless surface. Because you are on the moon, there is no air resistance. Would the slide of the beer glass ever end?

If your aim is not parallel to the length of the counter, then at the point where the beer glass slides off the edge, how would its motion change? And if you accept Aristotle's belief that an object in motion must be sustained in motion by a force, how can the motion continue once the beer glass has left your hand? Besides correct mathematics, correct answers required a philosophical revolution. In order for Galileo's impetus to end centuries of intellectual inertia enjoyed by the Aristotelian dictum *"quod movetur ab alio movetur"*,[15] he must articulate convincingly the inertial principle he calls *impeto*. In the *Discourses*, he takes up that challenge and offers an insight into Newton's later understanding of how objects move anywhere in the universe: [16]

"... whatever degree of speed is found in the moveable, this is by its nature [suapte natura] *indelibly impressed on it when external causes of acceleration or retardation are removed, which occurs only on the horizontal plane; for on declining planes there is cause of more* [maioris] *acceleration, and on rising planes, of retardation. From this it likewise follows that motion in the horizontal is also eternal since if it is indeed equable it is not* [even] *weakened or remitted, much less removed.*

Furthermore, one must consider the existing degree of speed acquired by the moveable in natural descent to be naturally indelible and eternal; but if after descent along a declining plane it is diverted through another upward plane, a cause of retardation presents itself there, for on such a plane the same moveable would naturally descend. Wherefore a certain mixture of contrary influences [affectionum] *arises — that of the degree of speed acquired in the preceding descent, which by itself would carry the moveable away uniformly* in infinitum, *and* [that of] *a natural propensity to downward motion according to the same ratio of acceleration in which it is always moved. Whence it is seen to be quite reasonable if, in inquiring what events take place when a moveable is diverted through some rise after descent through some inclined plane, we assume that that maximum degree acquired in descent is in itself perpetually kept* [servari] *in the ascending plane, but* [that] *in the ascent there supervenes the natural tendency downward;*

---

15. "Motion requires a mover".
16. from Galileo Galilei, *Two New Sciences*, translated by Stillman Drake, Wall & Thompson, Toronto, 1989.

that is, to a motion from rest accelerated in the ratio always assumed. "

Galileo appears to attribute a conservative tendency to motion along a plane: a straight unaccelerated motion. In this view, "horizontal" refers to a tangent plane;[17] otherwise the velocity would be non-uniform and the "plane" non-planar. Noting that this is not the conservation of geocentric speed described in the *Dialogo*,[18] we see that under this interpretation, Galileo has deflected his theory to an upgrade: his mature description of *impeto* has now all the elements of what Kepler called (in Latin) *"vis inertiae"*.[19] Like "inertia" in the modern sense, it is natural, perpetual, rectilinear, and not preferential as to velocity or direction.[20] But it is also possible to interpret "horizontal" motion as occurring, consistently with conservation of geocentric speed, along the infinity of planes tangent to the horizon as the object moves. In either case, now that Galileo has explained how inertial motion and the downward acceleration of a falling object combine to produce its actual path, we see that to predict its trajectory depends on knowing how an object accelerates when it falls. How might Galileo have determined the manner in which an object accelerates when it falls? We can ponder this question as we mop up beer and shards of broken glass from the floor of the lunar pub.

Falling Bodies Revisited   Do falling objects accelerate? Aristotle embraced a hazy view of acceleration, but could not answer the question: *how does the change in speed of a falling object vary with time?* This is the question that Galileo set out to answer. He decided actually to measure this change, but measuring it required slowing the rate of fall and coupling the falling object to a distance reference frame. In his workshop at Padova, Galileo had solved these problems by using an inclined plane and a water clock that together allowed him to

---

17. That Galileo considered a horizon plane to be a tangent plane is clear in the following passage from his *De Motu*: "... a plane cannot actually be parallel to the horizon, since the surface of the earth is spherical and a plane cannot be parallel to such a surface. Hence, since the plane touches the sphere in only one point, if we move away from that point we shall be moving upward." Stillman Drake, *Galileo at Work*, p. 24.

18. Drake uses the term "conservation of geocentric speed" (which, in his notes at the Thomas Fisher Rare Book Library, he attributes to Jed Buchwald) to refer to Galileo's notion that a body not subject to external forces can be moved by any minimum force along an infinity of horizon planes (cf. Ch. 5, "Greasing the Skids"). In the *Dialogo*, Einaudi edition, pp. 182-183, Simplicio used *"impeto"* and *"un moto indelebilmente impressole"* (an indelibly impressed motion) to refer to the tendency of a stone having fallen from the mast of a ship to follow the ship; Salviati, replying, refers to two causes of the stone's total motion: *"la gravità"* as the cause of tending to the center (of Earth), and *"la virtù impressa"* as the cause of its being led around the center; hence, *virtù impressa* may correspond to what some call "circular inertia". By the time of the *Discourses*, Galileo's concept of inertia, here called *"impeto"*, has been refined to the perpetual tendency of an object to sustain its velocity.

19. See the Oxford English Dictionary, Clarendon Press, Oxford, 1933.

20. These properties of the modern inertia principle are enumerated by Dudley Shapere in *Galileo, A Philosophical Study*, U. of Chicago Press, 1974, p. 122. For an exposition relating Galileo's mature concept of inertia to the modern principle, see MacLachlan, James, "Drake Against the Philosophers", in *Nature, Experiment, and the Sciences* (eds. Trevor H. Levere and William R. Shea), Kluwer Academic Publishers, Netherlands, 1990.

measure the time of the fall. He designed the water clock to let water out at a controlled rate, so that the amount of water released measures the time duration of the observation period for any given experiment.

Instantaneous velocity cannot be measured directly with a water clock. To measure it indirectly, he sought and found a new experimental technique consistent with his kinematic approach. Galileo hints at this discovery in his comment *"For there seems to me to be no doubt that the heavy body coming from a height of six braccia has, and strikes with, double the impetus that it would have from falling three braccia, ..."*.[21] From the distance that a bowstring (of calibrated strength) is distended when coupled to a falling object, for example, one may measure the impetus of fall from a given height. In the percussive effect of free fall, he has found a striking approach to the problem of instantaneous speed measurement.

In the *De Motu*,[22] Galileo had rejected acceleration, except initially; in the *Discourses*, he presents it as continuous and constant.[23] Here he treats the medium not as a causal agent for motion, but as an inhibitor that detracts from the natural motion of free fall. Like the magician who drops a pigeon into a hat and makes it disappear before our eyes, he makes the medium vanish by transubstantiating the falling object (lead to ebony to tin) and rarefying the medium (fluid to vapor to vacuum) over a smooth continuum of evanescence. As the density of the medium approaches zero,[24] he shows that in the limiting case, weight and density of a falling object have no measurable effect. Similarly, he shows that free fall is a limiting case of a bronze ball rolling down an inclined plane, as the angle of inclination to the vertical vanishes.

While a bronze ball accelerates down an inclined plane, is the speed of fall proportional to the distance fallen? In the *Discourses*, Galileo admits that he once held this mistaken belief.[25] Aristotle would probably have conducted experiments to measure the increase of speed with time if he knew how. Galileo, circumventing the difficulty of measuring acceleration directly, measured instead the distance fallen with time — an equivalent test.[26] The path to that measurement was a long and thorny one. Avoiding algebra and predating the calculus, Galileo worked in proportion theory, dealing in ratios of distance to time. In 1604, he wanted to know how the speed of a falling object varies with time and with distance at every instant

---

21. Galileo's *Two New Sciences*, Drake trans., p. 159.
22. Cf. Chapter 4, "The Pisan *De Motu*".
23. *Discourses*, op. cit., p. 74 ff.
24. This is a thought experiment that was used by Aristotle (rdc); *Physics* IV, 215 a and b.
25. Professor Drake concluded that Galileo's mistake relates to his early interpretation of experiments on the impact effects of equal weights dropped from varying heights; cf. *History of Free Fall*, op. cit., p. 54.
26. Galileo measured $s = v_0 t + c \cdot t^2$. Here, v = speed of fall, t = elapsed time, s = distance fallen, c = coefficient of acceleration due to gravity. The initial downward speed $v_0 = 0$ if the ball starts from rest.

of fall, and he guessed wrong that speed is proportional to distance. It took him several years to distinguish the behavior of the velocity function from its square in relation to distance and time. By 1608, he had assured himself that, whatever the inclination through which they fall, for two objects falling through distances s and s′ in times t and t′ with terminal velocities v and v′, respectively,

$$\frac{s}{s'} = \frac{t^2}{(t')^2} \text{ and } \frac{v}{v'} = \frac{t}{t'} \text{ and he could deduce that } \frac{v^2}{(v')^2} = \frac{s}{s'}.$$

The distance that a ball rolls down a plane he found to be proportional to the square of elapsed time for an arbitrary angle of inclination. Combining kinematic and percussive experimental techniques with his substitution of an equivalent variable, and using the inclined plane to render that variable measurable, he discovered a law of nature which corrected his earlier mistake. From his measured relationship of distance to time, it is derivable that speed is proportional to time elapsed. The unroyal road to dynamics is not complete with this measurement; it is paved with a compound of experimental results imbedded in a theory. What Galileo must do with these results to create the theory is to explain them from first principles that predict all other observed results.

This he does in the *Discourses.* Introducing what he calls steady or uniform motion, Galileo observes the inadequacy of the classical definition: equal distances are traversed in equal times. You can walk from B to A in the time that you walked from A to B, but if on the return trip you stop at the local pub for a drink and then walk faster to recover your lost time, is your speed uniform? Refining the concept to eliminate this difficulty, he defines uniform motion as "*... one in which the distances traversed by the moving particle during <u>any</u> equal intervals of time, are themselves equal.*" [27] That is, no matter how small.

He follows with axioms on uniform motion. Time and distance are monotonically related. Speed and distance are monotonically related. It is on this humble but rugged foundation that he builds his kinematics. In the transitional discussion from uniform to accelerated motion, we find a very succinct and beautiful example of Galileo's characteristic scientific thinking collapsed to a few short paragraphs. Let's listen to his thoughts and the elegant language he uses to express them: [28]

"... In the investigation of naturally accelerated motion we were led, by hand as it were, in following the habit and custom of nature herself, in all her various other processes, to employ only those means which are most common, simple and easy.

---

27. *Discourses*, op. cit., p. 154.
28. Ibid., pp. 160-161.

For I think no one believes that swimming or flying can be accomplished in a manner simpler or easier than that instinctively employed by fishes and birds.

When, therefore, I observe a stone initially at rest falling from an elevated position and continually acquiring new increments of speed, why should I not believe that such increases take place in a manner which is exceedingly simple and rather obvious to everybody? If now we examine the matter carefully we find no addition or increment more simple than that which repeats itself always in the same manner. ... thus we may picture to our mind a motion as uniformly and continuously accelerated when, during any equal intervals of time whatever, equal increments of speed are given to it."

Here we see the genesis of a hypothesis, the one with which this passage ends. It is not simply an observation of nature. It is an abstraction about the simplicity and comprehensibility of nature derived from a multitude of observations.

In a variation of race between the slow and uniform tortoise versus the swift and accelerating hare, Galileo considers a body falling from rest and uniformly accelerated, versus one moving through the same distance with uniform speed equal to half the terminal speed of the falling object. Which would you put your money on? It doesn't matter, he shows us: they will finish simultaneously. This demonstration leads him to:

> *"Theorem II, Proposition II: the spaces described by a body falling from rest with a uniformly accelerated motion are to each other as the squares of the time intervals employed in traversing these distances."*

Theorem II not only completes the deductive argument from geometry underlying his law of falling bodies, but proves a relationship that he used to verify the constancy of free-fall acceleration by recourse to the method and tools of his workshop. The theory of falling bodies here emerges in its mature form. From Theorem II, he proceeds to derive:

GALILEO'S LAW OF FALLING BODIES

The distance through which an object freely falls is proportional to the square of the time elapsed; the rate of fall is proportional to the time elapsed; the acceleration is uniform.

This law refers to two proportionalities. The proportionality coefficient for the rate of fall is known as the acceleration due to gravity. That for the distance fallen is one half that value. In practice, Galileo usually stated distances and times as ratios to those of a baseline measurement and did not concern himself with the numeric value of the acceleration due to gravity, other than the one he apparently fabricated in working out the counter-example to Locher's cannonball drop. A value within about 80% of the modern accepted value (of 980.665 cm/sec^2 at sea level on the Earth's equator) was determined experimentally by Marin Mersenne.[29]

"Who is God? A spiritual circle whose center is everywhere
and circumference nowhere." [30]

... Ficino, *Theologica Platonica* 18.3

SIMPLICIO: I mean by "center," that of the universe; that of the world; that of the stellar sphere; that of the heavens.
SALVIATI: I might very reasonably dispute whether there is in nature such a center, seeing that neither you nor anyone else has so far proved whether the universe is finite and has a shape, or whether it is infinite and unbounded.

... Galileo (*Dialogo*, Third Day, trans. by Drake)

From Parabolic Fall to Planetary Periods

That "troublemaker" Galileo had made waves in the *Dialogo* — arming the astronomy of Copernicus with near-Earth physics which lifted up the long-abandoned Aristarchan vision against the current prevailing cosmology. Now in the *Discourses* he is poised to ride his own Renaissance waves. Unlike the anticlimactic Fourth Day of the *Dialogo*, the *Discourses* skate through Day Four and into their final pages along the crest of one of his scientific waves, pipelining from parabolic fall into a discussion of the properties of projectile paths. The wave breaks and rushes onto the shore of the receptive reader's mind as he invokes a Proposition from the Second Book of Euclid and two elemental theorems from the *Conic Sections* of Apollonios [31] in order to sketch out the superimposition of uniform inertial motion and accelerated fall:

"A projectile which is carried by a uniform horizontal motion compounded with a naturally accelerated vertical motion describes a path which is a semi-parabola."

So much for objects projected near the Earth's surface. What if God dropped the planets? He had posed this question to Ingoli, claiming that Plato had motivated it.

---

29. Mersenne's estimates imply g = 24 Parisian feet/sec^2 = 780 cm/sec^2. Mersenne, a contemporary of Galileo, reported his results after seeing Galileo's *Dialogo*. Source: James MacLachlan, "Mersenne and Galileo: Ideas in Motion", thesis at Harvard University, 1971.
30. quoted by Hallyn, *The Poetic Structure of the World*, MIT Press, 1990.
31. Here Galileo reformulates new proofs of these theorems so that the argument is self-contained and does not require all the development that Apollonios gave it.

Here are the salient words of the passage, spoken in the persona of Sagredo: [32]

> *"Plato thought that God, after having created the heavenly bodies, ... made them start from rest and move over definite distances under a natural and rectilinear acceleration such as governs the motion of terrestrial bodies. He added that once these bodies had gained their proper and permanent speed, their rectilinear motion was converted into a circular one, the only motion capable of maintaining uniformity, a motion in which the body revolves without either receding from or approaching its desired goal. This conception is truly worthy of Plato; ..."*

Smoldering frustration smokes through Galileo's guarded prose as Salviati recalls recondite calculations relating the speed of a planet to the size of its orbit:

> *"I think I remember his having told me that he once made the computation and found a satisfactory correspondence with observation. But he did not wish to speak of it, lest in view of the odium which his many new discoveries had already brought upon him, this might be adding fuel to the fire. But if anyone desires such information he can obtain it for himself from the theory set forth in the present treatment."*

Were his papers with these derivations removed by friends from his house at the time of his trial, and then lost? To reveal them here, would he have to reconstruct them? In any case, he did not wish to invite additional abuse from the Inquisition.

The Lunar Discus: Galileo's Parabolic Trajectories

The last quoted passage, inserted unobtrusively within his discussion of the parabolic paths of projectiles, discloses a dominant purpose of the book: to allow someone less vulnerable to theocratic thuggery to demonstrate rigorously the motion of Earth. That he concludes the *Discourses* with projectile paths is understandable; they are the visible outcome of the laws of inertia and fall. By reference to them, his new science of motion becomes verifiable.

So let's verify it. We can use it to predict something measurable: the range of a ballistic projectile. (Readers who do not enjoy mathematical derivations may skip this and the next section.) To establish how its range depends on the launch speed v and angle of projection ϑ to the horizon plane, recall first Galileo's discovery that inertial motion is unaccelerated. Unlike the lunar pub counter, which curves around the moon, the positive x axis can be considered directed along the horizon plane tangent to the launch point.

---

32. *Discourses*, op. cit., p. 261.

From this and from Galileo's other discovery that the
acceleration of gravity is locally constant, it follows that
the vertical distance y relates to the horizontal x in the
form $ax - bx^2$, which is immediately seen by inspection to
be a parabolic segment.[33] Not having enjoyed algebra or
calculus, Galileo infers a parabola by a brilliant (and
tedious) geometric argument.[34] Under the locality
assumption, the complex path of a ballistic projectile traces out a parabola. From a
post-Newtonian perspective, if we drop the locality constraint then Galileo's result
must be a special case of Kepler's First Law, mustn't it? [35] But within its domain of
validity, near the Earth's surface, Galileo's law of parabolic fall gives good results for
the trajectory and range of projectiles.

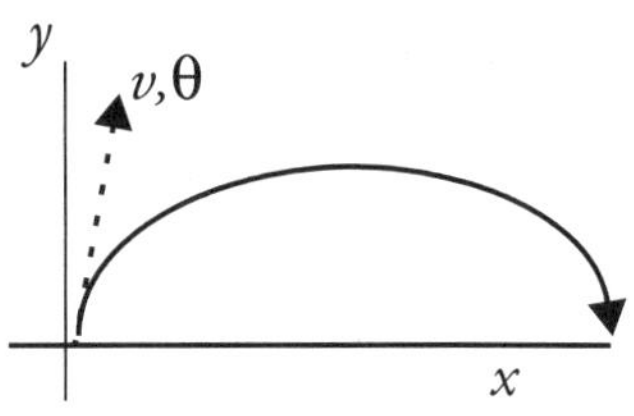

Maximum Range of a Projectile

Suppose that, wishing to throw a discus as far as possible on
the moon, one gives it an oblique upward thrust. Some
value of the elevation angle will yield the maximum range.
A reader conversant with the calculus may wish to show from the foregoing that the
horizontal displacement is $v^2 \cdot \sin(2\vartheta)/g$.[36] Differentiating with respect to the
elevation angle, we can find a necessary condition for a maximum:[37]

$$2\frac{v^2}{g} \cdot \cos 2\vartheta = 0$$

---

33. As the upwardly thrust projectile continues under Galilean inertia, then $\dfrac{d^2x}{dt^2} = 0$ horizontally.

Recalling his other discovery of the acceleration of gravity, g, along the vertical y axis, we may by convention

prefix a minus sign to denote the downward direction assigned by definition to fall: $\dfrac{d^2y}{dt^2} = -g$.

Integrating these, we see that $\dfrac{dx}{dt} = c_1$, while $\dfrac{dy}{dt} = -gt + c_2$, where c_1 and c_2 are the initial x and y

components of v; namely, $v \cdot \cos\vartheta$ and $v \cdot \sin\vartheta$, respectively. At time t = 0, let x = y = 0, then dx/dt =

$v \cdot \cos\vartheta$ and dy/dt = $v \cdot \sin\vartheta - gt$, and at any time t (integrating again):

$$x = (v \cdot \cos\vartheta)t \quad \text{and} \quad y = (v \cdot \sin\vartheta)t - \frac{g \cdot t^2}{2}$$

from which it follows by eliminating the variable t, that

$$y = x \cdot \tan\vartheta - x^2[g/(2v^2 \cdot \cos^2\vartheta)] \quad .$$

34. *Discourses*, op. cit., p. 217-222; recommended reading as a lesson in algebra and calculus appreciation.
35. The author recalls a book he read more than 30 years ago (*The Conquest of Space,* Viking Press, 1952) in
    which Willy Ley explained that projectiles trace out an ellipse with one focus at the center of Earth.
36. Hint: the range of the projectile is the product of its horizontal speed and the time to return to the ground.
    It will rise until the vertical component of its speed $v \cdot \sin\vartheta - gt = 0$. Solving this equation for t yields
    half the total path time. Use the identity $2\sin\vartheta\cos\vartheta = \sin 2\vartheta$.
37. The reader may wish to show that in this case it is also a sufficient condition.

(a condition satisfied when $\cos 2\vartheta = 0$) hence, $\vartheta = \pi/4$ or $3\pi/4$, etc. Galileo thus found that a 45° angle of launch maximizes the range of a projectile. And indeed, his tables #1 and #2 of Favaro's Opere VIII (304) [38] show the maximum amplitude of the semi-parabola at this angle.

By a Newtonian extension of Galileo's research we can explain his result at the Leaning Tower; namely, that he found no difference in the rate of

Explanation of the Leaning Tower Phenomenon

fall of heavy and light objects, at least none that is measurable and attributable to their difference in mass. If, as Newton declared,[39] the Earth and a falling object attract each other proportionally to the product of their masses, then why doesn't the heavier object, being attracted with greater force, fall faster? [40] Suppose one equates the mutual gravitational attraction between the Earth and either of Galileo's weights with the force required to overcome the inertia of its rest mass and produce the observed acceleration, a. Then, if G = Newton's gravitation constant, m_f the falling mass, m_e the mass of Earth, and r the nonzero distance of the falling mass from the gravity center of the Earth/mass system:

$$G \cdot m_f \cdot m_e / r^2 = m_f \cdot a \text{ , therefore } G \cdot m_e / r^2 = a = g \text{ .}$$

Equating the two physical laws governing inertial and gravitational force, we see that the value m_f will "drop out", with the result that the acceleration of a freely falling object in a vacuum does not depend on its mass, contrary to the belief of "Aristotelians" among Galileo's colleagues at Pisa.

From Newton's later work it is clear that the acceleration g is not a constant but a function $g(r)$ inversely proportional to the square of the distance, r, from the center of Earth. The value of g

The Gravity of His Error

in Galileo's law of falling bodies is subject to this correction if the fall takes place over astronomical distances. Without the correction, the deviation $g(r)$-g is a source of significant error. This fact recalls the passage from the English translation of the letter which the Church had sent around to all the Bishops of Italy and Papal Nuncios of Europe requiring that his sentence and abjuration be published to them and to all professors of philosophy and mathematics:

---

38. in Drake's trans. of Galileo's *Two New Sciences*, p. 251.
39. Newton's Law is discussed in Chapter 8, "Universal Gravitation (Centripetal Acceleration)".
40. A few years ago, I put this question to a young man who had recently graduated from the University of California with a degree in physics. He said that he would have to think about it. I still don't have his answer. Are mathematics and physics being taught today by a method substantially better than that of the universities in Galileo's time?

"... that they may know why we proceeded against the said Galileo, and recognize the gravity of his error, in order that they may avoid it, ... " [41]

Galileo's gravity law remains adequate for terrestrial physics. It was Isaac Newton and not any member of the Roman Catholic Church who corrected Galileo's law for use at more general altitudes. These same churchmen who presumed to judge the gravity of his error were incapable of judging the error of his gravity law.

The Theory of Mechanical Stresses

In his *Discourses*, Galileo constructs the foundation of construction engineering. Some of his conclusions are counter-intuitive. As weights are continually added on both sides of a fulcrum, the lever is subject to fracture — but if the fulcrum is not centered between the applied forces, a force that would fracture the lever at its center may not fracture it even if multiplied a thousandfold. This observation leads him to distinguish validation of thought from discovery of truths:

"I quite understood the action [facoltà] of the lever, and how by increasing or reducing its length, the moment of its force and of the resistance grew or diminished; yet, for all that, I was mistaken in the solution of the present problem, ... It seems to me that logic teaches how to know whether or not the reasonings and demonstrations already discovered are conclusive, but I do not believe that it teaches how to find conclusive reasonings and demonstrations." [42]

Apart from Aristotle's pioneering of the science of logic, it would be a serious mistake to assume that all of Aristotle's physics was wrong or overturned by Galileo. To create the theory of mechanical stresses, Galileo turns to Aristotle and to Archimedes, whose Law of the Lever he takes as fundamental:

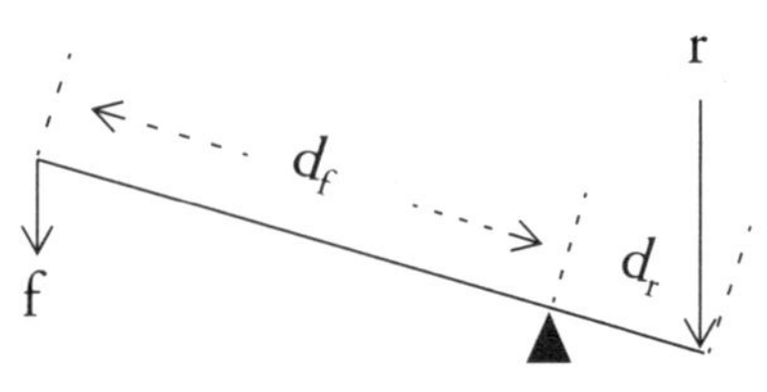

"that in using a lever, the force is to the resistance in the inverse ratio of the distances from the fulcrum to the force and to the resistance." [43]

This gives the equilibrium condition: $\dfrac{f}{r} = \dfrac{d_r}{d_f}$. While acknowledging the temporal priority of Aristotle, he credits Archimedes with the proof of the law:

"Yes, I am willing to concede him [Aristotle] priority in point of time; but as regards

41. in A. D. White, op. cit., p. 144.
42. Galileo's *Two New Sciences*, p. 133 [Drake's trans. from Favaro's VIII (175)]
43. Ibid., p. 109 [Drake's trans. from Favaro's VIII (152)]

rigor of demonstration the first place must be given to Archimedes, since upon a single proposition proved in his book on Equilibrium *depends not only the law of the lever but also those of most other mechanical devices.*"[44]

Galileo gives a proof of this law different than that advanced by Archimedes. Then he uses the law to derive the resistance R to fracture of a cantilever beam of span s and radius r under load L, a result he modifies to take into account the

Galileo's Laws of Resistance to Fracture

weight of the cantilever itself. Similar beams of different size, he shows, cannot have the same resistance to fracture; hence, there is a limit to the stable size of a natural form. Though intrinsically interesting, the discussion has an another purpose. It is an appetizer, a lead-in to introduce a law of his new science:

> "*Proposition V: Prisms and cylinders which differ in both length and thickness offer resistances to fracture (i.e., can support at their ends loads) which are directly proportional to the cubes of the diameters of their bases and inversely proportional to their lengths.*"[45]

A question of interest to designers of machines is how to increase strength without adding to weight. Answering it, he calculates the strength of solids cut from a prism by means of a parabolic section: a technique that can diminish the weight of a beam by as much as 33% without reducing its resistance to fracture.[46] Concluding this topic he explains the (even more surprising) resistance of hollow solids to fracture. His discussion leads to a somewhat counter-intuitive result:

> "*The robustness of the* [hollow] *tube therefore gains over the robustness of the solid cylinder in proportion to the diameters, provided always that both are of the same material, weight and length.*"[47]

These principles are applied in construction engineering and hold implications for

---

44. Galileo's *Two New Sciences*, Crew and de Salvio trans. from Favaro's (152). Drake observes in a note what Crew and de Salvio defaulted comment on; viz., "What Galileo calls a single proposition made up two for Archimedes, who proved separately the cases of commensurable and incommensurable distances."

45. Galileo's *Two New Sciences*, Crew and de Salvio trans. from Favaro's VIII (163). Again, Drake notes something important and missed by Crew and de Salvio: "This is probably the first expression of s [*sic*] strictly physical property in terms of two independent variables. Archimedes had used the compounding of ratios in a similar way, but only for mathematical relationships. Cf. Heath, *Archimedes*, p. clixxix." **Here illustrated is the importance of translator's notes** and the probable immense loss to students caused by the suppression of Drake's revised notes to the Third Edition of the *Dialogo* (See Epilogue).

46. In a style of argument reminiscent of Archimedes' application of Eudoxian exhaustion to the area of a circle $A = \pi r^2$ (see Chapter 1), and anticipatory of Cavalieri's principles, Galileo invokes Archimedes Prop. 10 from *On Spiral Lines* as a lemma to show that the portion of the interior of a rectangle contained within a parabolic section connecting its opposite vertices is two thirds in area of the rectangle.

47. Galileo's *Two New Sciences*, Drake trans. from Favaro's VIII (188).

architecture. Galileo formulated them in a few months while recovering from the torment and rigors of a judicial trial, and while burdened by sickness, grief and hostile suppression that could have crushed ordinary men. Galileo's greatest discoveries occurred after the age of 40, and some of his best works, including his masterpiece, *Discourses and Mathematical Demonstrations Concerning Two New Sciences,*[48] were completed after the age of 70, a stunning counter-example to the overly emphasized generalization that scientists do their best work at a young age. Here is a man whose purpose, determination and intellect utterly demolished every obstacle thrown in his path, including but not limited to age.

Tragedy, Discovery and Invention in the Final Years

"... and though
We are not now that strength which in old days
Moved earth and heaven, that which we are, we are —
One equal temper of heroic hearts,
Made weak by time and fate, but strong in will:
To strive, to seek, to find, and not to yield."

... from *Ulysses* by Alfred Lord Tennyson

Theory and Reckoning of Longitude

In his final years, Galileo revisits a project on the reckoning of longitude. Back in 1617, as we saw in Chapter 5, he had designed a scheme to deliver ships from the clutches of pirates. Let's flashback to that year to follow the ensuing events. Aware of reports that European governments were willing to pay huge prizes for a solution to the determination of longitude at sea, Galileo sent a proposal to a Spanish dignitary who offered to assist in presenting it to the King. What is Galileo's solution?

He looks at Jupiter with its bright moons, their orbital periods ranging from 1.8 to 16.7 days, and sees in his mind's eye a giant clock in the sky. The eclipses of these satellites occur with a regularity exceeding that of the most accurate clocks made at this time. Observing Jupiter from shipboard, a sailor can determine the time at any reference longitude, such as that of Greenwich, by reference to the "hands" of a giant longitude-independent clock. Measuring the hour angle or right ascension of any star transiting the meridian local to the ship would allow the sailor to reckon longitude as a function of the difference between this measurable local time (corrected for the

---

48. This book is a spectacular example that no work can be evaluated by its immediate reception. Though the demand for Galileo's writings will last as long as civilization, *Two New Sciences* "sold so poorly that Galileo's publishers abandoned their project of reprinting his collected works when he was old and blind". Stillman Drake, *Cause, Experiment and Science*, U. of Chicago Press, 1981, p. xi.

"equation of time" [49]) and the time at Greenwich. It was to make this calculation easy for persons with no mathematical training that Galileo invented the *giovilabio*. But his patient explanations of this invention to the Spaniards do not prevent the Spanish navy from arguing endlessly about his proposal. The King of Spain takes no action.

The Golden Boomerang

Returning to the year 1636, we find Galileo bringing similar proposals to the Dutch government, and patiently answering their questions as he did for the Spaniards who royally wasted his time. This time his invention garners a more favorable response. The Dutch government even sends Galileo a gold chain for his contributions to the problem of longitude determination. But Galileo is blocked from supplying updated tables on the positions of Jupiter and abandons the project. Why? He is advised that the Church at Rome would disapprove his acceptance of any remuneration from a *Protestant* government. Having been prevented from finishing the project, Galileo returns the gold chain with a polite refusal. Had he accepted it, it would have been a rare time in his life when he would have been paid for inventing a technique or process. Here is a case where he who has the gold does not make the rules.

Continuing on the project, he runs into a problem nearly intractable given the technology of his time: the unsteadiness of support for a telescope on shipboard. Shipborne clocks adequate to determine longitude directly from time will not be invented until after 1740 when Benjamin Huntsman begins to use steel, tempered in crucibles, to form the springs of shipborne chronometers.[50] Once the longitude problem is solved, the accuracy of determining latitude on a moving ship by sighting the elevation of a star is improved in 1757 by the use of John Dollond's achromatic lens in John Hadley's sextant.[51]

Galileo Elucidates the Phenomenon of Light

The man who enlightened the world about the nature of curious points of light in the night sky now becomes curious about the nature of light itself. Given a finite speed, it should be measurable, at least in principle. At his hilltop villa, Galileo conceives an experiment to time light signals exchanged between distant mountain peaks by humans equipped with lanterns. The experiment is correct in principle but fails by assuming human reaction time insignificant in relation to the travel time of a beam between the peaks. After nearly three centuries, in 1926 that is, Albert Michelson will redesign and repeat Galileo's experiment with more sophisticated equipment. Using a spinning stroboscope — an octagonal mirror — to

---

49. The equation of time is a function of calendar time expressing the difference of local apparent time and local mean time; i.e., the difference between a sundial and a mechanical clock at any time of the year.
50. Specimens have been on display at the Greenwich Royal Observatory (at least during my 1987 visit there).
51. See James Burke, *Connections*, Little, Brown & Co., Boston, 1978. Burke spells the name "Dolland".

synchronize the return of a light beam reflected from Mount San Antonio back to Mount Wilson over a precisely surveyed path, Michelson obtains to four significant places, rounded, the accepted value of 299.8 kilometers per second.

In the meantime, Olaus Roemer chose two remote points of Earth's orbit, sufficiently distant to allow accurate measurement of light travel time using Galileo's sky clock. In 1676 he measured the difference in eclipse times of Medicean satellites between the two points and calculated about ¾ of the modern accepted value for the speed of light.[52] Galileo aided Roemer's experiment by discovering the satellites in question, by noting and determining the accurate periodicity of their orbits and, though Roemer's astronomical experiment differed fundamentally, by designing the terrestrial prototype for measuring the speed of light.

Galileo did not confine his curiosity about light to its speed. As we have seen,[53] he theorized how light is produced and arrived at a description that might have raised the eyebrows of his successors Newton and Huygens, who each formulated contrasting explanations of the nature of light.

The Discovery of Lunar Libration

Lunar libration, not to be confused with libation (something that one might enjoy in a lunar pub if there were one), refers to that familiar old "man in the moon" not always showing us the same face; over time, Galileo discovered, we see more than half of it. Galileo first describes his discovery, apparently one of his last before the onset of blindness, in his 7 November 1637 letter to Micanzio:

"I have discovered a very marvelous observation in the face of the moon, ... it changes its aspect with all three possible variations, making for us those changes that are made by one who shows to our eyes his full face, head on so to speak, and then goes changing this in all possible ways, that is, turning now a bit to the right, and then a bit to the left, or else raising and lowering [his face], or finally, tilting his left shoulder to right and left. ... "[54]

Having cited Galileo's discovery, Henry Norris Russell [55] distinguishes three kinds of lunar librations: [56]

---

52. Roemer's estimate will not be improved until 1849. Fizeau then will repeat Galileo's experiment using a mirror in place of the second lantern. His result will be accurate within 5% of Michelson's very accurate determination in 1926.

53. Chapter 8, "The Atomic Genesis of Light".

54. Trans. by Stillman Drake in *Galileo at Work*, Bibliography.

55. American astronomer, regarded as co-discoverer with Herzsprung of the stellar luminosity relationship to spectral class. (*Astronomy* co-authored with Dugan and Stewart, Ginn and Co., 1945, pp. 169-170. The author purchased his copy at the age of ten; although probably out of print, it is a timeless classic.)

56. Lunar librations exclude the "diurnal" effect at moonrise and moonset, when we can in effect "peer over the top" (the western edge at moonrise, eastern at moonset) of the moon. The diurnal effect is thus a libration of the observer, and not of the moon.

"... *libration in latitude* is due to the fact that the moon's equator does not coincide with the plane of its orbit, but makes with it an angle of about 6½°. The inclination of the moon's equator causes its north pole at one time in the month to be tipped 6½° toward the earth, while a fortnight later the south pole is similarly inclined ..."

... *libration in longitude* depends on the fact that the moon's angular motion in its elliptical orbit is *variable*, while the motion of the rotation is *uniform*, like that of any other undisturbed body; the two motions, therefore, do not keep pace exactly during the month, and we see alternately a few degrees around the eastern edge and around the western edge of the lunar globe. This libration amounts to about 7¾° each way.

... Physical libration. Besides these geometrical librations, which arise from the lack of uniformity of the motion of the observer relatively to the moon, there is a minute *physical libration* due to real irregularities in its rotation."

This last effect, undetectible by Galileo's equipment, is a more recent discovery. These effects combine to show Earthlings 59% of the moon's surface over time, a constant 41% and a time-dependent 18%. Before Galileo, nobody noticed.

✦✦✦✦✦

Of major importance to Galileo in the few precious months after his return to Arcetri is his relationship with his elder daughter Virginia, who took the name Sister Maria Celeste.

Maria Celeste Galilei

Back in 1610, Galileo, having left Padova for Florence, separated amicably from his family. The mother of his children married another, and his daughters entered a convent. Thanks to the research of Desmond O'Grady, we can sample 134 letters which Maria Celeste wrote to Galileo between 1623 and 1633. In 1623, Maria Celeste wrote suggesting that Galileo ask Pope Urban VIII to *"send the convent a religious priest as confessor because there was no money to pay the diocesan confessors who 'often come to dine and become too friendly with the nuns' and 'are more used to hunting hares than guiding souls' "*.[57] On December 28 of that same year "she told him she was sending the candied limes he had requested, *'two baked pears for the days of the (Christmas) vigil' and a rare winter rose"*.

In 1628 she wrote to her father: *'I love you more than myself'*. At another time, she

---

57. Desmond O'Grady, "Candied limes for Galileo", *The Tablet*, 9 January 1993; thanks are due to Rev. Fr. Vincent M. O'Brien, S.J., for bringing this study to my attention. All quotations in this section are from this article; double quotes enclose O'Grady's words, single quotes, Maria Celeste's. The article refers to Urban "VI", but Urban VIII is intended. (Cf. also Allan-Olney, *The Private Life of Galileo*, and Morandini, *Suor Maria Celeste Galilei: Lettere al Padre*, La Rosa, Torino, 1983.)

invited him to *'send any collars you need starched'*. Concerned about his overwork and lack of sleep, she wrote: *'I would not want that, trying to immortalize your fame, you should shorten your life'*. Later, she sent him a preparation of dried figs, walnuts, rue and salt, mixed with honey, as a protection against the plague.[58] And indeed, Florence was racked by the plague in 1630. Here, a Board of Health was set up by the ruling Medici to treat plague victims, as Young tells us:[59]

"The pestilence raged for thirteen dismal months, during which time in and around the city twelve thousand people died. Ferdinand established a Board of Health, and this body issued many wise regulations, while they also forced the inmates of the immense number of monasteries and convents with which the city was crowded both to obey sanitary rules, and also to bear their share in receiving and helping those who were convalescent. But Ferdinand's sound sanitary regulations were denounced by the priests as impious; the Pope [Urban VIII] demanded that the Board of Health should be censured, and required that a severe penance should be exacted from its members; ... with the result that the Board of Health was made to do penance for having adopted measures which were in every way right and desirable."

Meanwhile, Grand Duke Ferdinand and his brothers Giovanni Carlo, Mattia and Francesco endangered themselves by going daily into the city and administering comfort to the stricken sufferers.[60] Thus did the secular state exceed Urban VIII and his Church in courage, compassion, and humanitarianism, all of which are considered Christian virtues and, in the case of these Medici family members, appeared to originate from something more deeply rooted in the heart of a man than the tenets of a particular religion.[61]

When the Inquisition arrested Galileo, Maria Celeste advised him not to ruin his health by worrying too much. After his condemnation, she wrote: *'You should be fully aware of the falsity and fickleness of all things in this rotten world, you should not pay much attention to these events.'* A month later, all the nuns in her convent *"had been delighted to hear that Galileo, to avoid the plague in Florence, was staying with his friend Archbishop Ascanio Piccolomini of Siena."* Later, she wrote: *'God knows how much I want your return, but I think it's a good idea that you stay some months more ...'*

---

58. to be taken in the morning with a little Greek wine. If the figs are baked too much, they become lumpy (she confesses this mistake). Recipe from Bonelli and Shea, *Galileo's Florentine Residences*, Istituto e Museo di Storia della Scienza (undated; author purchased his copy in 1984).

59. G. F. Young, *The Medici*, p. 668.

60. related by G. F. Young , ibid., based on the testimony of an eyewitness (Rondinelli).

61. Dante Alighieri had encapsulated certain Popes (who were for the most part accused of accepting money for indulgences) upside down in baptismal urns with their feet on fire. As Dante predated Urban VIII, there was no compartment of the *Inferno* to accommodate Urban's unique kind of behavior. Instead, history has remembered Urban VIII and his Medieval Church, not for the good and holy exercise of his office, but for his persecution of a man who brought science and philosophy out of the Dark Ages.

Maria Celeste undoubtedly sensed the importance of her father's work in the advancement of civilization, but seems to have underestimated her own great value as a provider of emotional support to him. She wrote: *'If I were sick or even taken from the world, it wouldn't matter as I'm not very good at anything, whereas in your case the opposite is true.'* She even offered to say for Galileo the weekly penitential psalms prescribed by the Inquisition. Four months later, Maria Celeste died, while Galileo lived on in house arrest another eight years. During the remainder of his life, Galileo heard his daughter constantly calling him.

As we read her words we, too, hear her voice in those letters, and from them we learn that Archbishop Piccolomini "told visitors the condemnation was a mistake and that Galileo was the most important man in the world, whose writings would live forever.

"I followed him. He showed me,
Looking along his outstretched hand, a star,
A point of light above our olive-trees.
It was the star called Jupiter. And then
He bade me look again, but through his glass.
I feared to look at first, lest I should see
Some wonder never meant for mortal eyes.
He too, had felt the same, not fear, but awe
As if his hand were laid upon the veil
Between this world and heaven.

Then ... I, too, saw,
Small as the smallest bead of mist that clings
To a spider's thread at dawn, the floating disk
Of what had been a star, a planet now,
And near it, with no disk that eyes could see,
Four needle-points of light, unseen before."

... Alfred Noyes

excerpted from *Watchers of the Sky*, pub. by Frederick A. Stokes Company, NY, 1922. Mr. Noyes assigns the narrator of this passage to the persona of Maria Celeste Galilei. He relates in his introduction that "this poem began to take definite shape during what was to me an unforgettable experience" — the first trial of the Hooker 2.5 meter telescope on Mount Wilson, which Mr. Noyes attended. I am grateful to Mr. Eklund for bringing Noyes' work to my attention.

"Come, fill the cup, and in the fire of Spring,
The Winter garment of repentance fling:
The Bird of Time has but a little way
To fly - and Lo! The bird is on the wing.

...

Then of thee in me who works behind
The veil of universe I cried to find
A lamp to guide me through the darkness; and
Something then said - "An understanding blind."

... Omar Khayyam, *Rubaiyat,* VII and XXXVII

Galileo, the Poet Milton, and Blindness

Anguished by the condemnation of a Church he loved and served, yet determined to lift the veil from his still suppressed discoveries in the precious time that remains, Galileo presses on with his mission to reveal those discoveries. In 1634, his request to visit a physician in Florence for treatment of his painful hernia condition is denied with the warning that he may be cast into the dungeons of the Inquisition should he make future requests of that nature. This charming response arrives on the very day when Galileo is told that his daughter, Sister Maria Celeste, is terminally ill and will probably not survive the night.[62] His period of enforced solitude ends briefly later in that year when his sister-in-law, Chiara Galilei, and her four children come to live with him. All of them die of the plague shortly after their arrival; his nephew Alberto joins him for a few months, then leaves to marry.[63] Alone again, Galileo plunges wholeheartedly into expanding the *Discourses*, occasionally receiving visitors, among them the English philosopher Thomas Hobbes.[64]

In 1638, now 74 years old, solitary, and blind, he is visited by John Milton, then in his youth. Galileo lost the sight of both eyes to inflammation in 1637, within a year after publishing the *Discourses*. "A few years later", says Professor MacLachlan, "Milton mentioned the visit in his *Areopagitica*, a defense of freedom of speech. He wrote eloquently about the plight of Italian scholars who felt tyrannized by the Inquisition". Of these Italian scholars, Milton said:

"they did nothing but bemoan the servile condition into which learning

---

62. See Drake, *Galileo at Work*, Ch. 19.
63. See Bonelli and Shea, *Galileo's Florentine Residences*, op. cit.
64. "... Thomas Hobbes of Malmsbury, during his travels abroad, 1634-37, spent some time in Florence, *circa* 1635-1636, and often met his brother philosopher for whom he conceived and ever retained the warmest admiration." Fahie, p. 390. Fahie cites Galileo to Micanzio, 1 Dec. 1635.

amongst them was brought; that this it was which had damped the glory of Italian wits; that nothing had been there written now these many years but flattery and flummery. There it was that I found and visited the famous Galileo, grown old, a prisoner to the Inquisition for thinking in astronomy otherwise than the Franciscan and Dominican licensers thought." [65]

Of his own blindness, Galileo says in a 1638 letter to Diodati:

"... this heaven, this earth, this universe, which by my remarkable observations and clear demonstrations I have enlarged a hundred, nay, a thousand fold beyond the limits universally accepted by the learned men of all previous ages, are now shrivelled up for me into such a narrow compass as is filled by my own bodily sensations." [66]

Incipient blindness does not prevent Galileo from making important additions to the *Discourses*, which he does by dictation. He is never able to see the final printed copy. In 1639, Urban VIII refuses a request to free him from house arrest.

Galileo is to outlive his eyes but not his magnificent innovativeness. While becoming imprisoned in the dark closet of a sightless body, he widens that optical window to the universe that he himself opened; he writes a treatise called *Astronomical Operations* full of astrometric techniques, like the measurement of sidereal time by the count of pendulum ticks between consecutive culminations of a star seen crossing the field of a telescope aligned to the meridian.[67] As we have seen,[68] he later (mid-1641) applies his lifelong pendulum researches to the design of a pendulum-driven clock, a design that he dictates to his son Vincenzio. It is a fitting close to his career, for the pendulum has been central to many of Galileo's discoveries, as eloquently noted by Signor Antinori:

The Pendulum Revisited

"The pendulum, as is already known, was the result of the first observations of our philosopher in Pisa; it was the spark which kindled his genius, the instrument by which he tested the conceptions of his mind, the torch which led him along the path of his discoveries. The pendulum, by proving the resistance of air, served to confirm him in his theory of gravitation; it likewise illustrated his theory of music by the intersection of waves of sound. The pendulum suspended to a fixed centre suggested to him the motion of the earth, with the moon, round the sun. And it is singular to reflect how the two

---

65. as quoted by James MacLachlan, *Children of Prometheus*, p. 115. Professor Drake notes this passage, having quoted the last sentence of it in his translation of the *Dialogo*, p. xxv.
66. trans. by J. Bronowski, *The Ascent of Man*, Little, Brown, & Co., Boston, 1973.
67. See *Galileo at Work*, pp. 384-385. As told to me many years ago by Dr. Woolard, the United States Naval Observatory (USNO) used a related technique as the national standard of timekeeping prior to the use of atomic clocks. In the USNO timekeeping, a quartz crystal oscillator substituted for Galileo's pendulum.
68. Ch. 5.

marvellous discoveries with which he so happily commenced his glorious career, the isochronism of the pendulum and gravitation, should have occupied him at its close." [69]

Galileo, Torricelli, and the Barometer

With advanced students, Galileo does not teach by giving rote solutions to standard problems, but by suggesting real problems for them to solve. In at least one case, that of Torricelli, this method had spectacular consequences. Late in life Galileo retains some Aristotelian concepts, one of which is that nature abhors a vacuum. In 1630 he ponders Giovanni Baliani's observation that water cannot be siphoned beyond a certain height and conjectures that air has weight, a hypothesis he then teaches to his last students.[70] Being now ill and totally blind, and wishing to develop the abilities of his student, Evangelista Torricelli, Galileo assigns him the problem. While working on the problem, Torricelli invents the barometer, measures the variability of air pressure and discovers the inverse proportionality of air pressure to altitude. We will resume that story in the next chapter.

The Race With Death: Galileo's Analysis of the Continuum

Galileo's method of dealing with tragedy of all sorts was simply to outrun it. In his last year we find him confined to bed with terminal illness, blind but intent on an expansion of the *Discourses* into number theory, strengthening the bridge between mathematics and physics that Archimedes had begun and that Galileo had continued. In this, he has given us perhaps the greatest lesson of a well lived life. From our perspective four centuries beyond him, we still feel the tragedy of his oppression by the Church; yet however severe his setbacks may have seemed at close range, his achievements survive and tower above them.

Galileo, Mathematician to the End

Those misguided commentators who are fond of suggesting that Galileo is a physicist and not a mathematician, and those who embrace the all-too-common belief that scientists can only do good work at a young age, might better pay attention to what the permanently bedridden Galileo accomplished in his last days. However much distracted by pain, for the alleviation of which he was not even permitted to drink wine, Galileo directed his concentration to formulating a rigorous analysis of the continuum, and in particular, to the question: "how a rigorous theory of irrational magnitudes can be built on the natural numbers by

---

69. Signor Vincenzo Antinori, Director of the Scientific Museum of Florence, as quoted in G. F. Young, *The Medici*, op. cit.

70. " 'Nature's horror of a vacuum' ... was questioned by Galileo, who taught that air has weight. Torricelli proved (this) by his famous experiment ..."; Thomas R. Baker, PhD., *Elements of Natural Philosophy*, Porter & Coates, Philadelphia, 1881.

means of equimultiples, a term left undefined in Euclid's presentation." For his discussion of uniform motion in the *Discourses*, Galileo had made use of the fifth definition of Book Five of Euclid, generally attributed to Eudoxos. I now ask the reader: is the meaning of the following definition immediately obvious to you?

> "Magnitudes are said to be in the same ratio, the first to the second and the third to the fourth, when, if any equimultiples whatever be taken of the first and third, and any equimultiples whatever of the second and fourth, the former equimultiples alike exceed, are alike equal to, or alike fall short of, the latter equimultiples respectively taken in the corresponding order." [71]

If you think that this definition is not self-evident, you are in good company. Galileo didn't think so either; indeed, he pointed out that it is difficult to comprehend. Dictating an extension of his *Discourses* to Torricelli, who remained by his bedside in the final days, Galileo takes it as a theorem to be proved, rather than as a definition. He begins by proposing a definition of a more fundamental concept: proportionality among four magnitudes. In the dialogue, he develops the theory of proportions so as to replace Euclid's original definition.[72] This restated definition underlies the Dedekind cut, the rigorous basis on which real numbers are defined in modern mathematics. Galileo's physics, and indeed the bulk of modern physics, will require a well-defined mathematical continuum. Here, Galileo has supplied it. Surely there can be no more fitting close to the lifelong career of the Chief *Philosopher and Mathematician* to the Grand Duke of Toscana.

Beginning of the Failed Campaign To Stifle Galileo's Legacy

As if it were of value to science, Galileo's Sentence and Abjuration were read to assembled professors of mathematics and astronomy in the Papal States and everywhere in Europe where the long arm of the Vatican reached. Faculties of major universities were instructed to suppress evidence for the Copernican system from their teachings, and the Inquisitors were ordered not to permit the publication of a new edition of *any of Galileo's works*.[73] Though the Inquisition stifled an immediate school of Galilean science, and shoved science with all its sinew out of Florence and the Papal States, Galileo had two students at his bedside in the final days. Torricelli, besides being first to create a sustained vacuum and inventing the barometer, wrote a treatise on mechanics including his combined findings on fluid and projectile motion. Here we find at last answered Leonardo da Vinci's question for which he had compiled data but lacked a general solution: "Torricelli's

---

71. Euclid, *The Elements*, trans. by Sir Thomas Heath, Dover, NY, 1956, vol. II p. 114.
72. See Galileo's dialogue on Euclid in Stillman Drake, *Galileo at Work*, p. 422 ff.
73. See A. D. White, op. cit. Even *Sidereus Nuncius*, which the Jesuits had supported, was banned! Evidently, the Collegio Romano had approved a dangerous and heretical book.

Theorem" $v = \sqrt{2gh}$, where v is the speed of liquid exiting an orifice h units of vertical distance below the water's surface. Viviani, 20 years of age when Galileo died, went on to write a scientific biography of Galileo, and to publish an Italian textbook of Euclidean geometry that included Galileo's treatise on Euclid's definitions of ratios.[74]

	Prince Leopold de' Medici appointed Viviani to the

The Academy of Experiment

Prince Leopold de' Medici appointed Viviani to the *Accademia del Cimento*. Young's account of it so vividly illustrates the failure of the campaign against Galileo's science in his own home country that it deserves to be quoted at length:[75]

"In the year 1657, when Ferdinand was forty-seven, there was formed under his patronage by his talented brother Prince Leopold the celebrated *Accademia del Cimento* (Academy of Experiment), *the first society for experiments in natural science ever formed in Europe*, and one which became the model for all those subsequently established in England, France, and other countries; and this new Academy held its first meeting on the 16th June 1657 in the Grand Ducal palace, presided over by its founder, Prince Leopold de' Medici, then forty years old. Truly the Pitti Palace, honored as it is by all artists for its magnificent picture-gallery, should be no less honored by all scientists as the building in which originated this notable event in the world of Science. The Royal Society of England was not incorporated until 1663, and the French Academy of Science not until 1666; so that Florence in this matter also, as in former days it had done in Learning, and as it had done in Art, led the way. And prominent as had been the leadership of the Medici as to Learning, and as to Art, in neither of these was it so directly marked as in this case of Science. Prince Leopold, both as an earnest pupil of Galileo, and on account of his own proficiency in science, was chosen by the new society as the proper man to lead it as its President.

And very ably he did so. At its first meeting the society ruled that its fundamental law should be that no special school of philosophy or system of science should be adopted by it, and that it bound itself 'to investigate nature by the pure light of experimental facts'; also that the society should be open to all talent, and that the privilege of selecting the experiments to be made should lie with the President. It adopted as its motto, *Provando e Riprovando*.[76] Magalotti was chosen as its secretary; and on the walls of the entrance hall of the present National Library (in the Uffizi building) are to be seen the portraits of the distinguished men who were the first members of this famous society.

Thus took place the first case on record of the formation of a society purely for the pursuit of inductive science, and for the furtherance of that new philosophy which Galileo had inaugurated and of which Bacon was to be the chief exponent. Ferdinand took the greatest interest in the work of the new society, and devised several of the experiments, among others the suggestion of the use of the expansion of liquids for thermometric purposes, instead of the air of Galileo's thermoscope. The results of the experiments ...

---

74. Additionally, Galileo had an earlier student Cavalieri, who had developed the pre-calculus theory of indivisibles, used to find areas and volumes of curved spaces.

75. *The Medici*, p. 695-696.

76. "Testing and retesting."

were published in Florence in 1667 ..."

Research by the Academy went beyond that of Galileo, its inspirational founder. After Marin Mersenne and Pierre Gassendi made the earliest measurements of the speed of sound, for example, the Academy confirmed the constancy of that speed and its independence of pitch. In at least two notable areas, the Academy also directly extended Galileo's own work. It members used a working model of a vibration counter, made to Galileo's 1637 design by Archduke Ferdinand II's clockmaker J. P. Treffler, for recording variations in moisture collecting on a cooling glass under varied conditions as noted in its 1662 minutes. And around 1650, the first sealed liquid-in-glass thermometer, a prototype of modern thermometers, was developed there. At that time, these were pioneering studies in transition of state between two of the three recognized states of matter: liquid and gas.[77] (Nowadays we recognize a fourth state: plasma.) Regrettably, the Academy broke up after ten years when Leopold was made a Cardinal in place of his deceased brother Giovanni Carlo; its Secretary, Magalotti, then joined the newly formed Royal Society of England.[78] This breakup introduced a disconnection between Renaissance and modern studies, not only in the physics of state transitions, but in all the diverse disciplines to which the Academy had applied groundbreaking research extending the work of Galileo.

Galileo's Papers: Wrappings for a Pork Butcher

Viviani inherited Galileo's papers and through his life added to this collection documents pertaining to Galileo's life and work. Unfortunately, he never published the whole collection. It was still dangerous to publish anything of Galilean science in a country controlled by the Vatican and policed by the Inquisition. And so we have as one of the worst tragedies of this disease called the Inquisition that many of Galileo's papers were forever lost. Between 1703 and 1750, some were lost to students and others were sold:

"During this period, students frequently borrowed from the collection without (so it is said) bothering to return what they had borrowed. In 1744, a somewhat fuller edition appeared (four volumes; Padua), which for the first time included the *Dialogues* and also many letters. A few years later (1750), the Viviani heirs were discovered to have been selling the Galileo papers to a pork-butcher for wrapping paper. (A scholar happened to notice that his sausage was wrapped in a letter signed by Galileo!) It would appear that a considerable part of the papers perished in this and other ways ..."[79]

Despite these losses, Favaro, aided by the Viviani archives, compiled over 4,000 letters from, to, or about Galileo in his National Edition of Galileo's works. What a

---

77. (rdc)
78. Ibid. p. 696, and footnote 70, p. 817.

tribute to Galileo's influence on others of his day! That historians, scientific scholars, and wealthy amateurs did not negotiate with the heirs for the preservation of these papers reflects the effectiveness with which the Inquisition drove the scientific Renaissance out of cisalpine territory for a century after his death.

End of the Failed Campaign To Stifle Galileo's Legacy

Across the Alps, the vehicle bearing the Copernican evidence, the *Dialogo*, had spread across the continent, and to England. There the still-unproved Copernican system had been introduced for purposes of calculation and prediction by John Field [80] in his ephemeris of 1557. Galileo's heroic arguments for it, published in London in Salusbury's 1665 English translation of the *Dialogo* and the *Two New Sciences* "which contains practically all that Galileo has to say on the subject of physics" [81] awaited the genius of Isaac Newton. Still smoldering in the ashes of the London fire of 1666 were English renderings of Galileo's great works, which Newton had read in derivative form as a student at Trinity College, Cambridge. Having learned Galilean science from these derivative works (by Anderson, Charleton, Digby and Gregory),[82] Newton would then connect Galileo's dynamics to Kepler's ellipses in a heliocentric planetary scheme.

Through all its attempts to strangle Galileo's intellectual legacy, the Church succeeded only in calling attention to its blunder — when in 1687, Newton publicly credited Galileo's contributions to his System of the World, Galileo's remains still lay hidden in an unmarked tomb and his *Dialogo* was still under the ban. Evidently, the same philosophers and clerics who noticed Galileo's "heresy" in the *Dialogo* missed this admonition to Simplicio:

> *"So you see that the power of truth is such that when you try to attack it, your very assaults reinforce and validate it."* [83]

---

79. This tragic story is related by Michael J. Crowe in a short note buried within his annotated list of Galileo's works appended to McMullin's *Galileo, Man of Science*, op. cit., pp. lxxxv, lxxxvi. Crowe's account does not say whether the papers sold by the Viviani heirs included papers on matters of intellectual interest; however, it is likely that most of Galileo's papers were of such a nature.
80. See *Field Genealogy*, Hammond Press, W. B. Conkey Company, Chicago, 1901.
81. In the preface to their translation of *Two New Sciences*, Crew and de Salvio utter these significant (if only approximate) words and add that Salusbury's edition (the one that Drake tells us was tragically destroyed only a year later in the great London fire) was "worthily printed in two handsome quarto volumes".
82. For having informed me about these sources of Newton on Galilean physics, I now thank Professor Cohen, curiously also the creator of the "Enormous Gulf" myth (see Epilogue).
83. from the *Dialogo*, Second Day; see Ch. 9 "The Resilience of Truth".

Reflections on Galileo's Life, His Death and Rebirth of His Work

"... We hail
Thy sunny slope, Arcetri, sung of old
For its green vine; dearer to me, to most,
As dwelt on by the great astronomer;
... Sacred be
His villa (justly was it called the Gem),
Sacred the lawn, where many a cypress threw
Its length of shadow, while he watched the stars." [84]

... Giacomo Zanella

In 1984 there circulated a 2000 lire note depicting Galileo on its face and the Arcetri Astrophysical Observatory on the reverse side. To the right of Galileo's portrait, a blank space floats like a little white cloud. If you raise it to light, letting the background light shine through the paper, there appears an image of Torricelli, whose achievement shines in the aftermath of Galileo's work. On the streets of Rome, in the late afternoon shadow of Vatican offices where Galileo's *Dialogo* had been condemned, the numismatic shops were selling a silver coin commemorating that same *Dialogo*. The philatelic shops offered stamps depicting Galileo teaching, lecturing, and using the telescope, as well as other stamps honoring scientists whom he influenced: Marconi, Galvani and Volta. For all of this superficial public attention, one may well wonder why the teaching of science and mathematics pays too little heed to Galileo or to his methods. The contrast between the memory of the man and the neglect of his significance for the future of discovery is a contradiction that will require the rest of this book to address.

A Souvenir: Galileo's Unwritten Signature

A street vendor in Florence from whom I purchased a souvenir seemed perplexed that I was so intrigued by it. In the center of the photograph mounted on it, the *Ponte Vecchio* spans the Arno River adjacent to the museum where Galileo's telescopes reside. There is an elliptical border that intimates a generalization of his laws of fall — a Keplerian consequence of Newtonian gravitation interacting with Galilean inertia. And there is an evolute of his invention: a real thermometer. Though not labelled with his name, it suggests that the memory of Galileo is stamped in some invisible but indelible way on the consciousness of the artisan who created it.

---

84. quoted in Young's *Medici* and attributed there to Rogers' *Italy*. Fahie identifies the poet as Giacomo Zanella ("Versi", Florence, 1868).

◆◆◆◆◆

"... we Italians are making ourselves look like ignoramuses and are a laughingstock for foreigners, especially for those who have broken with our religion; ..." [85]

Stranger in His Home Town?

Twentieth century visitors to the Galilei Airport in Pisa can see a multicolored globe circumscribed by the words *"Nel cielo, nel mondo, nel universo ..."* with which a celebrated comment from Galileo to Diodati begins.[86] Visitors should exercise caution; the local police have been known to search the pockets and camera bags of a visitor photographing the monument. Do they not realize that this is a monument to Galileo? Or, is photographing a public monument to Galileo a strange and suspicious thing to do in Galileo's own home town? Other signs exist that Galileo and Fibonacci are like strangers where they lived. Apparently educated natives of Pisa within one kilometer of the monument to Leonardo (Fibonacci) were unable to direct me to it. In fact, they were unaware of its existence; I found it on foot in a random walk. At Florence, a taxi driver and his dispatcher had not heard of *Il Gioiello*, Galileo's villa at Arcetri. In 1993, I found his tomb in the great Cathedral of Santa Croce enshrouded in scaffolding, cordoned off, and unapproachable — like his reputation, still eclipsed in relative darkness, and under repair.

The artifacts of the Academy, like those of Galileo, are archived by the Institute that inherited the surviving instrument collection of this society and manages the museum. Applications of condensation studies and other works by the Accademia del Cimento are either esoteric or have been abandoned. One easily wonders if Galileo's modern day countrymen have frozen the dynamic activity that he and his students fostered into statues of brass and glass, imprisoned in a museum, in order to make them more perfect than they were.[87]

Notwithstanding some contrary indications, sustained efforts are being made to preserve Galileo's memory and his artifacts, if not also the direction of his work.

---

85. spoken by Salviati in Galileo's *Dialogo*, Third Day, Drake's translation, p. 279.

86. See above: "Galileo, the Poet Milton, and Blindness", the quotation beginning " ... this heaven, this earth, this universe, which by my remarkable observations and clear demonstrations I have enlarged ..."

87. Recalling a remark by Galileo: *"It is scarcity and plenty that make the vulgar take things to be precious or worthless; they call a diamond very beautiful because it is like pure water, and then would not exchange one for ten barrels of water. Those who so greatly exalt incorruptibility, inalterability, etc., are reduced to talking this way, I believe, by their great desire to go on living, and by the terror they have of death. They do not reflect that if men were immortal, they themselves would never have come into the world. Such men really deserve to encounter a Medusa's head which would transmute them into statues of jasper or of diamond, and thus make them more perfect than they are"*; trans. by Drake in the *Dialogo*, op. cit., p. 59.

"Non si puo compartir quanto sia lungo,
Si smisuratamente e tutto grosso"

... Ludovico Ariosto [88]

A Size Beyond Measure

Pausing here to reflect on the scope and compass of Galileo's achievements, we see laws of nature, a methodology of discovery, a slew of inventions, and original contributions to fields as diverse as mathematics and the development of literature. Among them was the geometric compass, a general-purpose calculator comparable in effect to that of the pocket computer today. Among them too are the optical devices for which he is well known, and with which he at once shrunk the vast heavens and opened a window into that miniature microcosm where the building blocks of our physical world had lain locked from sight. Look around you at everything that does not come from nature, and, upon analysis, it will be hard not to see the stamp that this individual put upon it. In versatility, in the sheer pervasiveness of his effect across the whole spectrum of human endeavor, he was, as Viviani put it, comparable to no one in his time.

Those achievements are all the more remarkable in view of certain natural and man-made burdens. There was blindness, ill health, and the death of his devoted daughter; there was suppression of his work, cruel prosecution, the ignominy of house arrest, and the disfavor of a Church he loved and served faithfully. Despite these hardships, he was well supported by the Medici, and under their aegis Galileo created his magnificent legacy, the *Discourses*. Though he missed the opportunity to unite fully the mechanics of celestial and terrestrial events, his work combined with that of Kepler made it possible for Newton to do just that.

The Death of Galileo and the Rebirth of His Work

On 8 January 1642, Galileo died. He reached 78 years of age, but died too young, still productive. Pope Urban VIII, unfortunately still alive, forbade Grand Duke Ferdinand to erect a monument in Galileo's honor lest any word on it "offend the reputation of the Holy Office". His remains[89] were interred in an unmarked tomb in an obscure part of the Cathedral of Santa Croce, a small closet by the Capella del Noviziato, where they remained for nearly a century. Ultimately, the injustice could not endure. Vincenzo Viviani, Galileo's

---

88. "One cannot reckon his height; yes, his size is beyond measure" from *Orlando Furioso*, XVII, 30 (by Ariosto, probably Galileo's favorite poet) quoted by Galileo in the *Discourses*, Second Day, trans. by Crew and de Salvio. What Galileo has quoted in another context applies equally well to him in the sense of intellectual stature.

89. except two fingerbones; one is on display in the museum. Where's the other one?

student, willed property for the purpose of erecting a proper monument to Galileo in the Cathedral at Santa Croce whenever permission could be obtained.

In 1737 Giovan de Nelli, trustee of Viviani's estate, used this bequest to commission the mausoleum at the front of the Cathedral; the then-current Pope (Clement XII) sanctioned the relocation of Galileo's remains to this magnificent tomb, designed by Giovanni Battista del Foggini, assisted by his son Vincenzio and Girolamo Ticciati. On the 12th of March that year, the transfer of Galileo's remains occurred with approval and in the presence of Church authorities.[90] By this time, the physics of Kepler/Galilei/Newton had so successfully validated the Aristarchan/Copernican solar system, that it had altered the *Weltanschauung* of people in general. The Latin inscription[91] on the tomb begins with Viviani's fitting epitaph, which I translate as follows:

"GALILEO GALILEI FLORENTINE NOBLEMAN
Most Eminent Restorer of Geometry, Astronomy and Philosophy
Comparable to no one in his time
May he repose well here"

Santa Croce is situated in the Flood District which was inundated by the 1966 overflow of the Arno River. As if to remind us that Galileo had begun his mathematical career studying Archimedes and the principles of floating and submerged bodies, the water rose up above the tomb, and the surrounding frescoes had to be restored.[92] The raging water current subjected everything in its path — the Cathedral, museums, libraries, irreplaceable historical and artistic monuments — to the laws of hydrodynamics. An intellectual current from which these laws flowed, the hydrostatics of Archimedes with which Galileo began his career in physics, culminated in the magnificent law of parabolic fall.

I was reminded of that law when first I visited St. Louis, Missouri and gazed upon the elegant parabolic arch designed by the architect Eero Saarinen. Then there came to my mind the haunting words of the poet Tennyson, words I had committed to memory on a windy street corner while commuting to a physics lecture, the

---

90. See Bonelli and Shea, *Galileo's Florentine Residences*, op. cit.
91. "GALILAEUS GALILEIUS PATRIC. FLOR. / Geometriae, Astronomiae, Philosophiae Maximum Restitutor / Nulli aetatis suis comparandus / Hic Bene Quiescat"
92. Santa Croce monks, assisted by restoration experts, erased the effect of oily water that filled the cathedral to a depth of 5 meters. They gave the statues a coat of talcum to draw out moisture and oil, and treated the stains with powerful detergents; Champ Clark, *Flood*, Time-Life Books, 1982.

content of which was influenced by Galileo. They seem to capture a semblance of his spirit upon his return to Arcetri after his intellectual energy had been "stored and hoarded" in the Inquisition trial and its aftermath:

> *"... Yet all experience is an arch wherethro'*
> *Gleams that untravell'd world whose margin fades*
> *For ever and for ever, when I move.*
> *How dull it is to pause, to make an end,*
> *To rust unburnish'd, not to shine in use!*
> *As tho' to breathe were life. Life piled on life*
> *Were all too little, and of one to me*
> *Little remains; but every hour is saved*
> *From that eternal silence, something more,*
> *A bringer of new things; and vile it were*
> *For some three suns to store and hoard myself,*
> *And this gray spirit yearning in desire*
> *To follow knowledge like a sinking star*
> *Beyond the utmost bound of human thought. ..."*

... from *Ulysses*[93] by Alfred Tennyson [94]

What the poet has said of all experience applies conspicuously to that of the great Galileo in his heroic harnessing and redirection of forces that would have had him rust unburnished. The margin of the universe, opened up by Galileo's escape from the dark labyrinth of the Middle Ages, continually advances as our tools for studying the universe advance. Galileo's work continues, today and as far into the indefinite future as the civilization of man may survive. For there is no progress that does not build on his work. And with no progress against the annihilating forces of nature and of man, there is ultimately nothing.

---

93. Re the use of italics for the title of this poem: Galileo's example of the incomplete dichotomy (see Ch. 8, "The Language of Natural Philosophy") applies to meaningless advice like the following that originates from a committee: "Quotation marks are used ... to enclose titles of ... short poems ..."; *A Manual of Style*, Gramercy Publishing Co., NY 1986. How short is a short poem?

94. *The Complete Works of Alfred Tennyson*, Richard Worthington, NY, undated.

View of Light-Beam Path in Michelson's Measurement:
Mount San Antonio from Michelson Plaque on Mount Wilson
(photo by author)

Marker at Mt. Wilson Site of Michelson's Light-Speed Measurement
(photo by author)

Duemila (2000 Lire Currency)
Depicting Galileo and Hidden Torricelli
(photo by author)

<u>Notes</u>
1. Portrait of Galileo appears to derive from the 1638 portrait by Sustermans.
2. This currency is marked with the name L. Lazzarini inv. — apparently an abbreviation for "inventore" (designer) — in letters too small to read without a magnifying glass. The signatures of Banca d'Italia officials are, however, quite prominent.
3. Note Leaning Tower and Cathedral of Pisa on front.
4. Note Arcetri Astrophysical Observatory on obverse side.
5. Portrait of Torricelli appears in blank space when held up to light.
6. Mint dates: 10 Sept. 1973, 9 Jan. 1976, 22 Oct. 1976.

Visitor Laying White Flower,
Symbolic of Hope, at Mausoleum
of Galileo

(photo by F. Obena Di Canzio)

Monument to
Galileo's "*Nel
Cielo, Nel Mondo,
Nel Universo ...*" at
the Galileo Galilei
Airport, Pisa

(photo by author)

Chapter 12. Edge of a Conic Slice
The Significance of Galileo in the History of Ideas, Part I

Parabola, edge of a conic slice
Traced out by the wind blown fall of an acorn,
Borders between capture and eternal nights,
Orbital path of those not yet reborn.

Who draws depth from shadows pries open the crypt,
Anesthetized to pain by the flight from the dark;
Children of the living are alive and equipped
To remodel the cosmos alone in the park.

Molding form to function they climb, pausing at the top;
Flowing they descend in rhyme
Then stop,
Freezing an instant of time.[1]

In an old childhood memory, some children intent on finishing their models before sunset were hastily digging for clay while steep cones of sunlight streamed down past the pinhole spaces formed by dense clusters of tree leaves. Like those children, sculpting physical clay models in wide-eyed scrutiny of sense impressions and unburdened by assumptions in which adults are prone to dress their uncertainty, Galileo and Kepler were reshaping the abstract clay of Aristotelian logic and natural philosophy, each man making a scientific mosaic of new and interrelated truths, each mosaic adaptable to an encompassing mathematical model of the cosmos. In the work of Galileo and Kepler, the operational rebirths of Eudoxos, Apollonios, Aristotle and Archimedes and of Thales, their common parent, awaited their culmination in that comprehensive mosaic that would be assembled by the not-yet-born Newton.

Parabola, the Unique Conic Shape

Of four distinct conic sections — the circle, the ellipse, the parabola, and the hyperbola — two have unique shapes: the circle and the parabola. An ellipse can be more or less eccentric or "squashed" in appearance depending on which of infinitely many possible angles the cutting plane makes with the generating cone; as the eccentricity tends to zero, it approximates a circle. Similarly, there is an infinity of hyperbolas corresponding to the angle of incidence of the cutting plane. Defined as the locus of points equidistant from a fixed point and a fixed line,[2] the parabola is unique because of a restriction: the cutting plane which produces it is

---

1. ... from *"Parabola"* © *1993* by Albert Di Canzio.

parallel to an element of its generating cone.

The parabola is the first conic form beyond the closed path of the ellipse; it lies on the boundary between the ellipse and hyperbola, and is open. Like the hyperbola it branches away from its focus to infinity; unlike the hyperbola, it is a unique individual form.

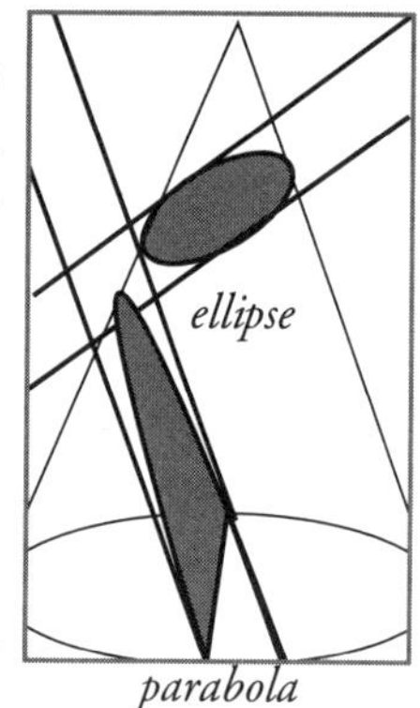

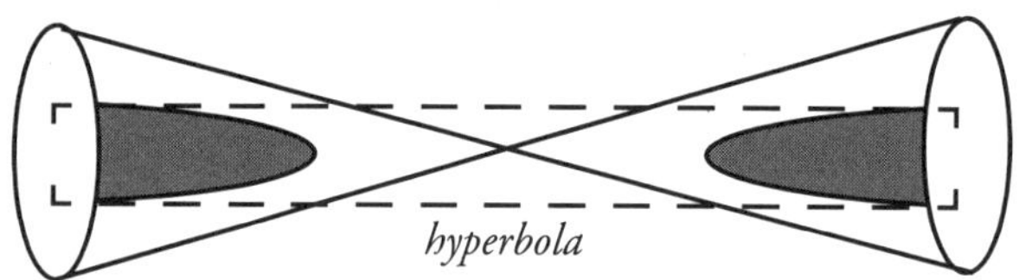

The Path of a Scientific Modeler

A dynamic course traversed by someone like Kepler or Galileo, who models processes in nature with mathematical clay, is analogous to a parabola. Each is a unique individual whose work propagates through the minds of participants now passed away and yet unborn. He produces in his model an image of the natural world conditioned by available knowledge. The model is an encapsulation not only of a process, but also of the state of scientific understanding of that time. Having made his closest approach to the focus of his work, he "freezes" an instant of time and instills into his model a set of assumptions about the natural world peculiar to that instant. Beyond his physical life, the vital presence of his work may recede from its focus, like a comet in a parabolic path, but at the instant of the rebirth of that work in the mind of a capable successor, it returns as if captured into an elliptical orbit by the interposition of a third gravitational source.

Legacies of Kepler and Galileo

Time and time again discarding shapes that did not fit the data, laboriously exhausting the 5! = 120 permutations of the five regular solids, Johannes Kepler reluctantly relinquished his quest to model the solar system with those Pythagorean solids. Having resumed with fresh insight, he came at last upon the secret of planetary orbits, recognizing in Tycho's data on the motion of Mars the form of an ellipse, a conic slice. Galileo and Kepler left behind significant unconnected thoughts. There they remained like rhyming lines of a verse not yet formed in the mind of a poet who would link them in an act of semantic synthesis. Before we see Newton's

---

2. This is a modern form of definition. Such definitions as Menaechmos and Euclid may have given have tragically been destroyed, but Archimedes described it in his greeting to Dositheos preceding the *Quadrature of the Parabola* as "a section of a right-angled cone"; cf. Sir Thomas L. Heath's translation reprinted in *Great Books of the Western World*, Brittanica, 1952, vol. 11, p. 527. In more modern terms of analytic geometry, if a locus of points satisfies the general equation of the second degree in which the coefficient of the mixed term and that of one of the second-order terms are each zero, then it is a parabola.

assembly of the puzzle, let's look at a seemingly unrelated interlocking piece that he picked up, one that Galileo left behind in the Fourth Day of his final masterpiece, the *Discourses.*

After considerable groping, Galileo had recognized that an object thrown horizontally from the Leaning Tower would have a complex motion described by what we now call the **Galileo's Law of Parabolic Fall** vector sum of an accelerated downward component and a uniform horizontal component. Lacking air resistance, the latter would hold constant in accordance with his inertial principle. As the addition vector of these components traces out the parabolic curve in which we observe that locally projected objects fall, the composition of his two models for downward and horizontal motion is known as *Galileo's Law of Parabolic Fall.* The verse introducing this chapter is a poetic application of his discovery in which it is assumed that an instantaneous gust of wind detaches an acorn with only an initial horizontal force. Rooted in the geometry of Thales, this dynamical concept fused inertial and gravitational conditions into a single and smooth curvilinear motion. The fusion created one of Newton's smoother pebbles:

> *"I do not know what I may appear to the world;*
> *But to myself I seem to have been only like a boy playing on the seashore,*
> *and diverting myself in now and then finding*
> *a smoother pebble or a prettier shell than ordinary,*
> *whilst the great ocean of truth lay all undiscovered before me."*

... Attributed to Isaac Newton [3]

The Law of Parabolic Fall marks a milestone in kinematics, the study of unperturbed motion, and in dynamics, the study of how objects move under the **Galileo's Gateway to Gravitation** action of forces, by treating them simultaneously. Galileo did not have a motion picture camera that would record successive positions of an object in free fall. He could not practically derive his law by fitting the data to geometric shapes, as Kepler succeeded in doing with real data, collected at Tycho's Uraniborg observatory, on successive positions of Mars. It is a tribute to Galileo's theoretical genius that, roadblocked on the path of empirical generalization, he arrived deductively at this composite law describing instantaneous behavior of a projected object. If you would like to guess how he did it, try to visualize a curve fitting the superimposition of its free forward motion (as it obeys his inertial principle of kinematics) onto its forced motion (as it obeys his odd-number rule of acceleration), and then turn the page.

------

3. Brewster's *Memoirs of Newton*, vol. II, chap. XXVII, per *Bartlett's Familiar Quotations, 1951.*

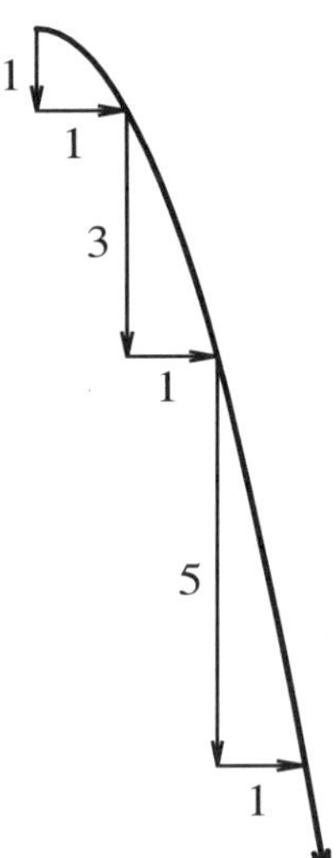

The same Galileo who would show the world how to effect structural weight reduction by means of parabolic sectioning of load-bearing beams built a parabolic arch leading to the Newtonian synthesis.[4]

Newton's Extension of Galilean and Keplerian Laws

Crossing the graveyard of the old cathedral[5] adjacent to the King's school at Grantham, young Isaac must have seen at the fringes of shadows of tombstones the final resting place of light rays hurled by the sun across the vast void of the solar system. Thinking that light consists of little particles that he would later call "corpuscles", he must have wondered about the path of anything that is hurled with so great a speed. Isaac Newton (1642-1727)[6] had been born the year Galileo died, and took up parts of his and Kepler's works. By performing a thought experiment, seeing in his mind's eye a projectile being hurled repeatedly and with increasing initial thrust, Newton eventually seized on the relationship of Galileo's parabolas to Kepler's ellipses, and described the behavior of planets and projectiles in a single system of equations.

As the projectile path in Newton's three-dimensional space continuum approaches its limiting case, an infinite straight line tangent to Earth's surface at the launch point,[7] it moves through sections of a larger and larger ellipse until the ellipse circumscribes the entire Earth, as in the case of a satellite orbit, and then, escaping Earth altogether, on through a parabola to an open hyperbolic branch.

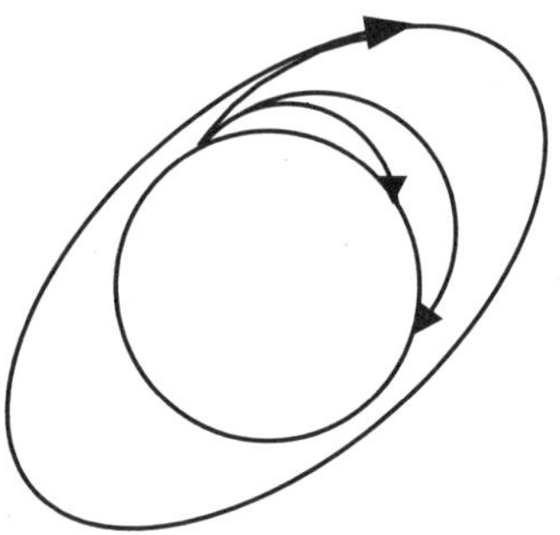

---

4. The ambitious reader might wish to relate Galileo's demonstration of parabolic sectioning to the moment diagrams used in beam and girder analysis, which show a parabolic shape under uniform load; cf. Linton E. Grinter, *Theory of Modern Steel Structures*, MacMillan Co., NY, 1942.

5. the 12th century Church of St. Wulfram (Archbishop of Sens, d. 720 A.D.)

6. These are Newton's biographical dates by the old calendar to which England clung during his lifetime. If England had, with Italy, gone to the Gregorian calendar in that time, then Newton's birthday (December 25 by the old calendar) would have been adjusted to 4 January 1643. By the current internationally accepted Gregorian calendar, Newton was born in 1643, the year after Galileo died.

7. I know of no reason to suppose that a light beam, moving at finite speed tangentially away from a gravitational source, would not move in a hyperbolic path that approximates rectilinearity.

Visualizing gravitational centripetal force and inertial
centrifugal force interacting to produce conic-
sectional motion, Newton's thought experiment

Huygens' Centrifugal Principle

invokes Galileo's principle of the composition of motions. At the same time, it
depends on a relationship discovered by Huygens. In 1659, a time between Galileo's
Two New Sciences and Newton's *Principia*, Christian Huygens extended Galilean
kinematics to describe the centrifugal force of an object moving in a curved path as
proportional directly to the square of velocity and inversely to the radius of
curvature. That extension is one piece of the puzzle that Newton would assemble,
among several pieces whose relationship we explore in Appendix C. Thus it was
that the same Huygens who explained Galileo's tripartite planet as Saturn with
rings, who disproved Galileo's conjecture that the catenary — the curve of a chain
hanging freely from its endpoints — is a parabola, and who turned into a working
reality the pendulum clock design that neither Galileo nor his son Vincenzo had
lived long enough to complete,[8] had brought Galilean kinematics to the threshold of
Newtonian dynamics.

Adding his own principle of centripetal force[9] to
that centrifugal principle due to Huygens,
Newton related the projectile path both to

Universal Gravitation and Conic Sectional Motion

Galileo's law of falling bodies and to Kepler's First Law, which might better be
called Kepler's Law of Elliptical Accelerated Motion. Kepler's laws are
traditionally called laws of planetary motion, but this name conceals their
wonderful and beautiful generality. They appear in the dance of distant double
stars, in the swirling internal motions of giant galaxies, and even in the
submicroscopic world where electrons orbit their nuclei elliptically.[10]

Newton demonstrated Galileo's and Kepler's laws as consequences of a still more
general law of gravitation. Galileo's law of parabolic fall is an approximation to
Newton's universal law over a region local enough that the gravitational force on
the projectile can be assumed equal in magnitude and direction. In Newton's
mechanics, a falling object describes a path fitting a conic section determined by
the masses of the attracting bodies and the initial conditions of projection. On the
question of the path of a projectile launched on a planetary surface, Newton's
answer is this: when a projectile encounters a cutoff of its path, its trajectory has
actually traced out part of an ellipse with the center of the planet[11] at one focus.

---

8. Huygens also showed that the pendulum whose bob moves in a cycloidal path is isochronous for arcs of all
    sizes. He tried to apply pendula to determine longitude, as had Galileo before him.
9. described in his *System of the World*, [3.], p. 551 of the Motte trans. revised by Cajori.
10. "In accordance with Kepler's First Law, the electron orbit is, in general, an ellipse"; John L. Powell and
    Bernd Crasemann, *Quantum Mechanics*, Addison-Wesley, Reading, Massachusetts, 1961.

And so it turns out that the knowledge of what really happens to a stone projected from the catapult of Archimedes has grown out of the revival of Archimedes' own mathematical physics.

Newton described gravitation algebraically as a field permeating a geometric space isomorphic to the familiar three-dimensional space of the natural world. From this description, he was able to generalize Kepler's planetary motion laws and Galileo's inertial principle in a way that explains simultaneously the motions of celestial bodies and of near-Earth objects acting under gravitational field forces.

Part of that generalization was his extension of celestial dynamics to three dimensions. Kepler found the shape of a planetary orbit within its orbital plane, but his model does not address the orientation of that plane in a space of three dimensions. Newton's mechanics supports calculation of the orbital orientation with equations that satisfy stated conditions in a three-dimensional space continuum. Einstein will later extend Newton's three-dimensional space to a "four-dimensional time-space continuum",[12] in which a planetary orbit can be represented as a locus of events in curved spacetime. The essential vehicle carrying coordinate geometry from a plane to higher dimensions is the Thalesian theorem "of Pythagoras" [13] — and here we see a fine specimen of the connectivity between ancient Greek and modern mathematical physics.

Galilean Units of Time and Distance

Still dormant in Galileo's work are intriguing questions, a fertile field for exploration by future historians and physicists. Was Kepler's Third Law of Planetary Motion motivated by analogy to Galileo's Law of the Pendulum, on account of the harmonic motion present in both situations? Was the discovery of the independence of electromagnetic wave frequency and amplitude similarly motivated, in that the information-bearing waves in the theories of Maxwell and Hertz are forms of harmonic motion? Would Galileo, if transported to the 20th century, be satisfied with the current uncertainty as to whether the Bode-Titius [14] relationship between planetary orbit radii and the astronomical unit is a consequence of physical theory? Having posed this last question to myself, I then came across a little monograph, one of Professor Stillman Drake's last works that

---

11. ... strictly speaking, the barycenter of the planet and the projectile.

12. Cf. Einstein and Infeld, *The Evolution of Physics*, Simon and Schuster, NY, 1958. This four-dimensional geometry was developed by Hermann Minkowski and later refined by H. Weyl to a "3+1" dimensional world; cf. A. S. Eddington, *The Mathematical Theory of Relativity*, Chelsea Publishing Company, NY, orig. pub. 1923.

13. Cf. Ch. 1 "Thales of Miletos, the Founder". See also K. O. Friedrichs, *From Pythagoras to Einstein*, Random House, 1965 for an exposition of the generalization of the distance function to higher dimensions.

had earlier escaped my attention, as it had been published locally in Toronto in 1989.[15] Imagine, then, my astonishment at reading Drake's monograph and finding that Drake himself had addressed the Bode-Titius question before I posed it, and that he did so using Galilean units!

Recalling[16] that the time of the pendulum clock moving through a small arc is $\pi/(2\sqrt{2})$ times the descent time through a distance equal to the length of the pendulum, we can convert this time relation to a distance relation by invoking Galileo's law of falling bodies. Since distances are proportional to the squares of times, a distance freely fallen is a multiple $\pi^2/8$ of the length of a pendulum swinging to the vertical in the same time.

Gravitation and Harmonic Motion

Drake has noted that Galileo chose his distance unit, the *punto (λ),* arbitrarily as ~0.94 mm, and his time unit, the *tempo (τ),* equal to ~1/92 second as determined by the function $16\sqrt{2L}$ mapping the lengths L in *punto* of his pendula into their time to the vertical in units for a certain flow rate of his water clock. He points out that calculations using λ and τ make the acceleration due to gravity:

$$\frac{\pi^2}{8} \cdot \frac{\lambda}{\tau^2}$$

In Galilean Units (G.U.), he notes, distance and time are "dimensionally commensurable" since the value for acceleration due to gravity is chosen such that the square of time expressed in units τ of fall through a distance expressed in units λ is numerically equivalent to the length of a pendulum timing the fall. This is not the case with arbitrary units such as the metric or British system.

Noting that the factor 10 is a constant in Bode's law, Drake observes that the ratio of semimajor axes of Venus to Mercury is, to six significant digits, equal to $\log 10e^2$. Assigning the Astronomical Unit (the mean sun-to-Earth distance of about 499 light seconds, used as a measuring unit by astronomers) the value $10/2\pi$, and expressing the planets' semimajor axes in *punti* and their sidereal periods in

---

14. Bode's "Law" was apparently (per Russell et. al. *Astronomy*, op. cit.) formulated by Johann Titius of Wittenberg in 1766 (later published by Bode). It is a relationship between the radius vectors of the planetary orbits in Astronomical Units (AU), and the sequence of positive integers. Number the planets n = 0, 1, 2, etc. in order of increasing distance from the sun. The mean distance of the nth planet from the sun in AU is approximately $(x+4)/10$, where x = 0 for planet 0 (Mercury) and $x = 3 \cdot 2^{n-1}$ for the other planets out to Neptune. The asteroid belt is considered (a fragmented) planet 4; I note that Pluto's present orbit, though not well conforming to the "law", is generally considered by theoretical astronomers least likely to have a common origin with those of other planets of the sun.
15. Stillman Drake, *History of Free Fall*, op. cit. (See Bibliography and Publisher's List of Other Titles.)
16. See Ch. 5 "Inclined Planes and Pendula".

tempi, he finds the Bode-Titius sequence visible in the resulting tabulated orbital circumferences. Though his analysis does not fully explain the Bode-Titius relationship, it offers a direction by which scientists may pursue a demonstration that the Bode-Titius "Law" is non-coincidental.

The Significance of Galileo in the Newtonian System of the World

"I do not doubt that in the course of time this new science [of Gilbert] will be improved with still further observations, and even more by true and conclusive demonstrations. But this need not diminish the glory of the first observer. I do not have a lesser regard for the original inventor of the harp because of the certainty that his instrument was very crudely constructed and more crudely played; rather, I admire him much more than a hundred artists who in ensuing centuries have brought this profession to the highest perfection."

... Galileo [17]

The Laws of Mechanics of Galilei-Newton

Three scientific masterpieces form a continuum on the science of motion: Galileo's *Discorsi*, Newton's *Principia*, and Einstein's *Relativity*. Those who teach, write, and think that Newton's Laws are entirely due to Newton haven't read the chapter of Newton's book in which these laws are stated: [18]

Lex I: "Every body continues in its state of rest, or of uniform motion in a right line, unless it is compelled to change that state by forces impressed upon it. ..."

Lex II: "The change of motion is proportional to the motive force impressed; and is made in the direction of the right line in which that force is impressed. ..."[19]

Lex III: "To every action there is always opposed an equal reaction: or, the mutual actions of two bodies upon each other are always equal, and directed to contrary parts. ..."

---

17. See Galileo's *Dialogo*, Third Day, p 406, in the Stillman Drake translation, op. cit. He is here referring to William Gilbert's new science of magnetism.

18. translated by Andrew Motte and revised by Florian Cajori (Newton wrote the *Principia* in Latin); cf. *Newton's Philosophy of Nature*, H. S. Thayer, ed., Hafner Press, NY 1953.

19. In modern terminology the foregoing Acceleration law can be restated:

    Lex II (restated): "Rate of change of momentum is proportional to the resultant force and is in the direction of this force" (Sears and Zemansky, *University Physics*, Addison-Wesley, Cambridge, Massachusetts, 1953) ; or symbolically:

$$F = k \cdot \frac{d}{dt}(m \cdot v)$$

    If k is eliminated by choice of units, and mass is invariant with time, then this law assumes the familiar form "*f = ma*".

Corollary I: "A body, acted upon by two forces simultaneously, will describe the diagonal of a parallelogram in the same time as it would describe the sides by those forces separately. ..."

Corollary II: "And hence is explained the composition of any one direct force ...out of any two oblique forces ...; and, on the contrary, the resolution of any one direct force ... into two oblique forces ... : which composition and resolution are abundantly confirmed from mechanics. ..."

Scholium: "Hitherto I have laid down such principles as have been received by mathematicians, and are confirmed by abundance of experiments. By the first two Laws and the first two Corollaries, *Galileo* discovered that the descent of bodies varied as the square of the time (*in duplicata ratione temporis*) and that the motion of projectiles was in the curve of a parabola; experience agreeing with both, unless so far as these motions are a little retarded by the resistance of the air. ..."

Here Newton explicitly credits Galileo with use of the first two of "Newton's Laws" and their first two corollaries in Galileo's discovery of the laws of falling bodies and of parabolic fall.

It is understandable that Newton did not extend credit for the Third Law; Galileo did not publish it. Still, Galileo should be remembered for anticipating that law in his unpublished *De Motu*. There, he says of weights arranged in a line (*ab*) along a supporting surface that "these bodies press down with a force equal to that with which the line *ab* presses upward", in reaction to which Professor Drake has said: "The equality of action and reaction (Newton's third law of motion) was by no means a commonplace in 16th-century mechanics; indeed, the idea of a table-top pressing upward when a weight is placed on it would have been ridiculed by Aristotelian physicists, and even today seems odd to many beginning students."[20]

Of Galileo's law of falling bodies, it can be said that Newton was so confident of its correctness and consistency with his theory, that he used it to analyze experimental errors in an extension of the Leaning Tower demonstration conducted by Hauksbee from the top of St. Paul's Church[21] in London in 1710.[22]

The Ponte Vecchio, built in 1345 as the first flat segmental arch bridge, was undoubtedly useful to Galileo in commuting between

Galileo and Newton in Modern Construction Engineering

downtown Florence and his villas south of the Arno. It is standing today, though it had to be extensively repaired after the 1966 flood. The next development in

---

20. Stillman Drake, *History of Free Fall*, op. cit.
21. "Christopher Wren designed many buildings to replace those destroyed in the London fire of 1666. His best known work is St. Paul's cathedral."; cf., Charles F. Linn, *The Golden Mean*, Doubleday, NY, 1974.
22. Newton's *Principia*, op. cit., Book II, p. 243.

bridge construction was the wood truss invented by the architect Andrea Palladio (1518-1560); his sketches "show structures practically of modern types".[23]

While Galileo used the flat segmental arch bridge for his trans-Arno commute, builders of bridges after him have used his principle of the composition and resolution of forces, expressed in the corollaries to the Laws of Mechanics of Galilei-Newton. This principle, recast in a Newtonian formulation, underlies the laws fundamental to modern construction engineering:

> "... for any set of coplanar forces in equilibrium, the algebraic sum of the components of these forces along any axis in this plane is equal to zero; ... the moment of these forces about any point in this plane is equal to zero."[24]

These are laws of statics due to Galileo and Newton. When the resultant of all the forces acting on an element of a structure is zero, that element is said to be in equilibrium, a way of saying that it is at rest in accordance with the inertial law of Galilei-Newton (First Law). By applying this law successively to the various elements of a structure, an engineer can "compute how much force each part must withstand and therefore how strong each girder, beam, or column must be."[25]

Newton's System of the World

Having synthesized Kepler's and Galileo's results, Newton created and articulated in geometric reasoning a universal theory and its diverse applications, called the System of the World.[26] Here is presented a gravitational theory of tides that at last realizes the grand intent of Day Four in Galileo's *Dialogo*; Newton calculates oceanic motions due to the solar and lunar influence, finding that the sun can raise the sea by nearly two feet, and the moon by as much as twelve and a half feet.[27] He published his theory in *Principia Mathematica* in 1687 under the auspices of Edmund Halley, a major innovator in his own right, whose correct prediction of the return of the comet that bears his name both confirmed and popularized Newton's achievement. As Newton's physics became increasingly perceived as able to predict a multiplicity of natural events from the behavior of tides to the return of comets, the perceived harmony of heavens and Earth began to hold sway over the imaginations of men; science had begotten a cultural revolution by removing cosmology from the exclusive domain of priests and pedants.

---

23. Linton Grinter, *Theory of Modern Steel Structures*, MacMillan Co., NY, 1942.
24. Ibid.
25. Sears and Zemansky, *University Physics*, op. cit.
26. Newton's theory exceeds the scope of this description. My focus is limited to his direct working relationship to Galileo; this is not an attempt to summarize Newton's book, the *Principia*.
27. These tidal forces operate on land masses as well, so that Newtonian mechanics provides a tool to calculate their effect on the stability of tectonic plates.

Galileo's conclusion, in his Padovan lectures on mechanics, that one cannot cheat nature with a machine, marked a milestone

Conservation Laws of Physics Formulated

on the road to the conservation laws of physics. In the mid-19th century, Thomson and Clausius independently extended Newton's Law II, a consequence of which is that *the acceleration of a body is determined by the forces which act upon it*, to a principle applicable to any displacement of a body by an external force:

> "...energy supplied from the outside is equal to the change in kinetic energy plus the change in gravitational potential energy plus the energy converted into heat. This is the *principle of conservation of energy* ... a consequence of Newton's second law." [28]

From a variant of this formulation emerges the First Law of Thermodynamics:

> "The *first law of thermodynamics* is the same as the *law of the conservation of energy*, which states that energy can be neither created nor destroyed. Thus, the quantity of energy in the universe is constant." [29]

This principle has important implications in cosmology. Any hypothesis concerning large-scale evolutionary change in the universe cannot be constructed in such a way as to add to or subtract from total energy without running afoul of this law. Cardinal Bellarmino's eternal incendiary mechanism was rendered unreal by scientific discovery more than a century before the Church declared, in its Jerusalem Bible, the symbolic nature of Revelation's hellfire.[30]

Where along the path leading to thermodynamics did dynamics, the science of accelerated motion, begin? A founder of celestial mechanics answers: [31]

Lagrange and Fourier on Galilean Dynamics

"Dynamics is a science due entirely to the moderns. Galileo laid its foundations. Before him philosophers considered the forces which act on bodies in a state of equilibrium only. Although they attributed in a vague way the acceleration of falling bodies and the curvilinear movement of projectiles to the constant action of gravity, nobody had yet succeeded in determining the laws of these phenomena. Galileo made the first important steps, and thereby opened a way, new and immense, to the advancement of mechanics as a science."

Joseph Louis Lagrange: *Mécanique Analytique,* 1788

---

28. Sears and Zemansky, op. cit.
29. Virgil Moring Faires, *Applied Thermodynamics*, MacMillan Company, NY, 1938 (the author is grateful to Geoffrey A. Masaki for the gift of his copy).
30. Cf. Ch 9, "Heaven Hotter than Hell".
31. as quoted in *Systems Analysis*, Harry J. White and Selmo Tauber, W. B. Saunders Co., 1969.

If dynamics is the first of the modern sciences, then by the opening sentences of Lagrange's statement, Galileo is the first modern scientist; the Archimedean origin and Newtonian extension of his work in mechanics make all the more visible his pivotal position in this regard. Later, another successor describes the effect of collaboration over time among Archimedes, Galileo and Newton:[32]

> "The knowledge of rational mechanics, which the most ancient nations had been able to acquire, has not come down to us, and the history of this science, if we except the first theorems in harmony, is not traced up beyond the discoveries of Archimedes. This great geometer explained the mathematical principles of the equilibrium of solids and fluids. About eighteen centuries elapsed before Galileo, the originator of dynamical theories, discovered the laws of motion of heavy bodies. Within this new science Newton comprised the whole system of the universe. The successors of these philosophers have extended these theories, and given them an admirable perfection: they have taught us that the most diverse phenomena are subject to a small number of fundamental laws which are reproduced in all the acts of nature. It is recognised that the same principles regulate all the movements of the stars, their form, the inequalities of their courses, the equilibrium and the oscillations of the seas, the harmonic vibrations of air and sonorous bodies, the transmission of light, capillary actions, the undulations of fluids, in fine the most complex effects of all the natural forces, and thus has the thought of Newton been confirmed: *quod tam paucis tam multa praestet geometria gloriatur.*"[33]

Joseph Fourier, *Analytical Theory of Heat*, 1822

Developments pioneered by Fourier led to the thermodynamic law that the total energy in the universe is constant. Fourier here quotes Newton expressing a useful concept of creativity that does not entail a variation in the total energy of the universe. Newton's concept of amplifying knowledge, from a minimal set of postulates to the understanding of diverse phenomena, is consistent with that thermodynamic law. In this way, it differs from the theological concept of creation from a void. Like the work of Galileo, offering as it does a scientific alternative to concepts whose origins are theocratic, Newton's dictum, found confirmed by Fourier, offers a rational alternative to the literal view of Genesis.

---

32. trans. by Alexander Freeman, *Great Books of the Western World*, op. cit.
33. "Geometry prides itself on enabling such a multitude of results from such few axioms." — my translation.

Extensions to Galileo's Work Between Newton and Einstein

Outstanding examples of Galilean mechanics and of the foundation laid by Newton[34] in orbital computations populate the work of Gauss, Laplace, Lagrange, and Euler in celestial mechanics. Closer to home are extensions to Galilean science visible in earthly phenomena. Here we can only sample the explosion of scientific activity into diverse areas that Galileo's new sciences ignited. Though I can afford them but scant coverage, I want to mention Galilean successors that may point you to trails of his work that lead beyond the scope of this book.

Strings, Lutes, Waves and Right Triangles

In acoustics, D'Alembert, Euler, and Daniel Bernoulli independently described in mathematical equations the concurrent contributions of simple harmonic oscillations. Transplanted far afield of Galileo's lute, these equations flourish in hydrodynamics, in the theory of elasticity, and in any application requiring roots for a function representable by superimposed waveforms. Later, Fourier discovered how to express arbitrary functions of a variable as an infinite series of trigonometric functions of that same variable. Fourier thus reduced any kind of curvature represented by a known integrable function to a polynomial expressed in ratios of sides of a right triangle. I think Galileo would have been particularly pleased to see that.

From Hydrostatics to Fluid Dynamics

You may recall that in 1612 Galileo published a *Discourse on Bodies in or on Water*.[35] Here he presented theorems on hydrostatics of objects floating in fluid containers and disproved the "Aristotelian" notion that an object will float or not depending on its shape. Anchored in calm waters, all ships float independently of their shape and of Church-sponsored medieval philosophy. Galileo's student Castelli calculated water flow in aqueducts with access to most of Galileo's results to that time.

In 1630 Ferdinando II called upon Galileo to evaluate two competitive plans for flood control of the Bisenzio River. This request led Galileo to consider how the behavior of water accelerating through a river channel differs from that of a solid body rolling down a slope. As a byproduct of his analysis, he enunciated the following principle by analogy to falling bodies:

> "... in two channels of equal total drop the speeds of motion will be equal, even though one channel is very long and the other short." [36]

---

34. Newton, for example, shows how to determine a parabolic comet orbit from three observations; cf. *Principia*, Book III, Proposition XLI.
35. Ch. 8 " Floating Bodies".

"Speeds of motion" are averaged over the drop. Initially, the shorter, steeper slope is traversed faster than a corresponding distance on the longer slope. Eventually, accelerating water on the long slope catches up with water on the steep slope. Steepening the river bed will not allow the water to drain faster.

While on a horizontal plane a metal ball will have no tendency to motion, a ball of fluid placed on such a plane will collapse and flow by virtue of the tendency of its molecules to seek the lowest level; hence, the pressure of a stream of water varies throughout its depth, and the flow rate of water depends on factors besides the tilt of a river bed. Galileo concludes in favor of the plan that addresses some of these factors by removing mud and silt, and against the plan that depends on straightening the river bed in order to shorten and steepen it.

Adding to hydrostatics the consideration of forces applied to a fluid (and the consequent acceleration of their molecules) led to hydrodynamics, the theory of fluids in motion. The design of ocean vessels profits from the ability to calculate and to measure such effects. After hydrodynamics, brigs and clippers, cutters and hulks, sloops and galleys, freighters and ocean liners plow their courses through the deep according to the improved designs of their shapes, though still independently of scholastic philosophy.

Because gases are also fluids, hydrodynamics treats of their behavior as well. Fundamental to this theory are the gas laws developed by Galileo's successors, laws founded on his pioneering recognition and proof that air has weight, that gases have measurable properties, and that it is worthwhile to measure such things.[37]

Solid Progress from Gases

Galileo's thermoscope and Torricelli's barometer, the first instruments that measure temperature and pressure of a gas, combined with their discoveries to enable new inventions and discoveries by others. For example, Galileo's friend Sagredo improved Galileo's thermoscope; the Academy of Experiment then made thermometers that were less sensitive to air pressure. Torricelli's discovery of atmospheric pressure led to the invention of the air pump by Otto von Guericke (1602-1686).[38] To a startled crowd of onlookers, von Guericke pumped out the air between two hemispheric plates and had two teams of horses unsuccessfully try to pull apart the plates, which then fell apart when he broke the vacuum.

---

36. trans. by Drake, *Galileo at Work*, p. 323.
37. Edwin Hubble summarizes Galileo's experimental technique described in *Two New Sciences*: "Water is forced into a vessel until the imprisoned air is under considerable pressure; the vessel is weighed; the imprisoned air is allowed to escape; the vessel is again weighed. Then the difference in weight is the weight of a volume of air equal to the volume of water in the vessel." *The Astrophysical Journal*, op. cit.
38. Cf. Virgil Moring Faires, op. cit.

One condition that fostered the development of gas laws is the knowledge of how pressure is transmitted. This knowledge began with an experiment conducted by Torricelli: when he upended a tube filled with mercury into a dish of this same substance, some of the mercury from the tube ran into the dish, but *stopped running into the dish* when the mercury column in the tube dropped to a height of about 76 cm. In one act he had done three things: (1) proved Galileo's conjecture that air has weight and that we live at the bottom of an ocean of air, (2) created a vacuum at the upper end of the tube, a phenomenon supposedly abhorrent to nature (and definitely so to philosophers like Cremonini), and (3) made an archetypal instrument to measure air pressure.

Through intermediaries including Viviani and Mersenne, Torricelli's result reached Blaise Pascal in Paris, who is said to have repeated this experiment using red wine, which required a tube over 14 meters long. I have yet to find an account of how Pascal disposed of all that wine after his experiment, but he was certainly inspired when he invented a law that is today the basis of hydraulic presses: [39]

> *"Pressure applied to an enclosed fluid is transmitted undiminished to every portion of the fluid and the walls of the containing vessel."* [40]

Of three related properties of a gas — pressure (p), volume (V) and temperature (T) — Galileo's and Torricelli's inventions measure two of them, the two that are not visible to the eye, and it was this measurability that led to general laws of gases; viz., the Boyle/Mariotte Law and the Charles/Gay-Lussac Law, which combine to the equation of state of an ideal gas (one that obeys the gas laws exactly):

$$pV = nRT,$$

where R is a proportionality coefficient depending on the units of the variables, and n is the number of moles of gas, Avogadro's number of molecules. If temperature and pressure are invariant between ideal gases, then *"equal volumes of* [those] *ideal gases contain the same number of molecules"* by a principle of Avogadro.[41] This is a law that we depend on every day as we move about in cars designed to harness exploding gases.

In diverse other ways, the effect of Galileo's science on invention and discovery

---

39. Asimov's *Biographical Encyclopedia of Science and Technology*, Doubleday, NY, 1982.

40. Sears and Zemansky, *University Physics*, op. cit. Modern aircraft make use of Pascal's law insofar as their control surfaces and brakes are hydraulically activated.

41. P. W. Atkins, *Physical Chemistry*, W. H. Freeman and Company, San Francisco, 1978. This work also contains a discussion of the equations of state for ideal and real gases.

reached beyond the post-Galilean Renaissance. His laws of mechanics and his experimental methods underlie the discovery and verification of the laws of physical chemistry and thermodynamics on which 20th century cosmology depends. Modern transportation systems are designed around the measured behavior of gases, beginning with a principle that we have just traced from Galileo and Torricelli to Pascal. A later principle due to Daniel Bernoulli explains the lift on an airplane wing as a pressure differential integrated over a wing surface area. So much for Urban's attempt to snuff out a Galilean school of science: his successors who fly about on jet airplanes to conduct their diplomatic missions are using Galilean science. Lucky for them that Urban failed, for the lift on an airfoil continually counterbalances the tendency of the aircraft to fall. And here is an even more direct influence of Galilean science on aviation technology, the knowledge of which begins with Galileo himself. Aviation technology requires taking into account how objects accelerate when they fall through the air.

Falling Through the Air

What caused Galileo finally to deny that falling objects continue to accelerate? It will be instructive to consider how objects denser than a resisting medium fall through it, and thereby gain insight into his change of mind on one of the issues that caused him to withhold the *De Motu* from publication. Newton covers this topic in Book II of the *Principia*,[42] showing geometrically how Galileo's free-fall parabola can be modified to a hyperbola in a resisting medium; Morris Kline has given a lucid presentation that is a lot easier to follow and uses only elementary calculus.[43]

> (Hint: If you aren't thrilled by the beauty of mathematical reasoning, you may now skip this and the next section.)

We would expect that the relative wind will increase with the velocity of the falling object, so let's assume that these variables are directly proportional in magnitude and opposite in direction. As a further simplifying assumption, we will consider any effects of buoyancy or viscosity in the atmospheric medium to be negligible. If we look through Professor Kline's discussion searching out a form of the vertical component of projected fall analogous to the form $v_y = gt + V \sin A$ that expresses the same component in fall through a vacuum according to the law of Galileo, we can find it in Kline's equation (40). Here he derives the downward velocity component of a projected fall through air in elapsed time t which, collecting terms, is equivalent to:

$$v_y = -g/k + (V \cdot \sin A + g/k) \cdot e^{-kt}$$

42. Prop. 3. Problem 1.
43. Morris Kline, *Calculus: An Intuitive and Physical Approach*, Wiley and Sons, NY, 1967, Part II, p. 22 ff.

for initial velocity V and angle A to the horizontal.[44] As in Galileo's law of fall through a vacuum, here v_y is a function of g and t. But notice that from its starting value ($v_y = V \cdot \sin A$ at time zero) the speed of fall approaches the terminal speed g/k while the term on the right vanishes with increasing time. This evanescence explains mathematically the eventual physical annihilation of an acceleration opposed by a drag effect. Air resistance at low speeds is nearly proportional to velocity but, at high speeds, to velocity squared. And so it turns out that Galileo's first instinct, to treat acceleration as a temporary effect during fall, is correct in a fluid medium, and that includes falling through air![45] A lucky guess? Remember that Galileo's observations of falling bodies were not made in a vacuum.

What if the medium *is* a viscous one? Using the Second Law of Mechanics as formulated by Newton, Euler had arrived at a general equation for continous

Falling Through a Viscous Fluid Medium

flow of a nonviscous fluid, a result that was extended for a viscous medium independently by Navier and Stokes. Sir George Gabriel Stokes, that is, a successor to Newton as Lucasian professor of Mathematics at Cambridge. The work of Stokes directly connected to Galileo's around 1845, when he investigated the slowing of pendula due to the viscosity of the fluid in which they are immersed; this investigation resulted in Stokes' Law for the resisting force R encountered by a spherical body of radius r falling with downward velocity v_y through a fluid (liquid or gas) of viscosity η :[46]

$$(12\text{-}1) \qquad\qquad R = 6\pi\eta r v$$

From this relationship, one can extend Galileo's law of falling bodies for a viscous fluid medium, and as long as we stick to certain simplifying assumptions,[47] the mathematics involved is not as sticky as one might think. The derivation, worked out in Appendix D, yields an expression for velocity of fall that is interesting to

---

44. k is here defined as a constant ratio of air resistance to velocity, divided by the mass of the projectile.

45. This was also Galileo's final conclusion: "My opinion is therefore, that under the circumstances which occur in nature, the acceleration of any body falling from rest reaches an end and that the resistance of the medium finally reduces its speed to a constant value which is thereafter maintained" ; spoken by Salviati, in the *Discourses*, First Day, op. cit., p. 94. But his conclusion in the *De Motu* of 1590 that "... even if it were conceded that the velocity of the body underwent continuous increase forever, the velocity still would not necessarily become indefinitely great" was drawn not from the resistance of the medium but from diminution of the original impressed force; cf. Drabkin's translation in *Galileo Galilei on Motion and on Mechanics*, U. of Wisconsin, 1960, p. 102.

46. Cf. Sears and Zemansky, *University Physics*, Addison-Wesley, 1970, 4th ed. The discussion in this section is equivalent to recombining relevant portions of their Chapters 5 and 14.

47. Here we will assume that the viscosity and density of the medium are invariant during the fall and that the speed of fall is not great enough to cause turbulence; an analysis of a long fall would need to take into account, for example, a gradual increase in the density of the medium.

compare with Kline's result given above for a nonviscous situation; viz., when the medium's viscosity η is greater than zero:

$$v_y = \frac{2}{9} \cdot \frac{r^2 g(\rho - \rho')}{\eta} \cdot \left[1 - e^{\frac{-6\pi\eta r}{m} \cdot t} \right]$$

Here, v_y is still a function of g and t, but notice that the bracketed function starts out at zero at time zero so that $v_y = t = 0$; as t increases toward infinity, the bracketed function approaches one and the unbracketed expression on the left becomes the terminal velocity. Notice also that this result is of the form (multiple of gravitational constant g) + (multiple of e^{-at}) for some coefficient a, making it analogous to Professor Kline's result for a nonviscous medium, with the addition of new variables to account for the viscous drag effect. Again, regardless of the (finite) amount of viscosity, an acceleration opposed by a drag effect that increases with time will eventually be annihilated. Also notice that when the two densities are equal, $v = 0$; e.g., air will not fall through air of the same density, nor water through water, nor will anything fall through an infinitely viscous medium. One might wonder whether this law may eventually pose a problem for aircraft attempting to land in Los Angeles.

Gravity and Gas

Gravity[48] and gas[49] are jointly as basic to cosmology as are Galileo's and Kepler's laws to Newtonian mechanics. Knowing the behavior of gas is essential to understanding large-scale change in the universe. The source of such change is the interaction of gravity and interstellar gas. From the perspective of current cosmological models, the existence of you and me can be traced back through a long chain of events linked by the imploding and exploding gases of stars at both endpoints of their "life cycle". In Chapter 13, we will use these gaseous events as a bridge to connect gravity and life.

Rocketry and Celestial Mechanics

Rocketry requires the application of thrust and guidance to a missile, and hence is an outgrowth of the technology of the ballistic missile, which (except at launch) has neither. After Archimedes' catapults, the engineering of launch vehicles resumes with determining the trajectory and range of a ballistic missile, in the ideas of Tartaglia[50] and in the Fourth Day of the *Discourses*. Galileo published

---

48. In this context, "gravity" refers to the force due to gravitation that is mutually exerted between a "parent" celestial body and one or more objects that are bound to the parent in this way.

49. "**gas**. A term (literally 'chaos') coined by J. B. van Helmont (1579-1644), describing a spiritual essence that could be released from substances but never isolated." Cf. *Dictionary of the History of Science*, ed. by Bynum, Brown, and Porter, Princeton U. Press, 1981.

there a novel solution to this problem, introducing the parabolic trajectory and tabulating range as a function of launch angle. The rocket thrust principle is Newton's Law III; from the foundation of the thrust principle and of the gas laws, the science of rocketry took off. The "rocket's red glare" that Francis Scott Key saw in the attack on Fort McHenry in 1814 came from rockets developed by William Congreve, an Englishman schooled in Newtonian science. But not all rockets were used in warfare: "In 1838 John Dennet, an English inventor, patented a rocket that would carry a rescue line to ships in distress."[51] The next improvement in rocketry came with fin stabilization by William Hale in 1846. By the early 20th century, the train of events moved into high gear with the work of K. E. Tsiolkovsky, Hermann Oberth, Robert H. Goddard and Fritz von Opel. That is a story we will leave for others to tell.

In a separate but related development, the transportation systems by means of which men travelled to the moon in our own time were made possible by a celestial mechanics that combines: (1) Galileo's principle that matter is "mindless" of the magnitude of motion and that the change of a moving object's velocity is determined by the forces which act upon it, (2) Kepler's laws and (3) the Newtonian world system. After Halley with conspicuous success applied (to the comet named for him) the method of orbit determination given by Newton in his Book III, celestial mechanics was ripe for development. Post-Newtonian developers of the science of celestial mechanics include: Euler, Clairaut, D'Alembert, Lambert, Lagrange, Laplace, Gauss, Gibbs, Jacobi and Poincaré.[52]

Extensions to Human Observational Domain

With the invention and application of the astronomical telescope and the compound microscope, Galileo opened up two new worlds that had been previously invisible to man. Together, his emphasis on measurement and the Galilean micrometer spurred the development of astrometric devices. In 1719, James Pound coupled a micrometer to his telescope and proceeded to measure variations in the polar and equatorial angular diameters of Jupiter over a 40 day period. These are stated to four significant digits in the Third Edition (1725-6) of the *Principia (Bk. III, Prop. XIX)*, where Newton uses them to confirm his theoretical calculations of how Jupiter's rotation affects its oblateness. But Galileo's optical devices operate at optical wavelengths, a miniscule portion of the

---

50. "In 1537 Tartaglia, as a mathematician, conceived of motion as continuous. He recognized that the path of even the swiftest cannonball must be curved downward from the instant at which it became free to fall, and assumed its two nearly-straight paths to be joined together by a circular arc." Stillman Drake, *History of Free Fall*, referring to S. Drake and I. E. Drabkin, *Mechanics in Sixteenth Century Italy*, Madison (1969).
51. Cf. *New Standard Encyclopaedia*, Standard Education Society, Chicago, 1965.
52. For a discussion of their individual contributions, cf. Forest Ray Moulton, *An Introduction to Celestial Mechanics*, Dover, NY, 1970, orig. pub. 1914.

full spectrum of information sources. Further crucial advances include spectroscopy, the radio telescope, and the electron scanning microscope, three fascinating stories that are beyond the scope of this synopsis.

Galileo's Posthumous Scientific Victories

Among the most exciting extensions to Galileo's work are those which remained unconsummated until after Newton. We begin here with Galileo's posthumous roles in the discovery of new planets, and we end with his role in completing the discovery that Earth itself is a planet. In the interim we'll cover results of modern experiments demonstrating the universality and exactness of his law of falling bodies. Consummating the proof that Earth is a planet are independent experiments which corroborate the hated heretical hypothesis of Copernicus and complete two lines of investigation started by Galileo: parallax and pendula.

Galileo's Role in the Discovery of Uranus

As illustrated dramatically at the 1933 Century of Progress Exposition in Chicago,[53] when his telescope is said to have been briefly recalled to active duty, Galileo is perhaps best remembered for pioneering application of the telescope to astronomy. Beyond his life and through his application of parallax measurements, Galileo played yet another role in astronomical discovery, this time of a major planet. It happened in 1781, when William Herschel discovered a major planet unknown to the ancients, and to Galileo himself:

"One of Herschel's subsidiary objectives was to determine the distances of some fixed stars. He was attracted to Galileo's suggestion for detecting stellar parallax by measuring periodic changes in the separation of close pairs of stars differing in brightness, and he began a methodical search for suitable star pairs. In the course of this search, on the night of March 13, 1781, he noticed a curious object in the neighborhood of H Geminorum. Increasing the power of his eyepiece he found that the object had a magnifiable disk; continued observation convinced him that it was in motion relative to nearby stars. Although he did not realize it at the time, Herschel had discovered the seventh planet."[54]

Herschel thought that he had discovered a new comet, and that is how he announced his discovery. But attempts to predict its position based on a parabolic

53. While newspaper accounts and other sources concur that Exposition lights were switched on as light from Arcturus passed through a telescope at Yerkes Observatory on May 27th, 1933, Asimov has come up with this story: "On October 2, 1933, the light of the moon was received through Galileo's 3½-centuries-old original telescope and was used to switch on the lights of the Century of Progress Exposition in Chicago. Photoelectric devices transformed the moonlight into electricity and the resulting current closed the illumination circuit of the exhibition"; *Isaac Asimov's Book of Facts*, Fawcett, 1979, p. 390 (author Larry Sessions mentioned this source in an e-mail note to me). Asimov may have described an independent event, though I've been unable to corroborate it. Here is a project for an ambitious historical researcher.

54. Morton Grosser, *The Discovery of Neptune*, Dover, NY, 1962, p. 18.

orbital model consistently failed; the true nature of Hershel's object was recognized after Anders Johann Lexell fitted its observed positions to a nearly circular orbit. Herschel had discovered a planet with his homemade 6.2 inch [55] reflector while using a parallax detection method suggested by Galileo.

At 3:45 am on the morning of 28 December 1612, Galileo sighted the planet Neptune, which he could not then have recognized as a planet, for it was paused in retrograde phase and had all the appearance of a fixed star. And so he referred to it in his journal drawing as "*stella fixa*", a fixed star. Largely because of bad weather over the next month, he made no further sightings until the night of 27 January, when he noted its position. At 11 pm on 28 January, Galileo observed Neptune again and found that it had moved relatively to a truly fixed star which has been identified by the modern designation SAO 119234; Neptune had drawn 2.5 Jovian radii closer to that star, an advance that is simply not possible in a single night for a fixed star. He drew the pair 3.75 Jovian radii apart on the night of 28 January and, labelling Neptune as fixed star "*a*" and SAO 119234 as fixed star "*b*", annotated his drawing:[56]

> "*Beyond fixed star* a *another followed in the same line, as [does]* b, *which also was observed on the preceding night; but they [then] seemed farther apart.*"

A slightly puzzled Galileo had not only sighted the yet-undiscovered planet Neptune, but noticed it wandering in planet-like fashion among the fixed stars. Unaware of the great significance of his observation, he left it to a successor to recognize the new planet. Like Herschel's discovery of Uranus, Galileo's Neptune sightings happened not in pursuit of a planet but of an astrometric goal. Galileo had invented a micrometer that superimposed a grid image in the field of view of his telescope, and was refining his tables of Jupiter's satellite motions.

Neptune was not discovered to be a planet until more than two centuries had passed. In October 1845, the English mathematician John Couch Adams, who had solved the problem of calculating the position of an undiscovered planet based on perturbations[57] of Uranus, had come to a dead end in his attempts to interest British astronomers to look for it. The Astronomer Royal's butler, it seems, was too protective of his employer's time, and failed to inform him of Adams' attempts

---

55. An "inch" is an archaic English unit equal to 2.54 centimeters.
56. Stillman Drake and Charles T. Kowal, "Galileo's Sighting of Neptune", *Scientific American*, December 1980, Vol. 243, No. 6, p.80. Charles T. Kowal is the discoverer of the 13th satellite of Jupiter in 1974 (source: Don Nicholson) and of asteroid 1977 UB.
57. A perturbation is a difference between an observed position and one correctly calculated from gravitational theory, and hence gives information about any gravitational source excluded from the calculation.

to see him.[58] Meanwhile in France, another mathematician, Urbain Leverrier (who had never heard of Adams) took up the same problem and completely independently arrived at a solution very close to that of Adams. Having failed to interest French astronomers in his calculations, he appealed to a German astronomer, Galle, who had sent him a dissertation. In September of 1846 Galle and d'Arrest were inspecting a small area of the sky with the Berlin Observatory's 22.86 cm Fraunhofer refractor where Leverrier's predictive calculations told them to look. As Galle called out the positions of observed stars he described an 8th-magnitude object at right ascension 22h 53m 25s.84, to which d'Arrest responded "That star is not on the map." It was Galileo's "*stella fixa 'a'*", not a fixed star at all, but the eighth major planet of the solar system.

The story of Neptune's discovery doesn't end here; otherwise, I wouldn't be including it as one of Galileo's posthumous scientific victories. In fact, Neptune is a planet still under examination — it has not quite completed one orbit since Galle spotted it — and its orbital elements are somewhat in doubt. Galileo's observations with the micrometer he invented were of such recognized precision, "often correct to better than 10 arc seconds and rarely wrong by as much as 20 arc seconds",[59] that, to refine the orbit of Neptune and investigate its perturbations, 20th century astronomers are taking into account his notebook entries showing its position relative to the star SAO 119234. The ninth planet, Pluto,[60] was found by Clyde Tombaugh in 1930 in a search inspired by Percival Lowell's then-estimated perturbations of Neptune — but in the end it was Tombaugh's incredibly exhaustive search (and not predictions from perturbation estimates) that enabled him to discover Pluto.[61]

---

58. The role of domestic servants in blocking the discovery did not stop there. Adams also wrote to Airy who, though unimpressed, showed the calculations to William Rutter Dawes. Realizing their importance, Dawes immediately sent Adams' calculations to Lassell, who had the largest telescope in England. Before Lassell could look for the planet, the letter was destroyed by a maid! Grosser concludes, "... owing to the pressure of other work, he never undertook the search." (Morton Grosser, op. cit., p. 92).

59. Stillman Drake and Charles T. Kowal, op. cit., p. 77.

60. There is currently a lot of quibbling among astronomers as to whether Pluto, the ninth planet, is a planet at all. If by a planet we mean a non-luminous solid body, gravitationally bound to a star, of stable mass large enough to be round, and not a satellite or fragment of such a body, then Pluto qualifies.

61. In a live presentation by Dr. Tombaugh which I attended in Orange County, California, on 16 October 1987, he mentioned that in the search he had examined photographic plates covering some 65% of the sky — and containing a total of 90 million star images and 29,000 galaxies down to 17th magnitude — with a "blink" comparator. (His search could have identified a planet the size of Earth at 2.5 times the distance of Pluto, but omitted an area south of Saggitarius inaccessible at the latitude of Flagstaff, Arizona.) Besides Percival Lowell's foresight and investment, important elements in his success included E. C. Slipher's search procedures and the Zeiss stereo-comparator that he used to "blink" thousands of pairs of time-separated exposures covering identical portions of the sky. The Zeiss stereo-comparator automates a method pioneered by Tycho and Galileo — the detection and measurement of parallax.

Pluto is not Lowell's Planet X — it doesn't explain Neptune's behavior. If there is no tenth planet, then it is left to a future mathematician to break new ground in celestial mechanics by explaining the perturbations in Galileo's data. But Tombaugh acknowledges holes in his optical search. And there may be dark matter, invisible to an optical search, perhaps even a black hole, that is perturbing Neptune. Just as perturbations of Uranus pointed astronomers to Neptune, data on perturbations of Neptune, supplied in part by Galileo's pioneering astrometric observations, are available to assist in the discovery of a tenth planet.

Already demonstrated at the Leaning Tower, Galileo's law of falling bodies has been dramatically corroborated at least

Modern Corroborations of Galileo's Law of Falling Bodies

three times in our generation. On the moon's surface, David Scott of the NASA Apollo 15 (July 1971) lunar landing mission dropped a falcon feather and a hammer simultaneously, showing that Galileo's Law holds in a vacuum. In the laboratory, Roll, Krotkov and Dicke reported in 1964 that aluminum and gold in free fall have the same acceleration, to an accuracy of 3 parts in one hundred billion ($3 \cdot 10^{-11}$). Recently, according to Lawrence Krauss of Yale University, Galileo's discovery that the speed of fall is independent of the weight of a falling body was verified when light and heavy particles from Supernova 1987A — 170,000 light years away — were received at Earth within an interval that astrophysical theory predicts for the delay between their release at the supernova.

In his *Dialogo*, Galileo suggested that Earth's motion would cause a distant star to appear, relative to a still-more-distant star, to trace out a tiny image of Earth's orbit. The tools for

Bessel's Demonstration

observing this phenomenon did not materialize for two centuries. In the 1830s, Friedrich Bessel made the first determination of the parallax of a "fixed" star, 61 Cygni: less than a half-second of arc. His data for that star showed the image of Earth's orbit predicted by Galileo.

Soon after, Foucault produced what he cited as a visible demonstration of Earth's rotation, using an iron pendulum bob suspended from the dome of the Pantheon in Paris. To

Foucault's Demonstration

create the demonstration, he applied his own discovery of the tendency of a pendulum to maintain its plane of oscillation. A pendulum, suspended so as to rotate freely, is set in motion and maintains its direction while the Earth rotates beneath it. To an observer, bound to Earth anywhere except at the equator, it appears to be slowly changing direction, and completes an apparent revolution in a time synchronous with Earth's rotation, according to the relationship:

$$T = \frac{d}{\sin \varphi}$$

where T is the time required for one apparent revolution, d is Earth's diurnal (rotational) period (about 23 hours, 56 minutes, 4 seconds), and φ is the latitude of the pendulum. Surely this demonstration was something that Galileo, whose first documented scientific observations were of a pendulum, would have loved to see. A re-enactment of Foucault's experiment, almost as dramatic as the original, underscored its clinching effect:

> "To make the matter complete, this experiment was publicly made in one of the Churches at Rome by the eminent astronomer, Father Secchi, of the Jesuits, in 1852 — just two hundred and twenty years after the Jesuits had done so much to secure Galileo's condemnation."[62]

The Pendulum, Falling Bodies, and Copernicus

We now know that Galileo developed his theory of falling bodies and of the pendulum in the same series of experiments. At the Griffith Observatory in Los Angeles, visitors can see a demonstration of the pendulum and the Copernican system in simultaneous interaction. In a continuous re-enactment of Foucault's 1851 demonstration, a 240 pound brass ball swings on a 40 foot long wire,[63] successively knocking down free standing markers and completing an apparent revolution in about 42.8 hours, which is one day divided by the sine of the latitude of Los Angeles. Here the rotation of the Earth under the pendulum carries the markers into its path, one by one, converting them into falling bodies. *Eppur si muove*!

Given a place to stand, namely on the mathematical physics of his hero Archimedes, Galileo "moved" the Earth by proving to Earthlings that the Earth moves — on its own *impeto*. As to moving Earth literally, he conceived kinematic laws that his successors have used in developing transfer-orbital maneuvers that can be applied to our planet as to ordinary spacecraft. Galileo's achievement thus realizes the phrase *SAPIENS NEMPE DOMINATUR ET ASTRIS*[64] that Vittorio Croster engraved on his lens mounting. But it did not end with the kinematics of slow-moving near-Earth bodies. As our story continues, we will see how the principle that Galileo used to counter the objections to Earth's motion based on the deception of the senses became a starting point for Einsteinian Relativity.

---

62. A. D. White, op. cit., p. 157
63. The pound is an obsolescent British unit equal to approximately 0.4536 kilogram; the foot is an obsolescent British unit equal to approximately 0.3048 meter.
64. "The intellect of a wise man will not be subordinated even to the stars" (translation mine).

Some Developments of Galilean Science up to Einstein

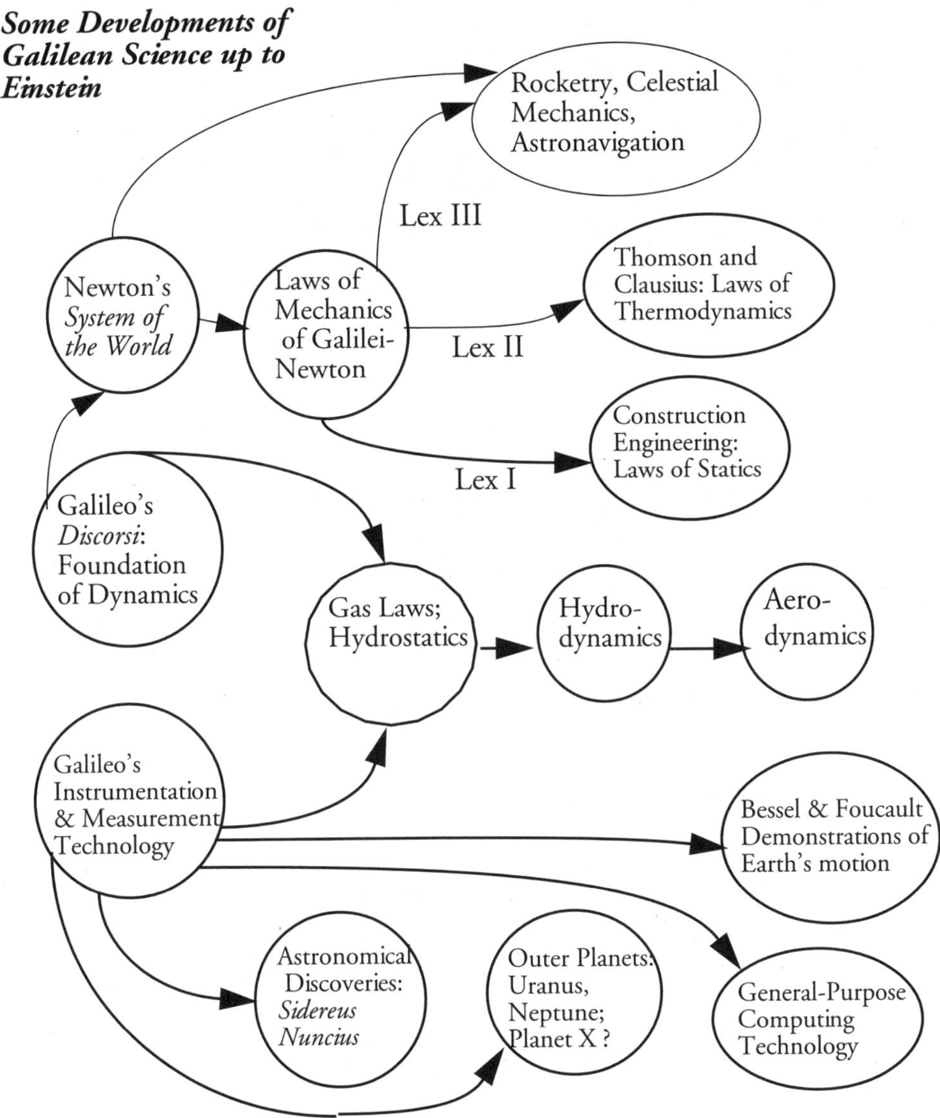

Tycho Brahe (1546-1601),
Reaper of the Information Harvest
Underlying Kepler's Great Discovery.
Monument at Weil der Stadt.
(photo by author)

Johannes Kepler (1571-1630),
Discoverer of the Laws of Planetary Motion, Founder of Geometric Optics,
Supreme Contemporary Augmenter of Galileo's *Message from the Stars.*
Monument at Weil der Stadt.
(photo by author)

Specimen of Galileo's Handwritten Notes
on the Medicean Satellites of Jupiter
(from Favaro's National Edition)

Isaac Newton

Creator of the Mathematical Principles of
Natural Philosophy and the System of the World,
a Universal Set of Dynamical Laws Unifying
Galileo's Terrestrial Kinematics and Kepler's
Celestial Mechanics into a Single Theory;
Inventor of the Reflecting Telescope.

Monument at Trinity College, Cambridge,
England

(photo by author)

Chapter 13. Path of a Pyramid
The Significance of Galileo in the History of Ideas, Part II

Galileo:
"For when the sun
draws up some vapors here,
or warms a plant there, it draws these
and warms this as if it had nothing else to do.
Even in ripening a bunch of grapes, or perhaps just a
single grape, it applies itself so effectively that it could not do more
even if the goal of all its affairs were just the ripening of this one grape." [1]

In the forests of the Tuscan morning, branches clad in red and brown leaves of autumn beckon a busy squirrel stashing acorns in his cache to make implausible leaps against the pull of Earth. Loose twigs dislodged by his acrobatics fall parabolically past parched and dying leaves that waft down into the river. Skating on ripples formed as the twigs hit the water, the leaves float to the riverbank where they join with other leaves and twigs in a decaying mulch that is nursery and home to myriad species of bugs and microscopic animals. Here in the forest, as in the deep caverns of the interstellar void, gravity is an engine of life. While centuries ago Galileo's coach made its way past countless colonies of microbes floating on a latticework of twigs by this same riverbank, this prober into the mysteries of moving, falling and floating could hardly have foreseen the extent to which his findings would prove essential to understanding life-sustaining processes everywhere. As that same coach scaled the hills of Bellosguardo to his laboratory, was he conceiving his next experiment or invention to pave the way for some successor? No doubt he was not. Yet multitudes of successors have extended developments he originated to results that he could hardly have imagined. In the origination of each development, he applied his passion for discovery and his post-Archimedean methods as effectively as if the goal of all his affairs were just the ripening of this one development.

Einstein's Continuation of Galileo's Work

Albert Einstein has improved on the traditional name "Newton's Laws of Motion" by calling them the "Laws of Mechanics of Galilei-Newton". His popular exposition of Relativity Theory begins with the Galilean principle by which Einstein formalizes the relative motion concepts discussed by Galileo in the *Dialogo* in the context of the tower, the ship's mast, and the shipborne aquarium.[2]

---

1. from translation by Stillman Drake of Galileo's *Dialogo*, Third Day, pp. 367-368.

Galilean Coordinate Systems

To accomplish this, Einstein introduces what he calls the "Galileian system of co-ordinates":

> "A system of co-ordinates of which the state of motion is such that the law of inertia holds relative to it is called a 'Galileian system of co-ordinates.' "

To this definition he immediately adds:

> "The laws of the mechanics of Galilei-Newton can be regarded as valid only for a Galileian system of co-ordinates."

He considers a raven flying so that its motion is uniform and rectilinear relative to an observer fixed on Earth's surface. Observed again from a moving train, the flight of the same raven might appear to differ in speed and direction from the first observation, but would still appear uniform in velocity. With this example he introduces the concept of uniform relative motion:

> "If K is a Galileian co-ordinate system, then every other co-ordinate system K$'$ is a Galileian one, when, in relation to K, it is in a condition of uniform motion of translation. Relative to K$'$, the mechanical laws of Galilei-Newton hold good exactly as they do with respect to K."

Remember Galileo's shipborne experiments and his principle that after the ship is set in motion relative to Earth's surface *"You will discover not the least change in all the effects named, nor could you tell from any of them whether the ship was moving or standing still"*? Einstein reformulates this "principle of relativity (in the restricted sense)":

> "If, relative to K, K$'$ is a uniformly moving co-ordinate system devoid of rotation, then natural phenomena run their course with respect to K$'$ according to exactly the same general laws as with respect to K."

This principle holds that the laws of nature are invariant for events observed by observers that are in uniform motion of translation relative to one another. Or, as Einstein later stated it in a form that he calls the *"Galilean relativity principle"*, [3]

> "If the laws of mechanics are valid in one co-ordinate system, then they are valid in any other co-ordinate system moving uniformly relative to the first."

A readily observable consequence of the principle is that when the attendant on an

---

2. Albert Einstein, *Relativity: The Special and the General Theory*, Crown Publishers, Inc., NY, 1961 (orig. pub. 1916). This topic is covered above in Chapter 9:"Licking the Tower", "Galilean Relativity and the Earth's Motion" and "Galileo's Close Brush with New Discovery".
3. Einstein and Infeld, *The Evolution of Physics*, op. cit. In this later publication, the spelling "Galileian" has been altered to "Galilean".

airplane flight pours hot coffee, it does not "fly" into your face at 500 knots.

Galilean and Lorentz Transformations

Ironically, though Galileo conjectured that light has a finite speed, c, of transmission and tried to measure it, Einstein shows that Galilean dynamics tacitly assumes an infinite speed of light. As his baseline, he introduces the Lorentz transformation showing the change of spacetime position that would be incurred by an object in transition between inertial frames; i.e., it answers with the following four transformation equations the question what is the position (x', y', z', t') relative to the coordinate system K' of the same object whose position is (x, y, z, t) relative to K, given that these two coordinate systems are moving uniformly with speed v along the x-axis relative to one another:

$$x' = \frac{x - vt}{\sqrt{1 - (v^2/c^2)}}; \qquad y' = y; \qquad z' = z; \qquad t' = \frac{t - (v/c^2)x}{\sqrt{1 - (v^2/c^2)}}.$$

Then, to the Galilean transformation he gives the following modern formulation: [4]

> "If in place of the law of transmission of light we had taken as our basis the tacit assumptions of the older mechanics as to the absolute character of times and lengths, then instead of the above we should have obtained the following equations:
>
> $$x' = x - vt$$
> $$y' = y$$
> $$z' = z$$
> $$t' = t."$$

I note that the Lorentz transformation would approach that of Galileo as $c \to \infty$. [5]

Despite what Einstein's observation has caused me to notice in retrospect that Galileo's own dynamics implied, Galileo declared that c is a finite and measurable quantity. Moreover, Galileo's primitive concepts of momentum and inertia underlie a consequence of Einstein's special relativity; viz., that the inertial mass of a particle measures its energy content, the two being related by a proportionality, the square of c. This relationship is famous in its algebraic form $e = m \cdot c^2$ and depends on the conservation of momentum of an atom, the product of its mass and its velocity. [6]

---

4. Einstein, *Relativity: The Special and the General Theory*, op. cit., p. 33.
5. My treatment of c as a variable here is only a mathematical exercise; physically, the theory (special theory of relativity) postulates that c is a constant.
6. For a complete discussion of this topic, see Baierlein, *Newton to Einstein: the trail of light*, Cambridge U. Press, 1992, chapter 11.

Einstein's Commentary on Galileo

In his "Foreword" to Drake's translation of the *Dialogo*, Einstein says of Galileo:

> "A man is here revealed who possesses the passionate will, the intelligence, and the courage to stand up as the representative of rational thinking against the host of those who, relying on the ignorance of the people and the indolence of teachers in priest's and scholar's garb, maintain and defend their positions of authority. His unusual literary gift enables him to address the educated men of his age in such clear and impressive language as to overcome the anthropocentric and mythical thinking of his contemporaries and to lead them back to an objective and causal attitude toward the cosmos, an attitude which had become lost to humanity with the decline of Greek culture." [7]

Of Galileo's motivation for boldly supporting the hypothesis of Aristarchos/ Copernicus, he says:

> "In advocating and fighting for the Copernican theory Galileo was not only motivated by a striving to simplify the representation of the celestial motions. His aim was to substitute for a petrified and barren system of ideas the unbiased and strenuous quest for a deeper and more consistent comprehension of the physical and astronomical facts."

Einstein characterizes Galileo's failure to make use of Kepler's planetary laws as "a grotesque illustration of the fact that creative individuals are often not receptive". [8]

And of Galileo's innovation in methodology of science, Einstein and Infeld say: [9]

> "The discovery and use of scientific reasoning by Galileo was one of the most important achievements in the history of human thought, and marks the real beginning of physics. This discovery taught us that intuitive conclusions based on immediate observation are not always to be trusted, for they sometimes lead to the wrong clews. ...

> Science connecting theory and experiment really began with the work of Galileo."

---

7. from German-to-English translation of Einstein's foreword by Sonja Bargmann.
8. Galileo had been quite receptive to Archimedes and other precursors, so the real issue is why his receptiveness did not extend to his contemporary, Kepler. This is one of the great unsolved mysteries in the history of science. We may never know the reason, though it was not lack of respect for Kepler. In the *Assayer* Galileo said: "Kepler has always been known to me as a man no less frank and honest than intelligent and learned"; trans. by Stillman Drake, *Discoveries and Opinions of Galileo*, op. cit., p. 262.
9. Einstein and Infeld, *The Evolution of Physics*, op. cit., p. 6 and p. 52.

Galileo, Computer Science, and System Methodology

To historians who specialize in the information-processing industry, scientific methodology explains why the history of computing machines begins with Galileo.[10] But there are other and equally important reasons for this industry to consider Galileo one of its founders, reasons that I have not come across in the literature on the history of information systems.

Prior to the Pascal-Leibniz calculator, analog calculators were invented for special purposes; for example, the lunar eclipse computer and planetary

General-purpose Computing Machines

computer of the Iranian astronomer/mathematician al-Kashi (1393-1449).[11] There was, of course, Galileo's jovilabe. Prior to that, he invented his geometric compass, which set an important precedent: it had general problem-solving capabilities.[12] Another milestone was set by Wilhelm Schickard's (1592-1635) calculator, a mechanical digital instrument that performed addition and subtraction with carry and borrow operations, and limited multiplication and division. Tragically, this machine and its inventor became casualties of the Thirty Years' War and the ensuing plague. Unaware of Schickard's instrument, Blaise Pascal developed his (somewhat less advanced) calculator about 1642, the year of Galileo's death and Newton's birth. This is the machine which Leibniz improved to give it multiplication and division capabilities; it became the prototype of modern mechanical calculators. In a sense, then, that pivotal year 1642 can also be regarded as the start of the era of general-purpose mechanical digital calculation.

There is an independent reason to regard Pascal along with Galileo as a "father" of the computer industry : it was Pascal who, in collaboration with Fermat, developed the foundations of probability theory. Modern computers are designed, modelled, and analyzed as flow networks; to the extent that these activities are done scientifically, they are made possible by queuing theory, a derivative of probability theory. Pascal's probability theory was extended by Laplace, who first applied it to astronomy. From al-Kashi's and Galileo's instruments to Laplace's probability theory, a strong case can be made that the computer technology of today has its origins in astronomy. Reciprocally, the fledgling computer industry is benefitting its parent, the astronomy industry, through computer-based astronomical instrumentation and automation of celestial mechanics calculations.[13]

---

10. "Prior to Galileo (1564-1642) there were of course intellectual giants, but his great contribution was to mathematize the physical sciences"; Herman H. Goldstine, *The Computer from Pascal to von Neumann*, Princeton U. Press, 1972. As Goldstine's account of the history of the computer begins with Galileo, he does not discuss the Greeks, but his statement would better say "re-mathematize" in view of Archimedes.

11. Ibid.

12. Ch. 5, "A Revolutionary General-purpose Computer".

Galileo's Groundwork in Computing Technology

Semi-automatic calculators are a step in the direction of programmable algorithmic machines. Galileo invented a prototype of analog calculators, the geometric proportional compass. An advance in analog calculators was the slide rule, which multiplies by addition of logarithms (invented by a Scottish contemporary of Galileo, John Napier) rather than by Galileo's scaling ratios. Yet the proportional compass was no less general in the scope of its utility than was the slide rule. Equally important in the development of computer engineering is the concept of a finite and measurable speed of light and, as we know from Maxwell's theory, of electromagnetic signals in general. An electromagnetic oscillation is like Galileo's pendulum timer with the added benefit of near-synchroneity over short distances between the clock and the timed events; i.e., it is a timing mechanism capable of synchronizing, over a distance, the operations of an algorithm embodied in a machine. When an event at one position in a logic circuit depends on the time of an event at another, the path length of a signal that traverses the pair of positions is a factor in the circuit layout. As computing machines shrink, circuits must be designed according to the speed of signal propagation, which must be known with increasing precision. Designers must know not only that light has a measurable speed, but what its exact value is. Galileo has made a fundamental contribution to computer technology through his hypothesis that light has a finite speed and his original design for later experiments that measured it.

New Industries from Two Galilean Ideas

Galileo's proportional compass is to modern general-purpose computing machines much as a Wright glider is to modern commercial aircraft. His idea of general-purpose calculations embodied in a machine was extended by Charles Babbage in 1833 when he undertook the development of a general-purpose programmable computing machine,[14] and by John Vincent Atanasoff in 1940 who designed the first such electronic machine. Galileo's concept of general-

---

13. Programmable computing machines are spectacularly useful in celestial mechanics. As a student, I was assigned to take three pairs of observed coordinates of a celestial object, and to calculate the orbital elements that determine the size and shape of the object's orbit and the orientation of its orbit in space. In those days, neither an electronic nor a mechanical calculator was available to me. After about thirty pages of calculations and many hours of work with pencil, paper and logarithms, I discovered that the method I had chosen was non-convergent for those coordinates. Nowadays, a test for convergence and the actual calculations can be carried out in a matter of seconds by a suitably programmed computing machine.

14. Cf. Arno Penzias, *Ideas and Information*, Simon and Schuster, 1989. Babbage's difference engine is described and shown in Doron D. Swade's article, *Le Scienze* #297 Maggio 1993, p. 40 ff. Babbage, who referred to Italy as "the country of Archimedes and Galileo", corresponded with a successor of Galileo's employer in 1828 and later. Babbage sent the Grand Duke of Toscana specimens of British manufactured goods, receiving in return a thermometer from Galileo's time. See Anthony Hyman, *Charles Babbage: Pioneer of the Computer*, Princeton U. Press, 1982, pp. 181-183.

purpose computing instruments has been applied to machines of the 20th century, but not uniformly. What are called "personal computers" used to feature access to the hardware instruction set, or when powered up they would vector to an interpretive computer language processor.[15] The present trend is to make machines featuring pre-packaged application software and a set of "icons", gates to special-purpose programs. Access to general-purpose capabilities present in the instruction set of the underlying hardware is usually considered an optional "enhancement" for which one must buy and install a compiler or other software language processor. As a market response to users who demand ease of use and are willing to forego general-purpose capabilities, the general-purpose machine has been downgraded to a set of special purposes at a cost to the user. But just like Galileo's proportional compass that offers ease of calculation to individuals, the advent of personal computing and networking can be seen as a favorable development.[16]

A variation of Galilean scientific method that lends itself to the production and use of computing machines has found a home in systems engineering, another industry growing from Galileo's ideas. A model in systems engineering is typically the incorporation, into a simulation or set of mathematical expressions, of a predictive hypothesis about how the system being modelled will perform. In part due to the requirement to model systems that do not yet exist and cannot be observed, Galilean scientific methodology has evolved into system methodology:[17]

"Simulation is based on a problem solving method that has been in use for many years, sometimes referred to as the model-building method or more commonly the scientific method. Thus when system simulation is used to solve a problem, the following time-tested steps, or stages, are applied.

1. **Observation of the System**
2. **Formulation of hypotheses or theories that account for the observed behavior**
3. **Prediction of the future behavior of the system based on the assumption that the hypotheses are correct.**
4. **Comparison of the predicted behavior with the actual behavior.**

... The scientific method's requirement for prior observation of the system has resulted in a slightly different approach to problem solving, called system methodology. This

---

15. The first personal computer I recall was the Altair 8800, which permitted the user to read, enter or modify machine instructions via a panel. Now that's general-purpose!

16. It should also be noted that Galileo's compass is the product of an individual; the modern personal computer is a complex aggregation of products from autonomous sources lacking rigorously defined interface constraints. This makes computers cheaper to produce and also lacking in robustness and reliability, as software products "step on each other" (contend for the same resources, like dynamic random-access memory). In case you think that this is some remote phenomenon not likely to affect you, you should be aware that it already has. It has, for example, delayed the availability of this book to you.

17. Cf. Graybeal and Pooch, *Simulation: Principles and Methods*, Winthrop Publishers, Inc., Cambridge, Mass., 1972, p. 5.

methodology consists of four phases: planning, modeling, validation, and application."

These steps can be compared with Galileo's as cited by MacLachlan:[18]

"Galileo brought physics into mathematics alongside astronomy, by modeling his way of studying motion on the methods in astronomy. In 1602, he wrote a memorandum on how to investigate nature, stating the procedure in four parts:

1. start with the phenomena, the sense observations we see every day;
2. then make hypotheses, suppositions about the underlying structure of things;
3. use geometry to demonstrate how various events can follow from the hypotheses;
4. finally, use the geometrical model to calculate particular values for the phenomena.

Although Galileo didn't say so, his own methods in following years used an additional step:

5. compare calculations from the model with measurements of the phenomena. And where they compare poorly, revise your hypotheses (step 2), and then repeat steps 3 and 4."

Theory of Statistics, a Tool of Mathematical Physics

The phases of the system methodology correspond to the steps of the scientific method; the system is planned rather than observed, its behavior is described in an analytic model rather than in a hypothesis, and the results may be validated by comparison with those produced by simulation of the real system, after which the model is applied to the real world. The validation requires statistical algorithms to reduce, interpret, and fit real data to the model; hence, system methodology creates demand for the development of mathematics. Outstanding examples include the "Student's t" function formulated by W. S. Gosset, and the Kolmogorov-Smirnov algorithm.[19] Indeed, the growth of system methodology spurred the entire new science of queuing theory, beginning with A. K. Erlang's pioneering investigations of the arrival and service

---

18. Here MacLachlan has revised Drake's translation (in *Galileo at Work*, p. 352) of three opening paragraphs in Galileo's manuscript for his course on Cosmography; see James MacLachlan, *Children of Prometheus*, pp. 121-122. MacLachlan tells us that "Galileo devised this procedure by elaborating on some statements that had appeared in the preface of Ptolemy's *Almagest*."

19. Gosset published his paper in 1908 under the pen name of "A. Student". "Gosset worked for the Guiness Brewery in Dublin which, at that time, did not allow its employees to publish research papers."; cf. Arnold O. Allen, *Probability, Statistics, and Queuing Theory*, Academic Press, NY, 1978. This excellent book discusses both Student-t random variables and the Kolmogorov-Smirnov test. The author was A Student of Dr. Allen.

distributions of telephone calls at a switching center.[20] And the value that Archimedes and Galileo created in applying mathematics to physics has returned to mathematics in the self-reinforcing cycle of theory and application of the theory through system methodology. Such an application today is becoming indispensable to astrophysical research: the computing technology of adaptive optics has erased the effect of atmospheric turbulence and light pollution,[21] algorithmically cleansing Galileo's window to the stars.

Misapplications of Mathematics

Any outrage can be justified by those who apply science while abandoning its foundation of rational principle. In particular, the superhuman computing power of modern machines can be misapplied to tasks for which the superautomatous thinking ability of a human is more important than speed.[22] Galileo said, "*... once you have denied the principles of the sciences and have cast doubt upon the most evident things, everybody knows that you may prove whatever you will, and maintain any paradox.*"[23] Paradoxically enough, as if its own economic growth has no relationship to that of the rest of the world, the United States imposes on its international borders a massive interference, aided by computer-based algorithms, in physical movement.[24] In the harbor she adorns at New York, the great Lady Liberty, an accepted gift from France, stands proudly beckoning the oppressed of other nations, an otherwise grand gesture spoilt by this hypocrisy. While computing machines blindly play software roulette to select tens of thousands for immigrant visas,[25] the act of employing already-productive resident "aliens" is outlawed, and these persons risk deportation. Are linear congruential random-number generators running in machines more informed or intelligent than the decision that a business owner in country X and a skilled contractor in country Y would make if they had the opportunity to meet each other? "*Let us not make random conjectures about matters of the highest*

---

20. Erlang first saw the importance to queuing theory of the parameter "traffic intensity", defined as the ratio of mean arrival rate to mean service rate; cf. E. Brockmeyer, et al., *The Life and Works of A. K. Erlang*, Acta Polytechnica Scandinavica, Mathematics and Computing Machinery Series No. 6, MA 6 (287/1960), Danish Academy of Technical Sciences, Copenhagen, Denmark, 1960, orig. pub. 1948.

21. The direct effect of adaptive optics is a sharpening of the image. As a point source of light becomes concentrated into a sharper image, the image narrows while its amplitude correspondingly increases. Hence, the amplitude boost can carry a signal above the threshold of background light pollution.

22. Not to mention admirable qualities of humans (like compassion, kindness and gratitude) that no artificial intelligence duplicates. Increases in computer power continue to mount, as massively parallel techniques complement manufacturing innovation; e.g., the AT&T-IBM-Motorola-Loral joint venture to use X-rays instead of ultraviolet light in semiconductor production. See the *Wall Street Journal*, 27 July 1994.

23. translation by Stillman Drake, *Dialogo*, First Day, op. cit., p. 41.

24. This conclusion is based on my research in 1977 leading to an application of economic principles of David Ricardo to immigration and tariff restrictions.

25. for example, the State Department's "DV-1 Lottery" of 1994.

importance", said a philosopher identified near the close of this chapter. There is no surer sign of a civilized people than an enlightened attitude toward personal mobility, a prerequisite of that freedom of inquiry for which Galileo won a long-term victory. Yet four centuries after him, the gates to the first and only stronghold of individual freedom on our planet are policed in emulation of the gatekeepers at Arcetri. There are other inappropriate applications of machine-based mathematical algorithms, whose detection I leave as an exercise for interested readers.

Operations Research What is the relevance of mathematics to everyday problems? Galileo had been a consultant to the Venetian arsenal on practical matters that required calculation, such as the optimum positioning of oars in a naval galley. The application of mathematics to military operations during World War II, and then to business operations after that war, became known as operations research. In modern terms, Galileo had been an operations research consultant to the Venetian arsenal.

At least two developments in mathematics led to modern operations research. First, because businessmen frequently have reason to quantify their uncertainty, operations research makes extensive application of the probability theory pioneered by Pascal and Fermat and extended by Laplace. Many operations research algorithms have been or are being embodied in computer software, the (correct) production of which is itself a mathematical discipline.[26] The other development is the calculus, formulated independently by Newton and Leibniz, partly in response to questions arising from Galileo's science of dynamics.

A small sample of questions tackled by modern operations researchers spans an enormous diversity of applications.[27] Airlines, for example, may use a probability distribution for the unavailability due to illness of crew members in order to decide how many pilots and flight attendants should be held in reserve as replacements.

Manufacturers can determine bidding strategies by keeping records of bids submitted on contracts with corresponding cost estimates, characterizing the difference in bids as a random variable with known mean and standard deviation. Retail managers can determine how many of each seasonal item to order from a density function characterizing the probability that demand for the item will lie in a given interval. Automobile dealers who know the function p(x) for the probabilities of a demand of x cars in a week, the one-time cost of ordering a new

---

26. "... correct [computer] programming involves only combinatorial selection, and not problems requiring perfect precision, on a continuous scale." Harlan D. Mills, "Mathematical Foundations for Structured Programming", IBM Document FSC72-6012, February 1972.
27. For discussion of these kinds of problems, cf. Sasieni, Yaspan, and Friedman, *Operations Research: Methods and Problems*, John Wiley and Sons, NY, 1959.

car, and the cost per unit time of holding a car in stock due to insurance, deterioration, interest on borrowed capital, etc., can determine the frequency of ordering and the optimum level of inventory. Bookbinders who record the distributions of times to print and bind each book based on manuscript size and other variables can use operations research to determine in what order to process the books so as to minimize the total time required to print all orders.

Even a barber shop owner can use operations research: if his customers arrive at a known average Poisson[28] rate, and he tracks the average time to complete one haircut, he can determine how many barbers to hire so that a customer has less than a 10% chance of having to wait more than ten minutes.

Astrophysical Research Extending Galileo's Discoveries

Can you identify the following activities in time and in place?

- Measurement of the speed of light;
- Measurement of the rate of a star's rotation;
- Using the telescope, expanding the visible universe;
- Moving ourselves away from its imagined center.

If your answer is California in the 20th century,[29] then you are as correct as if you had answered Italy in the 17th century. On Mount Wilson, using giant telescopes with up to $(254/1.135)^2$ or more than fifty thousand times[30] the light amplification of the one Galileo used to discover the Medicean satellites of Jupiter, astronomers have discovered that:

- We inhabit an island universe of stars — a galaxy;
- There are many other galaxies separated by vast distances from our own;
- They are receding from us in a way that suggests, not that Earth is the center of these motions, but that ...

---

28. The Poisson distribution, named for its discoverer Simeon-Denis Poisson (1781-1840), is fundamental to problems such as radiation and traffic flow, wherein interarrival times are exponentially distributed. If arrivals are Poisson-distributed, then the probability that the interarrival time does not exceed the value t is simply $1 - e^{-\lambda t}$, where λ represents the average arrival rate.

29. There is yet another parallel: the discovery of four new satellites of Jupiter in the 20th century in California by one person; namely, Seth Nicholson (source: Laura Eklund, as confirmed in my personal conversation with Don Nicholson, son of Seth Nicholson). Nicholson discovered satellite #9 in 1914 with the 0.914 meter Crossley reflector at Lick Observatory, and #10, #11 and #12 between 1938 and 1951 with the 2.54 meter telescope at Mount Wilson; see also Biographical Memoirs of NAS vol. XLII, Columbia U. Press, 1971. (Galileo still holds the record on satellites discovered per unit time.)

30. This calculation depends on my conjecture that Galileo's objective lens was stopped to a diameter of 1.135 cm based on the assumptions and calculation I make in Chapter 5 "Galileo's Handmade Lenses".

there is no center.

Galilean Astronomy at Mount Wilson

These discoveries emerged from explorations and measurements carried out by 20th century successors of Galileo, primarily at Mount Wilson. Indeed, Edwin Hubble, whose research there culminated in his 1929 announcement of the expanding universe, had 14 years earlier described his delight upon reading Galileo's *Discourses*. Mount Wilson Observatory, enshrouded in light pollution after Harlow Shapley had mapped our galaxy and Hubble had discovered the linear expansion rate of space (the "Hubble effect"), is making its comeback as a leading observatory due to adaptive optics technology and to interferometry.[31] Algorithmically cutting through the light pollution, an adaptive system installed on the 2.5-meter Hooker Telescope produces images with a resolution approaching the theoretical diffraction limit of its mirror: about 0.05 seconds of arc, or the angular diameter of a grapefruit in Los Angeles if it could be viewed from a high enough tower in Las Vegas. The CHARA array of five 1-meter telescopes, scheduled for initial operation in 1998, will operate as a single 400-meter telescope, and "resolve detail 100 times finer than the Hubble Space Telescope — the power equivalent of seeing the disc of a nickel located 10,000 miles away".[32] Using information in interference fringes that result from splitting a light beam, sending the parts along different paths, and recombining them in controlled phase relationships, the array will enable astronomers to detect planets around distant stars and the formation of proto-stars. Already today at Mount Wilson, astronomers are measuring stellar rotation rates by means of "starspots", phenomena that Galileo discovered on our own star and correctly predicted of other stars.[33]

Galilean Science: Unfinished and Unfound

Much hypothesizing and testing remains to be done in cosmology, using telescopes of varying wavebands and other space-mapping tools. The determination whether the mass of the universe is sufficient to reverse the Hubble effect, recollapsing itself, is important to understanding where we are going.[34] If,

---

31. Adaptive optics technology was pioneered by Horace Babcock; cf., *Unusual Telescopes*, Peter L. Manly, Cambridge U. Press, 1991. Interferometry depends on the discovery of diffraction by the Jesuit Father Grimaldi and was developed by Albert Michelson, who initially called it "interference refractometry"; Loyd S. Swenson, Jr., *The Ethereal Aether*, U. of Texas Press, Austin, 1972. (Source: Robert Eklund).

32. source: Center for High Angular Resolution Astronomy (CHARA) at Georgia State University. For the edification of future readers of this book, a "nickel" is an obsolescent U. S. coin about 2 cm in diameter.

33. "And if the generations and corruptions occurring on the very globe of the sun are so many, so great, and so frequent, while this can reasonably be called the noblest part of the heavens, then what argument remains that can dissuade us from believing that others take place on the other globes?" Galileo Galilei, *Dialogo,* Drake trans., The First Day, p. 58.

34. or, indeed, whether it is an absolute truth that expansion is the only explanation of the Hubble effect. (rdc)

as in the eternal oscillations of the Brahman spider,[35] there are cyclic nodes in the expansion/collapse sequence at which natural laws can be reshuffled, might there be a potential for internodal re-engineering?

Besides Galilean science not yet extended, there is other Galilean science that has simply been lost. If you are a historian of science, there is the unsolved problem of Galileo's lost papers: on the mathematical theory of continuity and indivisibles, and on light and color. While the Galileo papers were still in the custody of the Viviani estate there were, as we have seen, many loaned out and otherwise let loose that have never been recovered; very few of his letters during 1621 have survived. There are unresolved issues relating to his inventions. Little is known about the structure of the early compound microscopes that Galileo made. These problems do not become easier to solve as the events gain remoteness in time.

✦✦✦✦✦

"... And would it have been worth it, after all,
After the cups, the marmalade, the tea,
Among the porcelain, among some talk of you and me,
Would it have been worth while,
To have bitten off the matter with a smile,
To have squeezed the universe into a ball
To roll it toward some overwhelming question, ..."

... T. S. Eliot [36]

The Physics of Black Holes

Galileo's hypothesis of the finite speed of light is fundamental to the 20th-century concept of a spacetime locus from which light cannot escape due to the presence of a sufficiently compact mass. Pierre Simon Laplace (1749-1827) was a mathematician and astronomer who (although perhaps better known for his creation of modern probability theory) had applied the mechanics of Galilei-Newton to determining the long-term stability of the solar system, a problem to which his contemporary Lagrange also greatly contributed. Not only was Laplace aware of Galileo's hypothesis of finite light speed, but by his time, Roemer had already turned hypothesis to fact by measuring that speed.

In 1796 Laplace used this fact to express a distance from a point mass, at which the Newtonian escape velocity exceeds the velocity of light, as proportional to:[37]

---

35. "Just as the spider pours forth its thread from itself, and takes it back again, even so the Universe grows from the imperishable"; *Mundaka Upanishad* I, i, 7.

36. from "The Love Song of J. Alfred Prufrock", in *A Comprehensive Anthology of American Poetry*, op. cit.

$$G \cdot M / c^2$$

where G is Newton's universal constant of gravitation, M the mass of the object, and c the speed of light. Later, Karl Schwarzschild (1873-1916), who created an Einsteinian model of the solar system, determined the constant of proportionality to be 2. The imaginary sphere generated by the Laplace-Schwarzschild radius is called the "event horizon". As Newton demonstrated mathematically, every extended mass behaves gravitationally as would a point mass situated at its center of mass. What if an object were so densely massive that its event horizon encloses its own extension? Laplace conjectured the existence of objects sufficiently massive that light (or anything else) could not escape their gravity. Since each such entity would be an invisible dark mass to which any hapless passerby would be irretrievably extracted from the rest of the universe, John Archibald Wheeler named them "black holes". If light had an infinite speed, the radius around a mass within which light is trapped would be zero. Galileo's hypothesis of a finite light speed is required for us to understand how a black hole could exist.

An Invisible Star in the Swan

In our discussion so far, the black hole is a theoretical construct, a product of the mathematical representation of abstract events. Like Lagrange's L-points, which had no known representation in physical reality until Max Wolf in 1906 discovered asteroid 588 apparently "illegally parked" in Jupiter's orbit, 60° ahead of the giant planet,[38] the black hole began as a mathematical creation. From the time of Laplace's conjecture until astrophysicists began to estimate the masses of stars and calculate the conditions of their collapse, it remained so. Oppenheimer, working out the properties of neutron stars[39] in 1939, used Einstein's general relativity to predict what would happen to a sufficiently massive star, say greater than 3.2 solar masses, at the end of its life: it would continue its collapse toward a mathematical point, known as a singularity, at which all of its mass would be concentrated.[40] Then in 1971, astronomers discovered a flickering X-ray source, invisible at wavelengths detectible by the human eye. As its flicker interval is bounded by the travel time of light across its diameter, its flicker rate implies that it is smaller than

---

37. Cf. P. C. W. Davies, *Space and time in the modern universe*, Cambridge U. Press, 1977.

38. In 1774 Lagrange had published a purely abstract exercise in mathematical physics, proving that a 3-body system arranged in an equilateral triangle would form a stable gravitational system. After the discovery of 588 ("Achilles") and companions, and a similar cluster 60° behind Jupiter, these two parking positions 60° ahead and behind an orbiting planet became known, appropriately, as the Lagrangian points of its orbit.

39. In the inital collapse of a sufficiently massive star, atoms are crushed so that electrons and protons have coalesced into a super-dense "soup" of neutrons that support the star against further collapse. One teaspoonful of matter from the surface of such a star would outweigh a typical freight train, and an ocean liner dropped onto its surface would be crushed to about the size of a grain of rice; cf. P. C. W. Davies, *Space and time in the modern universe*, op. cit.

40. Cf. Isaac Asimov, *Asimov's New Guide to Science*, Basic Books, Inc., NY 1984.

the Earth. Cygnus X-1, as it is called, has a mass greater than 6 solar masses based on the perturbations of its visible companion, HDE 226868.[41] Taken together, these facts put its event horizon outside its mass: the defining property of a black hole. As evidence mounts for the physical reality of these mathematical constructs known as "black holes", the view of gravitation as a spatiotemporal property that harnesses planets kinematically in a Copernican system gains validity.

The "String" That Holds the Earth

How then is the Earth harnessed in orbit around the sun when, as a space traveller directly observed from his vantage point on the way to the moon, there is no string attached to it? Can a force act through a distance? Galileo developed a kinematic science of motion that did not refer to forces acting through distances. By contrast, Newton built a dynamic science of motion and postulated a gravitational force field co-extensive with physical space. In Einstein's general relativity, the force field of Newtonian gravitation expressed in Galilean coordinates becomes only a property of spatiotemporal geometry. In this Einsteinian view of gravitation, Newton's forces acting through distances are gone, and Galileo's vision of bodies moving in conformity to geometrical laws rather than to physical forces is realized.

With General Relativity, Einstein showed that objects in motion follow the geodesic of the spacetime through which they move, a curve affected by the proximity of masses. The "string" that holds Earth in its orbit, against its Galilean tendency to move forever away from the sun on a tangent line, is a geometric harness created by the proximity of the sun and the consequent curvature of spacetime in its vicinity. As the material of a star becomes more and more densely compacted, the curvature increases. And as a massive dying star collapses to a singularity, the curvature of spacetime surrounding it approaches infinity. Information about events internal to the black hole is lost to the outside world.

After Einstein, the understanding of the physical world refers back to the geometry of Thales of Miletos and his successors more than ever before.

Squeezing the Universe Into a Ball

What if an object of any mass whatever were to collapse or be squeezed into a ball smaller than its event horizon? For an object the mass of Earth, it would have to be compressed roughly to the size of a ping-pong ball. The 20th-century physicist Hawking, who collaborated with Penrose in developing a series of theorems dealing with the topology of spacetime, has suggested the existence of

---

41. The X-ray flickering was explained in 1984 by Smarr and Hawley as due to hot spots on the rotating inner edge of a disk of hot gaseous material ("accretion disk") orbiting and swirling into Cygnus X-1 from its visible companion; see William J. Kaufmann III, *Universe*, W. H. Freeman, NY, 1985, p. 451.

microscopic black holes, out of which energy can leak quantum-mechanically and thus visibly as light energy.[42] The model predicts that under certain conditions a black hole by this process can lose its entire mass to the surrounding space, like Dodgson's Cheshire Cat leaving behind only a smile formed by ripples of light.

The poet T. S. Eliot spoke of squeezing the universe into a ball and rolling it toward some overwhelming question of which he said: "Oh, do not ask, 'What is it?'" Ignoring his advice, I would ask: is it whether the universe of our experience is itself a black hole? The answer awaits the determination of the mass and radius of what we know as the universe. Now that we've asked "What is it?", we can pass through the doors of cosmology and make our visit to a related question: will that universe of our experience ultimately recollapse or expand to oblivion?

The way in which future evolutes of our species will react to large-scale changes in the universe will depend on developments that grow from the work of Galileo and of successors to whom he is operationally connected.

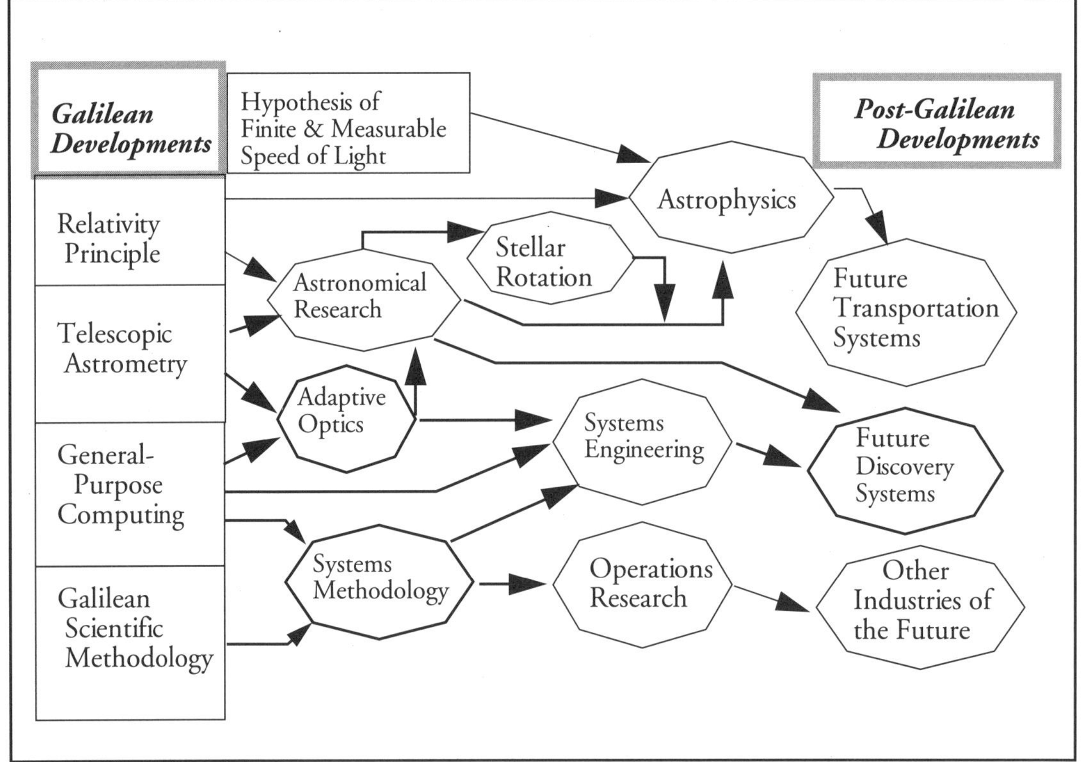

---

42. Ibid.

Star dust is more than a song writer's fascination:[43] it is the stuff of which planets are formed. The dust is mixed

Gravity and Life

with gas, and as we depend on gaseous explosions in internal combustion engines to move about in our automobiles, we depend on explosions of gas in the stars for our very lives. From the work of the astrophysicist Eddington, we can see that Galileo's law of falling bodies, extended by Newton to universal gravitation, takes on cosmic significance in a scenario in which planets are formed out of the gravitational collapse of vast clouds of gas and dust released by dying stars:

> "A star is born out of the condensation of clouds of gaseous hydrogen in outer space. As gravity pulls together the atoms of the cloud, its temperature rises until the hydrogen nuclei within it begin to fuse and burn in a series of reactions, forming helium first, and then all the remaining substances of the universe.[44] The elements of which our bodies are composed were manufactured in this way, in the interiors of stars now deceased, and distributed to space when these stars exploded. Subsequently, these elements were drawn together again in the cloud of gas out of which the sun and the earth condensed. If the sun explodes at the end of its life the planets will be consumed, and their substance once again distributed into space, to be reincarnated in another solar system as yet unborn."[45]

Traditionally, theology has concerned itself with the source of life. Not theology but astrophysics is the source of the concept that the energy required to sustain life on this planet is the energy emitted by a gravitational machine called a star. So much depends on loose dust. From its scattering at the base of the Leaning Tower to its coalescence in interstellar space, and on through the controlled collapse of a star that radiates the energy to nourish its orbiting life forms, Galileo's research is fundamental to the connection between falling bodies and the evolution of life.

Nor is the origin of life the only question which theology has been historically concerned to ask and astrophysics more recently concerned to answer.

---

43. The song "Stardust" is a composition of Hoagy Carmichael.
44. Dr. Jastrow has earlier explained that these reactions include supernovae in which the formation of elements higher than iron in the periodic table are manufactured "in the last gasp of a star's life".
45. Robert Jastrow, *Red Giants and White Dwarfs*, Warner Books, 1979.

Cosmology and Epistemology A paramount application of Galileo's work is to cosmological questions about the structure of the universe and its development cycle. Derivative questions, traditionally addressed by theologians, about whence we came and where we are headed as a species, are likewise susceptible to applications of Galilean science. Recognition of the Aristarchan/Copernican/Keplerian/Galilean revelation that Earth is not a special place around which the universe is designed is an initial step toward developing a cosmology by reference to the book of nature rather than the book of authority. Twentieth century discoveries such as Hubble's Law,[46] the cosmic microwave background,[47] and the physics of black holes — all outgrowths of the work of Galileo, Kepler, Newton, and their successors — supply the physical meaning absent in mythical allegories of creation, like Genesis taken literally. Such mythical accounts violate the dictum *ex nihilo nihil est*,[48] which paraphrases the later conservation-of-energy principle respected by accounts of the origin of the universe that arise from the exercise of Galilean scientific methodology.

The transition of operations research from military to business applications is a significant trend in the history of civilizations after Thales' application of science to the olive industry. When individuals the world over profit by unrestricted application of their skills to each other's demands and when economics is no longer perceived as a zero-sum game, what will be the perceived utility of an offensive military campaign?[49] It may be easier to populate such a campaign (for example, a religious "crusade") with men conditioned to believe on authority that death is only the loss of the less important physical aspect of an enduring self. It may be easier for anyone to confront loss of life with the expectation that this is true. But expectation does not equate to knowledge, and propositions about afterlife can be neither true nor false until they become testable. Bound up with the perennial question of assessing the meaning and value of the self is the method of testing and validating one's claims to knowledge; i.e., epistemology.

---

46. Edwin Hubble's discovery of the general proportionality to distance of the measured speed at which galaxies are receding from Earth has been variously called the Hubble effect, or Hubble's Law. For a detailed historical development of this topic, see Berendzen et al., *Man Discovers the Galaxies*, p. 197.

47. Knowledge of the cosmic microwave background is the product of two separate discoveries: (1) the discovery by Arno Penzias and Robert Wilson of AT&T in 1964 of an isotropic radiation matching a blackbody energy curve at 2.76° Kelvin; (2) the theoretical prediction by George Gamow in 1934, reconfirmed by Dicke and Peebles c. 1960, of a radiation remnant of the "Big Bang" fireball matching the characteristics received by the AT&T antenna and recognized by Penzias and Wilson. This discovery is a fine example of the application of experimental evidence to theory, a method pioneered by Galileo.

48. "Nothing comes from nothing." I recall hearing the dictum, probably from a Jesuit instructor, but have not been able to determine its origin.

49. See Ludwig von Mises, *Omnipotent Government*, Arlington House, New Rochelle, NY, 1969.

Of the painting reproduced at the end of this chapter, I do not know what intent may underlie the artist's vision in creating it. To me, it suggests old and new epistemologies. The foreground portraying a procession of ecclesiastics symbolizes claims to truth based on appeal to authority. The background depicting the cathedral and Leaning Tower, sites of Galileo's pendulum and gravity demonstrations, suggests approach to truth by integration of experiment and measurement into a consistent body of theory, pioneered by Galileo. At last it is possible to replace the creation and propagation of fables about the origin and destiny of the universe with a method of truth-seeking accessible to the individual. At last is is possible to replace prescriptive principles of morality with derivable ones. Since Heraclitos left us his Fragment 47, there has been an insight that better methods are required for handling these questions:

$$\text{ΜΗΕΙΚΗΠΕΡΙΤΩΝΜΕΓΙΣΤΩΝΣΥΜΒΑΛΛΩΜΕΘΑ}^{50}$$

And since Galileo, there is a better tool for deciding them.

What is the importance of cosmology to you and me, right here and now, at the end of the 20th century, the start of the Third Millenium? It is this.

Remember always that you are an individual far different from anything else or anyone else in the universe. If you could arrange all the free-standing particles and systems of particles in the universe in a pyramid whose base consists of the most common of these things, in terms of mass in the total volume of space, you would expect to find at the base highly abundant things like neutrinos and free-standing atoms that populate intergalactic space. Moving up the pyramid, you would find planets and stars, less common, but still composed of unorganized elements and still inanimate. Continuing your ascent, you would find on the planets molecules, complex crystalline structures and then (still higher and rarer) amino acids, the building blocks of life. Rarer still are the life forms themselves. At this point you may well pause to note that the ordering by commonality turns out to generate roughly the same pyramid as if you had ordered by structure; i.e, as you ascend the pyramid, you encounter things not only rarer but more highly organized.

Throughout the known universe, there are myriads of galaxies, each with billions of stars, separated by a vast expanse of nearly empty space. You live here on a

---

50. "Let us not make random conjectures about the most important matters"; trans. by T. M. Robinson, *Heraclitus: Fragments*, U. of Toronto Press, 1987.

relatively tiny and isolated planet that is teeming with rare and complex life forms, like the southernmost tree in the Black Forest. Among these life forms is the species of humans, the pinnacle of the pyramid. By smearing the pyramid along a fourth dimension of time in your imagination, you would generate a prism that contains the evolutionary path of the cosmos. In the aeons of time from the formation of our planet until our star is scheduled to exhaust its nuclear fuel, the lifetime of an individual human is like a brief candle flicker. The universe is enormous and unaccommodating to our species, the top edge of the prism. Worse than its neutrality is the destructive capacity of those man-made weapons wielded against reason by various institutions throughout human history: by the early creators of dogma, later by the Inquisition and, continuing into the present, by institutions whose product is indoctrination. Yet many individual humans, each in his or her own small or large way, have conquered and harnessed the otherwise destructive forces of nature, including those of our own species. Viewed along a fifth dimension, the dimension of human advancement, observe that this top edge of the prism differs from all the underlying layers in that its progress is self-directed by its own individual members, as Galileo's path of progress through formidable obstacles thrown up by collectively mindless institutions shows clearly. You could be one of those individuals.

Galileo's Baptistry: The Old and New Epistemologies
(painting by Carlini; purchased by author from
street vendor near Cathedral of Pisa)

Palomar Observatory (5-Meter Hale Telescope Dome)

Monument to Galileo, Inventor of Optically Aided Astronomy, and to George Ellery Hale, Patron of Astrophysics and Founder of Observatories. On Account of Galileo's Far-sighted Instrument and of Hale's Far-sighted Vision and Investment-Promotion of Optically Aided Astronomy, Astronomers Have Acquired Information about Man's Position in the Universe and Greatly Improved Galileo's Estimate of the Size of Interstellar Space, the Best Estimate Available in His Time (see p. 163 footnote 22)

(photo by author)

Epilogue: The Gatekeepers of *Il Gioiello*

"In Santa Croce's holy precincts lie
Ashes which make it holier, dust which is
Even in itself an immortality,
Though there were nothing save the past, and this,
The particle of those sublimities
Which have relapsed to chaos: - here repose
Angelo's, Alfieri's bones, and his,
The starry Galileo, with his woes; ..."

... George Gordon Byron
from *Childe Harold's Pilgrimage*, Canto IV, LIV
as quoted by Col. G. F. Young in *The Medici*

The Still-Unfinished Galilean Subversion of Indoctrination Systems

Will the memory of what happened to Galileo, whose physical form is relapsed to chaos but whose lessons are immortal, settle upon the consciousness of one who can devise means by which bringers of new things antithetical to authoritarian institutions will not be suppressed or oppressed by those institutions?[1] "New" does not mean "better", but the better requires the new. Many and varied are the institutional means for inhibiting innovation in literature, music, philosophy, science, art, mathematics and other pursuits that would sustain or improve life or reduce its hazards. The primal inhibitor surviving Galileo's campaign against it is indoctrination. Like the medieval variety of indoctrination that was more visibly enforced during the Inquisition, the modern variety impedes innovation just as well, but unlike its medieval counterpart, is openly sustained largely by deceit.

A subtle yet effective modern form of indoctrination is the bundling of Galileo's message from the stars with implicit denial of his personal example. In the year of publication of this book, a billboard has advertised the following message to travellers leaving the airport of a major city:

"Visit Our New Planetarium. You Tiny, Insignificant
Speck In The Universe."

This humorous advertising message presents an astronomical truth revealed to a large extent by Galileo, while it ignores his overturn of a false doctrine that, nurtured by authoritarian institutions, has lingered in Western civilization since pre-Renaissance days. That humans are physically tiny specks in the cosmos is an astronomical

---

1. "Bringer of new things" is a phrase coined by Tennyson; see excerpt from *Ulysses* at close of Chapter 11.

truth derived in large measure from Galileo's telescopic discoveries. The false doctrine, as tenacious as it is anachronistic, is: *Because you are tiny, you are insignificant.* Galileo's counter-example speaks to us: *An individual can rise above the limits of physical size and the brevity of life to affect the universe in ways that transcend these limits.*

A nine year old who has studied astronomy can read of Galileo's achievements and see this. But if even watchers of the skies have been diverted by doctrine from his message of human potential, how much more effective must be the sophistry of the indoctrinators among the general population of Earth? In its post-Renaissance form, this doctrine drove the campaign against an individual who dared to be more than a speck inside a vast whirling vault turned by angels.

The Institutional Suppression of Truth: 20th Century Case Studies

Case Study One: Classroom Indoctrination

The ecclesiastical campaign against Galileo moved into high gear once institutionally attached "philosophers" perceived as a threat to their institutions the truths this "mere individual" had been uncovering. You may ask: "What threat can there be in adducing evidence for Earth's motion?" A popular view of a moving Earth might suggest by analogy the kinetic energy of its inhabitants acting on their own decisions, a form of energy that authoritative institutions have sought to suppress:

"... in Europe the Kings and Popes, whose authority was derived from the belief that Authority controls a motionless universe, could not admit that this earth spins in space. If this earth does not rest upon something immovable, then there is energy.

The truly great intelligence of medieval Europeans saw at once that Energy operating in the universe may be *creative.* If there is Creative Energy, ... then this universe is not finished, changeless, static; and if it is not, then Authority is not its control.

The Europeans' whole world rested upon the pagan belief that Authority controls all things including men. When they heard that this earth spins in space, their whole minds and souls recoiled in horror ...

The Kings and Popes could not look for the East in the West; that would admit that the round earth spins in space; that would admit that the principle of the universe is Energy, and possibly Creativeness, Change, Progress. This is why they forced Galileo to recant his statement that the earth moves."[2]

---

2. Rose Wilder Lane, *The Discovery of Freedom*, Fox & Wilkes, San Francisco, 1993, pp. 117-118. As I demonstrate later in this Chapter (see "Myth Number Six: The Dishonor Myth"), Galileo "recanted" belief in Earth's motion in appearance but not in fact.

Having seen the threat, the indoctrinators and clergymen had to fabricate Galileo's indictment, his trial, and his conviction — not with truth but with smoke and mirrors. On this point Galileo portrays the intellectual inertia of the indoctrinators[3]:

"Please do not act the way most modern polemicists act: they first impress on their minds the conclusion without hearing other reasons and demonstrations; then, once having acquired the impression, they agree totally and very readily with every stupid and crude reason put forth in support of it; on the other hand, when faced with all kinds of evident and conclusive demonstrations to the contrary, they are motionless and immovable, having adopted the principle that perfect and true philosophizing consists of not letting oneself ever be convinced by any reason or experiment, however unequivocable it may be."

A Jesuit under whom I studied decades ago was a disciple of the 13th century Church philosopher Thomas Aquinas, whose followers even today continue, despite Galileo's inertial principle, to chant their mantra: *Quidquid movetur ab alio movetur.*[4] This man had such a severe case of what I'd call "anachronitis" that when I suggested to him that Aquinas' proofs of God's existence, resting on the denial that an infinite regression can be mapped into physical reality, suffer from a defect similar to that which invalidates Zeno's paradoxes, he reacted as if I were a heretic. I would have welcomed a rebuttal, but instead of being met with a thoughtful rejection, my original thought ended communication between us.

Being anchored in the thirteenth century, however, was not a phenomenon isolated to one priest.[5] A case in point is the textbook, *Answer Wisely,*[6] which Jesuits assigned me to read and, since almost everything known today about the Galileo affair was known when it was published, it is a reasonably current book with respect to its knowledge base. *Answer Wisely* is intended "for mature students who are about to go forth into the world as champions of Catholic Action". Champions of Catholic Action are defined to mean "educated laity who will cooperate in the

---

3. from Galileo's Reply to Ingoli, trans. by Finocchiaro, op. cit., p. 180

4. "Whatever changes is changed by something other than itself"; Maurice De Wulf, *The System of Thomas Aquinas*, Dover, NY, 1959, p. 75. "Newton's Law of Gravitation, the Law of the Equilibrium of Forces, The Principle of the Conservation of Energy", says De Wulf, "are all so many formulas which set forth in precise terms the influence of one being upon another." Does he think that inertia is a field force emanating from the Vatican? Does he think the Big Bang is impossible, since no law of nature would cause motion in an isolated singularity? Does he require assistance with his own movements?

5. On the other hand, I have personally known and studied under many 20th century Jesuits and can attest to the inclusion in their ranks of outstanding intellectuals and creative thinkers. It is not practical to mention them all here; they include the astronomer Father Francis P. Heyden, S. J. and Professor Father Vincent M. O'Brien, S. J., who, in my experience, is the outstanding living exemplar of how to teach and how to motivate students to learn and to think for themselves. The Jesuits have their un-indoctrinators, too.

6. Martin J. Scott, S. J., *Answer Wisely*, Loyola University Press, Chicago, 1943.

apostolate of the hierarchy"; i.e., who will explain to non-Catholic inquirers the position of the Church "on controverted points, and to make clear the Catholic side of historical and doctrinal questions", so as to "guide them to the truth". This book is especially important as an example of the product, not of one man, but of the Church as an institution. *Answer Wisely* had the highest approvals normally granted to a book by the Church at the time it was printed:

Imprimi Potest	Joseph A. Murphy, S. J.	Provincial of the NY-Maryland Province
Nihil Obstat	Austin G. Schmidt, S. J.	Censor Deputatus
Imprimatur	George Cardinal Mundelein	Archbishop of Chicago

As if "Catholic science" should differ from any other kind, its description of the Galileo affair begins with an impressive list of "Catholic scientists": Pasteur, Mendel, Fabre, Schwann, Mueller, Vesalius, Copernicus, Laplace, Lavoisier, Roentgen, Marconi, Galvani, Ampere, and Volta. Then it cites the membership of the Pontifical Academy of Sciences as the best answer to the "false charge" that "the Church is hostile to science". A comment is attributed to "Huxley" that in the Galileo case the Church had the better of it from the standpoint of scientific procedure. Another comment is attributed to Cardinal Newman that opponents of the Church could only find one case of Church opposition to science in three hundred years, one that proves to be no case at all.[7] These two comments are said to furnish "splendid evidence of the Church's truly scientific attitude and procedure". When such opinions are put forth as "splendid evidence", one begins to wonder how they square with the facts.

What the Church has presented as its view of the facts is contained in the following passage. As a specimen of Galileo's description of polemic methodology and as a case study showing how dispensers of doctrine can construct massive falsehoods that resemble reality in a superficial way, it is worth quoting in its entirety:[8]

1. "Briefly, the Galileo case was as follows. Fifty years before Galileo, Copernicus, the father of modern astronomy, questioned the then-accepted theory about the solar system, which was that the sun moved across the firmament and that the earth was stationary. Copernicus was one of the greatest astronomers of all time, and his theory that the sun was virtually still while it was the earth that moved, created no little discussion and investigation."

2. "Copernicus gave his view as a theory only, for as a true scientist he realized that he did not have sufficient evidence to substantiate it. He dedicated his monumental work to Pope Paul III, and was greatly honored by the Church. Galileo, without having sufficient evidence to corroborate his claim, asserted that it was a *fact* that the earth moved and that

---

7. Cf. A. D. White, op. cit., for examples to the contrary. As to the Galileo case, if it is "no case at all", that is because, as we have seen, the Church did not make its case. But it tried.
8. Paragraph numbering is mine and is supplied for ease of reference.

the sun stood still."

3. "It may be asked, What had the Church to do with the matter? The Church at that time was the leader in the scientific and religious world. When Galileo proclaimed as a fact something that revolutionized the accepted notion of the firmament, the Church asked him to give proof for his assertion. Galileo acquiesced, and a Church commission was designated to hear and weigh his arguments. This commission was not the Church, but only a body of very learned men. Its purpose was to pronounce on a debated astronomical question. On hearing the arguments this commission decided that Galileo had not substantiated his claim. Some of the greatest scientists of the day opposed Galileo's views, notably Francis Bacon and Tycho Brahe. On the recommendation of the commission, the Church forbade Galileo to teach his then-revolutionary and unproven doctrine. As he persisted in proclaiming his doctrine, the Church, after warning him, ordered him to be confined in the residence of one of the officials of the Holy Office."

4. "The question arises, Was not this condemnation of Galileo opposition to science? At the time of the condemnation the scientific world was divided on the matter, and since great scientists opposed Galileo, the Church, in the company of these scientists, cannot be said to be opposed to science. The condemnation of Galileo, however, has another phase, which it may be well to advert to briefly."

5. "The Holy Office not only decided against Galileo's claims, but, moreover, declared them heretical on the ground that they contradicted Scripture. In thus declaring Galileo's views heretical, the Holy Office fell into error, but that does not affect papal infallibility, since no Church tribunal, no matter how authoritative, has the unique gift of infallibility. But does not Galileo's theory contradict Scripture, since it is stated in the Bible that at Joshua's prayer the sun stood still, implying, of course, that it was in motion? To this question the very obvious answer is that Scripture, like every work of literature, may be understood in various ways. The statements of Scripture can be taken not only literally, but also figuratively or symbolically."

6. "It is because Scripture may be variously interpreted that the Church was divinely designated the sole authority for its interpretation. In point of fact, the Church has officially interpreted very few passages of Scripture, and until she does so interpret a passage, it is open to scholarly interpretation. The six days of creation, for instance, are interpreted by some to mean six periods or epochs. The Church has made no authoritative pronouncement in the matter. Until Galileo invented the telescope, and eventually substantiated his theory, the ordinary interpretation of the passage about the sun implied that it moved. When science demonstrated the contrary, the passage was interpreted figuratively, and that did away with apparent contradiction."

The opening paragraph of the quotation is hardly controversial. One might quibble about the attribution to Copernicus that the sun is virtually still (immobile with respect to the "fixed" stars), for this is not strictly implied by his having taken the sun as the center of planetary motions. But these opening statements are sufficiently conformable to the facts to produce immediate assent in the reader, and this is important in constructing such a polemic. The falsehoods must be slipped in after the reader is already in the habit of assenting to what he is reading, and we are scarcely into the second paragraph before a barrage of falsehoods, mixed with selective omission of relevant truths, commences. Copernicus is described as one of the greatest astronomers of all time, but nothing is said about the fact that, after Galileo, the Church censored and then banned this "monumental work" of the "father of modern astronomy" and all books treating of the motion of Earth. If having his book suspended for correction is what this author means by being "greatly honored by the Church" (I am aware of no other honor imputed by the Church to Copernicus), then why shouldn't Galileo be considered even more greatly honored by the Church since his person as well as his book was condemned? It is said here that Copernicus gave his view as a theory only. The Introduction to his *On the Revolutions of the Heavenly Spheres* described the content of the book as "new hypotheses" useful for calculating celestial motions, and from which no one can expect certainty "since astronomy can offer us nothing certain".[9] Nothing is said about this Introduction having been written by Osiander and not by Copernicus.

Moving along to paragraph (3), we are told that the Church at the time was the leader in the scientific world. Really? By virtue of what discovery? Perhaps the answer is Cardinal Bellarmino's perpetual hellfire-fanning energy source, so we will pass on this one. From context, it is clear that "When Galileo proclaimed as a fact something that revolutionized the accepted notion of the firmament ..." refers not to the *Dialogo*, but to a document written prior to 1616, probably the *Letter to Castelli*. The author continues: "... the Church asked him to give proof for his assertion. Galileo acquiesced, and a Church commission was designated to hear and weigh his arguments." Here is a superficial parallel to the correct official truth that a Church commission met behind his back, then without asking him any questions, told him to shut up, and he acquiesced. What happened according to official documents[10] is that the Cardinal warned Galileo that the opinion of Copernicus is erroneous and that he should abandon it; Galileo acquiesced and promised to obey; because Galileo acquiesced, the instructions which the Pope gave to the Inquisitors would have prohibited him not from teaching or defending the opinion but only from holding it. Paragraph (3) contains a triple discontinuity with the historical record:

The First Claim: A scientific commission (a body of "very learned men" whose purpose was "to pronounce on a debated

---

9.   from the Wallis translation of *De Revolutionibus* in *Great Books of the Western World*, op. cit.

10. Cf. Finocchiaro, *The Galileo Affair*, Inquisition Minutes 25 Feb 1616.

astronomical question") heard Galileo's arguments.

The Historical Record:

A group of consultants met, decided without questioning Galileo that the proposition of Earth's motion is "foolish and absurd in philosophy, and formally heretical" and made a report to the Inquisitors.

The Second Claim:

This commission decided that Galileo had not substantiated his claim.

The Historical Record:

One Consultant[11] had examined Lorini's defective copy of Galileo's *Letter to Castelli* and reported that it was "bad sounding" in three specific places but does not otherwise "diverge from the pathways of Catholic expression" — this report had no scientific content whatsoever.

The Third Claim:

The Church forbade Galileo to teach "his doctrine".

The Historical Record:

Pope Paul V gave instructions that Galileo was to be forbidden to teach "his doctrine" (and imprisoned) if he failed to acquiesce. He did acquiesce and did promise to obey and therefore no such prohibition is known to have been delivered to him.

Paragraph (3) ends with a compound falsehood. His "persistence" can only be a reference to the *Dialogo*, since that document was the basis of his confinement order. Here we are told that there was a "persistence", then a "warning", then a "confinement". What really happened was that the Church officially approved the *Dialogo*, after publication of which it tried, convicted, and sentenced him with no warning at all. His book was prohibited by public edict; he was sentenced to lifetime imprisonment, commuted to lifetime house arrest. That this is described as confinement in the residence of one of the officials of the Holy Office looks to me like an attempt to make the sentence appear temporary and even pleasant.

Paragraph (4) says in effect that the Church could condemn Galileo without being in opposition to science because it did so in the company of other scientists. Since when is science equivalent to "the company of scientists"? Three centuries after the condemnation, we see the Church clinging to the determination of "truth" by weight of opinion. Tycho Brahe (1546-1601) and Francis Bacon (1561-1626),[12] both of whom were dead before the *Dialogo* was published, are mentioned, with no supporting attribution or quotation, as having opposed Galileo. Tycho's alternative opinion is well known and conflicts, incidentally, with both that of the Church and that of the "father

---

11. Ibid., Consultant's Report on the Letter to Castelli, 1615.

of modern astronomy". Tycho may not have been in agreement with Copernicus or Galileo on the scientific issue, but would he have approved of this treatment of a fellow scientist? Would the judicial proceedings against Galileo be approved by the Francis Bacon who wrote as follows?[13]

> "Judges ought to remember that their office is *jus dicere*, and not *jus dare*; to interpret law, and not to make law, or give law; else will it be like the authority claimed by the Church of Rome, which, under pretext of exposition of Scripture, doth not stick[14] to add and alter, and to pronounce that which they do not find, and, by show of antiquity, to introduce novelty."

Surely the author of *Answer Wisely* would not allow us to say that the Church opposed science because it opposed a particular scientist, Galileo. Then neither should he allow himself to say that the Church in punishing Galileo acted on the side of science because two particular scientists may not have agreed with his conclusion. How consistent with Aristotelian logic is it to confuse science with scientists? Or to impute agreement with the Church's methods of suppressing a scientist to those who may only agree with its support of Ptolemaic astronomy? And by the way, Tycho did not agree with the system of Ptolemy.

In Paragraph (5), the author and the official Church censors who approved his textbook shift the blame for convicting Galileo of heresy to one of its "departments", calling this an error of the Holy Office that does not affect papal infallibility. But in 1664, Alexander VII issued a Papal Bull condemning all books that teach the movement of Earth and making adherence to the condemnation binding on all Roman Catholics. In retrospect, therefore, it appears that Galileo was, from the Church's later perspective, a heretic. You would think that if they are going to admit error on some point, they would at least get that right. But here they agree with Galileo that the Scriptures need not be interpreted only literally! From this embarrassing turnabout, Galileo himself had given them an escape by showing, in the *Letter to Castelli*, that the passage of Joshua they are citing could be taken literally by assuming the physical reality of the system of Copernicus.

Referring to "Galileo's theory" as the assertion of the physical reality of the hypothesis of Copernicus, Paragraph (6) states that after all this happened — the warning, the

---

12. In the Second Book of *Novum Organum*, Bacon gives a long list of conditions for validating that the diurnal motion belongs to the heavens and not the Earth. One of his conditions, that the east-to-west motion of celestial bodies be faster the greater their distance from Earth, was known to be false, and its falsity used by Galileo as a counter-argument to the Ptolemaic system. Bacon's position is hardly a condemnation of Galileo. Cf. *Great Books of the Western World*, Encyclopaedia Brittanica, 1952, vol. 30, p. 165.
13. Francis Bacon, *Essays and New Atlantis,* ed. by Gordon Haight, Walter J. Black Inc., Roslyn, NY, 1969.
14. The word "stick" in this context is synonymous with "hesitate".

condemnation, the "confinement" of Galileo — then Galileo invented the telescope and substantiated his theory! Well, hats off to the author, he got one thing right — Galileo proved the hypothesis of Copernicus. There's only one little problem with this statement. Galileo made his telescope in 1609 and by the end of 1610 had discovered most of the evidence for the Copernican hypothesis that was discovered in his own lifetime. That happened before he publicly espoused the Copernican hypothesis, before the warning, before the trial, before everything described in this incredibly convoluted discussion. And so if the Church had really wanted to examine Galileo's evidence before convicting him, it was all there.

Even today, it is difficult for me to accept that men of the cloth would deliberately mislead the young; but if this is not deliberate deception, then are we to believe that the Jesuits with all their erudition and access to Church archives could not have produced a more competent historical account than this? Conclusion: *Answer Wisely* is an indoctrination manual. But that was not obvious to me in student days. Young persons who have grown up in a loving family tend to be trusting of authority figures: their parents, their teachers. It is a sad thing to see the trust violated. The idea of teaching what to think, rather than how to think, was firmly entrenched in Galileo's time. *Answer Wisely* indicates that this notion of teaching what to think has persisted to the 20th century, confused with education.

The Baltimore catechism engraved upon the memory of Catholic students started with "Who made us?" The meaning of this question was not itself open to question, and the response ("God made us") was to be memorized. When a student asked, believing that the classroom is the place for understanding, "Who made God?", the result was a change of topic. But the mid-20th century was no time to change the topic; science had brought new information to bear on the question of the origin of our species and of its planetary home. Later, while probing the underlying meaning of the sharing of one nature by three persons in the doctrine of the Trinity, the same student was told that it is a mystery, not open to rational explanation, but a necessary article of faith. What bothered me about this response was not the attitude that some things might now lie outside comprehension, but rather that it is a closed issue, a matter of accepting conclusions on "faith". I could have accepted the notion of a metaphysics to be reduced by the advance of science, but certainly not of a static body of "mystery" that would forever remain beyond the boundaries of investigation. "Faith" could not be interpreted by me as anything but a collection of hypotheses. What possible reason can there be for failing to comprehend anything, except that the necessary examination of nature cannot be conducted with present tools or methods? What possible reason can there be for believing anything that cannot be fundamentally understood? Existing without thinking and understanding simulates nonexistence rather well.

Is not my being conscious of thinking proof of my existence? I was told that some chap named Descartes had thought of that (I had not yet heard of him and believed that my question was original), and that I should not entertain such questions unless I wished to be tagged as a solipsist or a doubting Thomas. I felt puzzled that the merit of a question should be judged by how it might reflect on the questioner.

Case Study Two: The Academic sup-Press

Confusing the teaching of "what to think" with education is not confined to indoctrination manuals and classrooms. A leading Galileo scholar, Stillman Drake, who has written new thoughts about the censored work of Galileo, has seen his thoughts suppressed by an academic press. Professor Drake's first translation of the *Dialogo* was published in 1953. Over the next 25 years he deepened his understanding of Galileo's work. In 1978, *Galileo at Work* was published. Realizing that his Translator's Preface to the *Dialogo* was misleading in the light of what he had learned in this intervening time, in 1979 he proposed to the publishers that they issue an edition of his translation with a new Introduction, revised notes and a revised index. It was his intent to provide readers with new insights derived from intensive study. His proposal was accepted and the publisher advertised the Third Edition in these words: [15]

> Galilei, Galileo. Dialogue concerning the Two Chief World Systems — Ptolemaic & Copernican. Third edition. Drake, Stillman, translator, notes, new introductory essay, & new analytical index by. Einstein, Albert, foreword by. 1980. (W) -04104-6 $30.00s.

Thus was the Third edition with a new introductory essay, new notes and a new analytical index advertised in 1980 as a "Book in Print". As of sixteen years later, it has not been published. Readers ordering the book today receive the Second Revised Edition (with the old index, old notes and old introduction) adorned with a notice that sounds as if it had been penned by Cardinal Bellarmino:

> "Alteration of the fore matter and notes to reflect substantial changes of viewpoint by the translator has been refused publication herein by the Editorial Committee of the University of California Press." [16]

In fact, the refused new Introduction contains valuable insights and clarifies issues that would assist readers in understanding Galileo's intentions in writing the *Dialogo*. In it, Drake cites unwarranted pre-conceptions in the literature about

---

15. University of California Press, *Books in Print, 1980-81, Author Index*, p. 16.

16. I am grateful to rdc for bringing to my attention the existence of this note in my own copy of Professor Drake's translation of the *Dialogo,* and to jm for facilitating my access to background information on which this discussion is based.

Galileo; for example, that Galileo's Copernican zeal affected his scientific judgment, and that his physics had the same basis and goals as Newton's. In rebutting the widely accepted view that Galileo was a Copernican zealot, Drake states that Galileo "was moved partly by zeal for the Church", pointing out that in the *Dialogo* he did present both sides of the issue.[17] He adds a point with which I enthusiastically concur: "It was not Galileo's fault if the preponderance of evidence lay on one side and could not be concealed when everything was set forth." Too bad for the Church that it failed to realize this.

Another excellent new point made by Drake in the refused Introduction, and one which I hold not only as generally true but of the highest importance in understanding both Galileo and much of the academic nonsense written about him, is the following: "Within the range of phenomena to which Galileo restricted [his] physics, what he said did not contradict experience. Outside that range his physics was not 'mistaken,' but inapplicable, viewed from his standpoint."

Against those who think that Galileo only upset the system of Ptolemy and offered no proof of the system of Copernicus, Drake cites Galileo's analysis of annual[18] variations in sunspot paths and his (widely and mistakenly held to be irrelevant) analysis of terrestrial tides. Drake's commentary on the tidal theory of the *Dialogo* revitalizes the Fourth Day and explains why Galileo himself believed that in terrestrial tides he had a significant argument for the system of Copernicus.

Though not necessarily in agreement with all that Drake has to say in his still-unpublished Introduction, I would ask: why not let him say it, and allow his readers to judge for themselves? One can only surmise why the publisher has not seen fit to do so. Does Drake's tone ring of self-assurance? So what? In light of his extensive research on Galileo, and of his having admirably translated the *Dialogo*, he has earned the right to be self-assured and his new views warrant the attention of his readership. Humility may be a virtue to academia as it is to the Church, but it does not educate readers. Does his view of Galileo's motivation in writing the *Dialogo* contradict the party line of academic philosophers that Galileo's thinking was warped by Copernican zeal? Perhaps, but is the function of a University to teach what to think? Why publish a book by Galileo, a central figure in the struggle for freedom of expression against authority-based institutions, and then adopt such an attitude toward the translator's expression of viewpoint? The woes of the starry Galileo, it seems, did not end in his own time, nor even after that of the poet Byron.

---

17. To which I would add that in the *Dialogo* Galileo developed the arguments for the Aristotelian/Ptolemaic side better than its own advocates.
18. The reader may well note that "annual" is linked to Earth — why would the sun have phenomena synchronized to Earth in the system of Ptolemy? On the other hand, such synchroneity is a consequence of the system of Copernicus.

Today we see that Galileo's book, to be annotated by the translator, requires the *imprimatur* of an editorial committee. Here one sees a modern "educational" institution withholding access to an innovative intellectual achievement. Through the courtesy of Mrs. Florence Drake, an analysis of Drake's refused Introduction with extensive excerpts from that document can be found in Appendix B.

✦✦✦✦✦

Case Study Three: Institutional Misnomenclature

A more subtle form of indoctrination can be found in institutional tolerance for improper nomenclature. Even scientific professionals should examine their behavior. Case in point: whose names do professional astronomers apply to the satellites discovered by Galileo? The names "Io, Europa, Ganymede and Callisto" are creations of Simon Mayr who had earlier sponsored the plagiarism of Galileo's proportional compass by Mayr's student, Capra.[19] Mayr claimed priority in the discovery of the moons of Jupiter, for the first time claiming in 1614 (!) that he had seen them in 1609. Let us make an improbable assumption in Mayr's favor: that he actually had seen them in 1609; still, it is undisputed that he did not publish before Galileo, and that, when he did, he used Galileo's data and conclusions without credit.[20] Yet the names accepted and used by astronomers today are Mayr's.[21] Galileo proposed individual names or pairs of alternate names for the Medicean satellites, as stated in the footnote. It is time to restore them. The name of Galileo's third Medicean moon, the most massive known satellite in the solar system (larger than the planet Mercury and

---

19. Simon Marius (Mayr) "claimed to have observed in his provocatively titled *Mundus jovialis anno 1609 detectus* (1614) the satellites in 1609, although his four-year period of silence added little to his credibility. A similar problem ocurred in 1606 when Mayr claimed to have pre-empted Galileo in the discovery of the geometrical and military compass. Since he did not succeed in this matter, Mayr exacted a minor revenge. Galileo had proposed to name Jupiter's satelites [sic] after the Medici children. .... Mayr proposed an alternative set, ... He also suggested using the names of four 'illicit loves' of Zeus or Jupiter as possibilities: Io, Europe [sic], Ganymede, and Callisto. ... These are the names that survived."; Derek Gjertsen, *The Classics of Science*, op. cit., p. 163, footnote 13. Willy Ley in *Watchers of the Skies*, op. cit., adds that Mayr's 1614 book "used much material from the *Sidereus Nuncius* without crediting his source", and that Galileo's own choices of names, reading outward from Jupiter were "Catharina or Franciscus, the next one Maria or Ferdinandus, the third one *Cosmus major* and the fourth one *Cosmus minor.*"
20. Mayr's account of his "sighting" matches Galileo's 8 January sighting even as to his mistake of missing the fourth satellite which was in plain view outside the narrow field of Galileo's telescope. Yet Mayr claimed to have been watching the satellites for weeks. (Cf. Drake, *The Unsung Journalist and the Origin of the Telescope*, op. cit., p. 16).
21. Galileo responded to Mayr's claim in *Il Saggiatore*, pointing out that Mayr had used the data on the Medicean satellites corresponding to Galileo's own dates of observation, beginning 7 January 1610, and then claimed priority on the basis that he had made his observations in 1609. *"But he neglects to warn the reader that he is a Protestant, and hence had not accepted the Gregorian calendar. Now the seventh day of January, 1610, for us Catholics, is the same as the twenty-eighth day of December, 1609, for those heretics. And so much for his pretended priority of observation."* Trans. by Stillman Drake.

more than twice as massive as Earth's moon) is Cosmus Major, not Ganymede.

Galileo has come to symbolize freedom of inquiry and of communication. His admonishment of polemicists speaks to educational systems of the future. Leaving the subject of forced miseducation entirely to others, I focus here on scientific reasoning in education, and counter-exemplify such reasoning by the approach to the history of science evident in various myths, rife in academia, about Galileo himself.

The Elimination of Mythology About Galileo

"I saw a man pursuing the horizon;
Round and round they sped.
I was disturbed at this; I accosted the man.
'It is futile,' I said, 'You can never — '
'You lie,' he cried.
And ran on."

... Stephen Crane, "I Saw a Man" [22]

Old myths usually die slowly and with great difficulty. To the extent that they do eventually die, an antidote to mythology is to counter new myths by application of individual scientific reasoning. Those myths by which various 20th century academics have understated or maligned Galileo's character and achievements are so numerous as to defy taxonomy. Many of them can be represented by the following specimens. In each case, I will illustrate my point by exposing some author's bias and will let that case stand for the similar biases of other authors. I will leave formal proofs of the falsity of each position to professional historians.

Myth Number One: The Anti-Copernican Relativity Myth

Few mythical messages could be more remarkable than that of a book which, purporting to be written about the achievement of Galileo, represents relativity theory as having invalidated the empirical support that Galileo's experiments gave to Copernicanism.[23] One must conclude, we are told, that "while twentieth-century science fully acknowledges its debt to Galileo for its mathematical-experimental method, it has unequivocally decided against him in his defense of the Copernican system."???!!! It is the duty of physicists, we are told, to acknowledge that "logic was on the side of Osiander, Bellarmine, and Urban VIII, and not on the side of Kepler and Galileo; that the

---

22. quoted from John Ciardi, *How Does a Poem Mean?*, Houghton Mifflin Company, Boston, 1959, p. 892.
23. Cf. Brophy and Paolucci ed., *The Achievement of Galileo*, Twayne, 1962., introduction by Paolucci, also containing excerpts from Cohen and Frank germane to this issue: see Morris Cohen, "Studies in Philosophy and Science", and Philipp Frank, "Modern Science and its Philosophy", for example.

former grasped the precise import of the experimental method and that the latter were mistaken in its regard."

Poor Osiander! Where is his prize for logic in distorting a man's work with a preface downgrading it to a hypothesis? Poor Bellarmino! Besides having protected us from Copernican heresy, he has hypothesized the infinite energy content of the universe from which a divine breath fans eternal flames of hell, and he didn't even get honorable mention in Fourier's *Theory of Heat* for this novel doctrine. And poor neglected Urban VIII! Where is his reward for punishing that vehemently suspected heretic, Galileo? The producers of this myth have given the title *The Achievement of Galileo* to a book in which they try to deny Galileo the achievement that consumed the lion's share of his career. To this end, they appeal to Einstein's theory as support for the claim that Kepler and Galileo were mistaken in defending Copernicus. What has Einstein to say about this matter?

Einstein has said of the Copernican theory that its support "over and above qualitative arguments", such as advanced by Galileo, "was possible only by determining the 'true orbits' of the planets — a problem of almost insurmountable difficulty, which, however, was solved by Kepler (during Galileo's lifetime) in a truly ingenious fashion."[24] If Einstein had overturned the Copernican system and disproved Galileo's qualitative arguments for it, would he have been able to say that Kepler, who showed that Earth and other planets move around the sun in elliptical paths, had solved the problem of determining their orbits? Einstein directly addressed the Copernican/Ptolemaic debate in a passage that holds it meaningless in the context of a relativistic physics valid in all coordinate systems:

> "Can we formulate physical laws so that they are valid for all CS [co-ordinate systems], not only those moving uniformly, but also those moving quite arbitrarily relative to each other? If this can be done, our difficulties will be over. We shall then be able to apply the laws of nature to any CS. The struggle, so violent in the early days of science, between the views of Ptolemy and Copernicus would then be quite meaningless. Either CS could be used with equal justification. The two sentences, 'the sun is at rest and the earth moves,' or 'the sun moves and the earth is at rest,' would simply mean two different conventions concerning two different CS.
>
> Could we build a real relativistic physics valid in all CS; a physics in which there would be no place for absolute, but only for relative motion? This is indeed possible!"[25]

Einstein's assertion that the Copernican/Ptolemaic debate is meaningless in the context of such a physics is not a denial of meaning in any context. Consider the context of

---

24. Sonja Bargmann's translation of Einstein's foreword to Drake's translation of Galileo's *Dialogue Concerning the Two Chief World Systems.*
25. Einstein and Infeld, *The Evolution of Physics*, Clarion (Simon & Schuster paperback), 1938, page 212.

Einstein's own description of Earth's motion around the sun:

> "Now in virtue of its motion *in orbit round the sun*, our earth is comparable to a railway carriage travelling with a velocity of about 30 kilometers per second." [26]

Relative to the "fixed" stars, which moves: the Earth or the sun? A first approximating answer is: from observation, one or both, but not neither. Ptolemy had chosen the sun, not Earth. Copernicus may have realized that the case of both cannot be excluded, but decided that choosing Earth and taking the sun as a stable reference point would simplify the whole matter. Galileo, who pioneered the concept of the relativity of motion, and discovered the rotation of the sun, likewise suspected both, and then adopted Copernicus' attitude.[27] But all we mean by saying that an object moves relative to another is that there is a change of separation or direction between the two, a change that can be described relative to either, equally validly. Just as from one airplane we see another airplane appear to be flying almost "sideways", we tend to attribute to another moving object the vector difference of its motion and ours. The senses can be deceptive! From a geosynchronous satellite, I would expect Earth to appear motionless and the universe to look as Ptolemy described it from his arbitrary geostatic viewpoint.

The point of the controversy is not just that using the sun as the stable origin of reference for the dynamic system of its planets simplifies the mathematical description of the apparent motions of the sun, moon, and planets.[28] It is this. Suppose you are a suitably equpped astronaut able to stand on the surface of 61 Cygni and look back at the solar system over the course of a Saturnian year. Would you see planets orbiting in Keplerian ellipses around our star, or the sun and planets moving around the Earth in deferents and epicycles? Which of the "two great world systems" would fit your description? It is futile to bring modern objections based on post-Einsteinian relativity theory, or philosophical objections about the physical meaning of a point in space, against this question; one of the two systems, the Ptolemaic or the Copernican, can be expected to fit your observation. This is true independently of whatever equations might be written to describe the relative motions of this entire system as geocentric or heliocentric. Unable to stand on a star, Galileo inverted the design of the experiment with no loss of generality. It is to Galileo's credit that he was able to suggest Bessel's

---

26. quoted from Einstein's *Relativity*, trans. by Robert W. Lawson, pub. Crown Publishers, Inc., N.Y. 1961. Emphasis mine.
27. The preceding two sentences are my conjecture.
28. Here, because I wish to construct an argument compatible with Galileo's pre-Newtonian view, I pass over the Newtonian proof that the entire solar system has a gravitational barycenter, **located in the sun and not in the Earth**. Apparently, however, Newton's proof of this heliocentricity is lost on Paolucci and other propagators of the Anti-Copernican Relativity Myth (rdc).

experiment and to foresee Bessel's result, though he was inattentive to Kepler and may have thought the orbits would be circular. That Galileo was bound to Earth did not prevent him from conceiving this earthbound experiment or from moving himself off Earth in a thought experiment incorporating the phases of Venus. Myth Number One is a misguided attempt to deny him these achievements using an "Einsteinian" argument with which Einstein's statements did not agree.

Myth Number Two: The Juvenile Vanity Myth

A Professor of the history of science[29] has accused Galileo of exhibiting pettiness and juvenile vanity, "hard to reconcile with the unquestioned greatness of his genius", he says. He cites the time and energy Galileo expended trying to prove the priority of his thoughts on various topics, like the development of the telescope. Besides the fact that Galileo never claimed to be the original inventor of the telescope, this Professor seems to overlook that before the telescope, Galileo had been plagiarized on a major invention, the proportional compass. Though the legal process that he had to use to clear his name showed some respect for intellectual property and Galileo received satisfaction for his efforts, the plagiarism forced Galileo to waste his valuable time on self-defense. But it seems this Professor's scorn of Galileo's efforts to protect his inventions renders him blind to the difference between a character defect and a reasoned reaction to the malicious behavior of a plagiarizer. Clearly, a psychological diagnosis is not within his field of competence, whatever that may be.

Myth Number Three: The Enormous Gulf Myth

And then we have the case of a science-history Professor who produces both a myth and splendid evidence against it. Following a mathematical discussion,[30] he concludes that on the question of whether, falling from a point in space, "all the planets could be let fall from rest toward the sun with the same uniform acceleration so as to arrive at their respective orbits with speeds of the magnitude of their observed orbital speeds", Galileo's claim was invalid and Newton's was valid. Of two men whose lives never overlapped, he says that

---

29. "An impartial appreciation of Galileo's motives must, however, take into account a rather disagreeable trait by which the nobility of his character was perceptibly impaired and obscured. I am thinking of a certain pettiness, which became evident time and again, and which apparently sprang from a kind of juvenile vanity, hard to reconcile with the unquestioned greatness of his genius. Thus whenever an important discovery was made by someone else, it grieved him not to have made it himself; he often wasted time and energy in trying to prove that he had at least *thought* of it before anyone else had. ... The most serious incident of this sort was, of course, his claim to be the discoverer of the telescope." Willy Hartner, "Galileo's Contribution to Astronomy", see McMullin et al., *Galileo: Man of Science*, pp. 178 ff.
30. In his paper "Galileo, Newton, and the divine order of the solar system" (McMullin ed., op. cit., pp. 212 ff.), he invokes Kepler's Third Law, unknown to Galileo but "a quite accurate representation", he says, "of the relation among planetary data as known to Galileo and other Copernicans."

a comparison of their arguments "reveals to us the enormous gulf separating these two great men of science of the seventeenth century."

Upon examination, one sees that the crucial difference in the two contrasted treatments lies in Galileo's application to this astronomical scenario of the constancy of free-fall acceleration, observed over distances not greater than the 48 meter height of the Leaning Tower. He says that Newton points to the (invalid) assumption that the bodies fall "with uniform and equal gravities (as Galileo supposes) ...".[31] Where this Professor is right he has agreed with Newton, and elsewhere he has been right about the relationship of Galileo and Newton. Following Galileo, and standing on Galileo's and Kepler's shoulders,[32] Newton saw that the force causing the acceleration of free fall depends on the separation of the falling body from the primary. Is this then a contest between the two parties? Far from showing a discontinuity of thought, the later theory builds on the earlier. Any analysis of planets falling through astronomical distances needs the gravitational theory of Newton, incorporating Galileo's *impeto* and extending his law of falling bodies. The "enormous gulf" is no big theoretical lacuna but rather a big deal made of a variable treated by Galileo as a constant.

Greatly compounding the difficulty of his position, the author of this myth says: "We would nevertheless tend to agree with Professor Koyré's general conclusion concerning the lack of importance of Galileo's writings in the development of Newton's thought." (Who is "we"? Only one author's name appears on the paper.) His justification for this statement is what he *failed* to find: "For there is only the barest of direct reference to Galileo in Newton's student notes and these appear to be to the *Dialogo*. I have found no evidence that Newton had ever read the *Discorsi* at all, at least prior to 1700." And though I have found no evidence in the Professor's paper that he ever read Newton's credit to Galileo for fundamental contributions to Newton's own mechanics (see Chapter 12), I would not attempt to claim on that basis that he did not read it; I suppose it a near certainty that he did, but that Newton's own statement of the Galilean basis of Newtonian theory bears much less convenience than relevance to the Professor's myth. On the other hand, we can be grateful to this same Professor for having informed us elsewhere of Newton's extensive sources on Galileo.[33]

---

31. Newton to Bentley, February 25, 1692/3; see Thayer, op. cit., p. 57.

32. The author of this myth has written a book whose cover depicts Newton aloft between the shoulders of Kepler and Galileo.

33. I am grateful to Professor Cohen for having related (in *Saggi Su Galileo Galilei*, Barbera, Firenze, 1972, p. 400 footnote 53) that Newton had read about Galileo's mechanics in other than the *Dialogo*, notably in separate works by Anderson, Charleton, Digby and Gregory.

Myth Number Four: The Wealthy-Kleptomaniac Myth

Unlike science, muckmaking is a negative-sum game. Muckmaking is the publication of unjustified attacks on men of high achievement. An organization that claims to represent the advancement of science reported in its journal a speculation by a history professor at a university in the United States that Galileo pirated an idea from a student. "On December 5, 1610, a former student sent Galileo a letter suggesting that if the planets orbit the sun — instead of the earth, as was the prevailing theory of the day — then Venus should appear as a crescent and gradually grow into a half, then full, disk. The student suggested that the astronomer use his newly invented telescope to look. On December 11 Galileo announced that he had discovered that Venus went through phases." [34] What motivated him to "steal the idea for one of his most important discoveries from a student?" The answer: "Appointed earlier that year to the court of the Grand Duke of Tuscany on the strength of his discovery of the moons of Jupiter, he was under pressure to produce more." What a compliment! Galileo had to be the fastest or the boldest astronomer in history. The orbital period of Venus is about 225 days; it takes about 56 days to go from one phase to the next. The assumption here is that he had not conceived of this experiment prior to December 5 *plus the delivery time of the letter,* and yet on December 11 recklessly announced a cycle of phases that he could hardly have verified.

The insinuation that Galileo should, like some wealthy kleptomaniac, feel compelled to pilfer an idea from a student is almost too absurd even to merit a rebuttal. Nevertheless, I wrote the publisher (certified with return receipt) a prompt counter-argument, asking for substantiation or retraction; I received the return receipt, but after twelve years, neither the publisher nor the author he paraphrased has responded — or else their mail delivery service is a lot slower than they think Galileo's was. Further, Professor Stillman Drake's rebuttal had not been included in this article, though it had been earlier quoted by a competing popular journal:

> "You show me any astronomer who has got a new instrument that others haven't got, and he's looked at everything he can find. And there's one planet, the most brilliant of all, that he can't see. Well, when the thing comes into sight, he doesn't put off observing it for three months, waiting for some bright student to tell him he ought to take a look. It doesn't make sense. That's not how astronomers behave." [35]

There is a difference between an explanation that fits all currently known facts, and the truth. Attacks like this one do not even come close to fitting all the currently known facts, yet they are susceptible of awards from historical societies.

---

34. *Science 84*, March 1984, p. 6.
35. quoted in *Science News*, November 26, 1983, p. 347. (The later article in the journal to which I responded was the one I saw first.)

As many lawyers seem to have discovered, it is relatively easy to build a semi-proof of any desired falsehood by selectively choosing facts to fit it. A semi-proof is not a proof (except by the barbarous rules of the Inquisition).

Myth Number Five: The Einstein-Galileo Dichotomy

The following footnote that I found in a physics textbook, referring to a Galilean frame of reference isotropic with respect to mechanical and optical experiments, is a particularly grotesque specimen of misinformation about both Galileo and Relativity Theory:

> "This [Galilean frame of reference] is the usual name, and not a very good one, for Galileo lived before Newton and of course had no idea of the theory of relativity. 'Einstein frame of reference' would be a better name." [36]

If these authors had bothered to read Galileo's writings, they would have discovered the very principle they are attempting to describe. But if we forgive them this, then what is their excuse for ignoring Einstein, since they want to call an "Einstein frame of reference" what Einstein calls a "Galilean coordinate system"?

Myth Number Six: The Dishonor Myth

This widespread and odious myth holds that Galileo "sold his honor" to the Inquisition when he "recanted"; i.e., that this act of "recanting" was not true to his belief. Let's confront an insane myth by first indulging in a bit of insanity ourselves; that is, by granting its nutty premise that the so-called "recantation" of Galileo is a meaningful set of statements and considering whether, taken at face value, they violate what the proponents of this myth assume to be his belief.

Consulting the official record of Galileo's "abjuration",[37] the first thing to be noted is this: all but one of the propositions it contains describe acts or promises of acts, not belief. He renounces, curses, and detests his heresies — not a statement of belief because his heresy has never been defined by the Church in terms of belief, but only by the Tribunal in terms of disobedience of an order. (His earlier "Catholic answer" referred to a belief, but played no role in the "recantation" recital because it was accepted.) He renounces, curses, and detests his errors; there were two: (1) he wrote a book in violation of a judicial instruction with injunction and (2) he published the same book. These are acts of disobedience, not beliefs. He promises to observe his penances, to fink on other suspected "heretics", blah, blah, blah. He refers to his "heresy" as follows: "I have been judged vehemently suspected of heresy, namely of having held and believed that the sun is the center

---

36. John L. Synge and Byron Griffith, *Principles of Mechanics*, McGraw Hill Book Co., NY, 1949, p. 478.
37. See Finocchiaro, *The Galileo Affair*, pp. 292-293. An "abjuration" is a recantation with an oath.

of the world and motionless and the earth is not the center and moves." Here he attests, not to a belief, but to the tribunal's judgment as to his belief. This statement is not an admission of heresy, or indeed of any belief of his. There was one, only one, statement of <u>his</u> belief in the "recantation" recital: "... *I have always believed, I believe now, and with God's help I will believe in the future all that the Holy Catholic and Apostolic Church holds, preaches, and teaches.*" [38] (Notice that such belief is assumed to be so difficult to sustain that it requires God's help.) There is no reason to believe that Galileo did not utter this statement of belief with perfect sincerity. To show this, it is necessary to show that his state of mind about the Copernican theory did not preclude simultaneous belief in "all that the Holy Catholic and Apostolic Church holds, preaches, and teaches".

What about that "Catholic answer" given earlier under threat of torture: Was that true to his belief? In its context, what is the meaning of his answer ("*I do not hold this opinion of Copernicus, and I have not held it after being ordered by injunction to abandon it*")? In ordinary terms, this is a question about Galileo's state of mind with respect to the opinion of Copernicus after 26 May 1616. But here we must realize that the meaning of a proposition depends on the method used to determine whether it is true or false. There is no doubt, from the context in which this response of his was solicited, that the ecclesiastical epistemology, not the scientific one, applied. The ecclesiastical test for the "truth" of an answer given by an accused in the Inquisition proceeding is whether it is sustained under torture, or in Galileo's case, under threat of torture. It is "validated" not by his actual state of mind, but by his behavior. As an ecclesiastical proposition, uttered in the language of the Church, his answer was meaningful and "true" because he upheld it under threat of torture and did not confess heresy. I say this not to nauseate the reader with the moral depravity of this epistemology, but to show that even from the perverted perspective of his persecutors, Galileo's answer was "true". His "Catholic answer" was accepted — by the Church.

Only in the context of a physical proposition, not applicable to his "Catholic answer", does it make sense to ask: What was his true state of mind with respect to the opinion? And here we may say that Galileo gave every indication of belief in the physical reality of the system of Copernicus; further, he stated that "*before a physical proposition is condemned, it must be shown to be not rigorously demonstrated — and this is to be done not by those who hold the proposition to be true, but by those who judge it to be false.*" [39] By his own standard, since neither the Decree of the Index nor the Bellarmino injunction nor the Sentence of the Tribunal against him showed that Copernicanism as a physical proposition had not been rigorously demonstrated,

---

38. translation by Finocchiaro, op. cit., p. 292.
39. Stillman Drake's translation; cf. *Discoveries and Opinions of Galileo*, op. cit., p. 195.

the proposition that he held as physically true in a scientific context had not been condemned. Moreover, as the Ptolemaic system was never part of the Church's official teaching, it did not belong to "all that the Holy Catholic and Apostolic Church holds, preaches, and teaches". The compound belief in the sun's stability (heliostasis) AND Earth's motion (geokinesis) was declared merely anti-scriptural by the Decree of the Index, which did not constitute official Church teaching. And by requiring Galileo, who did not teach heliostasis, to refer to the teaching of both heliostasis and geokinesis as a false and heretical doctrine, the tribunal blundered in its attempt to create the false impression that Galileo had violated official Church teaching.

In the foregoing rebuttal I granted the myth proponents their false assumption that the "recantation" statement was meaningful and could be taken at face value. But whatever sounds Galileo had been forced to utter, had they had been chicken scratchings, dolphin moanings, Neanderthal grunts or even an utterly unintelligible sequence of sounds, a rational person under equivalent duress would likely have uttered them without critically evaluating their meaning. The real disproof of this foul myth follows from adopting the reasonable premise: no statement made under duress can violate one's conscience or one's belief. Galileo uttered some words which, by virtue of being forced, were arbitrary with respect to his intent. That utterance is not a dishonor nor even a recantation. Galileo never recanted.[40]

Myth Number Seven: The Impotence Myth

This pernicious class of myth holds that with respect to some (fill in the blanks) accomplishment of Galileo, it had no effect on science or technology. If you want to write like an academic, this is the myth for you! You only need to fill in the blanks and come up with some sophistry to support your point. Then you can publish, secure tenure, go on the lecture circuit, etc. Case in point:

> " ... It is astonishing to our 20th century minds how little impact Galileo and his circle had upon the technology either of their own time or of the following two hundred years. ... We must conclude, in the present state of the evidence, that the mastery of steam power was a purely technological feat, not influenced by Galilean science." [41]

We *must* conclude? Who is "we"? There is only one author's name on the paper. What is astonishing to my "20th century mind" is the reason given for his sweeping generalization. Newcomen undertook his innovation on the steam engine without

---

40. according to the terminological convention adopted in Chapter 10 " To 'Recant' or Not To 'Recant'".
41. Lynn White Jr., "Pumps and Pendula: Galileo and Technology", included in *Galileo Reappraised*, Carlo Golino ed., U. of California Press, 1966. When this academic press decided that White's easily refuted anti-Galilean myth was fit to print, was it guided by that same Editorial Committee which later censored Drake's Third Edition notes on Galileo's perspective and achievements? At least they were consistent.

knowing, he says, that air dissolves in water. Apparently, he would have us believe that neither Galileo's having established the principle by which the temperature of a gas can be measured nor his student Torricelli's having measured air pressure was known to Newcomen or of help to him.[42] Would the author have us believe that Newcomen operated in a vacuum of science? Apparently, he thinks that designing a machine is "purely technological" and has nothing to do with science, or at least that hydraulic systems have nothing whatever to do with Galilean science. He forgets, or ignores, that Torricelli communicated his research through Viviani to Pascal, who then developed the basic hydraulic law.[43] This is a direct communication link from the "Galilean school" to the discoverer of a principle that Newcomen applied to the steam engine. The Inquisition drove Galilean science out of Italy for two centuries; what is astonishing is not how little but how much impact Galilean science had on the technology of the time, notwithstanding this fact.

I will rest with these "Terrible Seven". There are lots more. But to avoid spoiling the story of Galileo's positive achievements through disproportionate attention to his detractors, I'll end the parade of myths here.

The Galilean inversion of indoctrination systems is unfinished. We have seen enough examples of mythology about Galileo himself to conclude that the practice of mythologizing, antithetical to Galileo's example, is pervasive in our own time and not limited to the Church. Historians, philosophers and scientists in and out of the Church have contributed to Galilean mythology in the 20th century. If myths continue to be created about one of the greatest myth debunkers who ever walked this planet, then is the study of anything safe from this practice?

No, but fortunately, his influence is rebounding with a positive message for the future of man. In this last century of the millenium of Galileo, we have seen poignant examples of the contrast between positive benefits of science flowing from its application by creative individuals and the destruction arising from its misuse by collective institutions. In the aftermath of these contrary developments, the aim of science is visibly susceptible of redirection in the mix of Galileo's scientific method with his realization: *one tiny spark of reason in an individual who applies that method, or uses an application of it, can produce progress where a wildfire of collective opinion produces only chaos.* There is a renewal of Galilean epistemology emerging within our civilization. And one of the individuals within whose mind it is emerging might be considered one of the least likely to nurture it.

---

42. This is rdc's immediate reaction as I mention this particular author's argument to him, prior to being influenced by my (concurring) view on the matter.
43. Cf. Chapter 12 "Solid Progress From Gases".

The Galilean Revolution in the Relationship of Science and the Church[44]

In the four centuries since Bruno was burnt and Galileo suppressed, scientific epistemology has been confirmed by technological achievement; the decline of Church influence can be attributed to its reaction to this development, culminating in reinforcement of the posture that the Church and science ought to operate in irreconcilable spheres of faith and knowledge, respectively. But the ghost of Galileo is astir within the system of his persecutors; it stalks the halls of the Vatican in the person of an unusual and far-seeing Pope.[45]

Mathematics has evolved, sometimes by denying its own axioms, and exploring the consequences. Physics has evolved, often by exploring non-corroborations of existing laws, or by changing boundary conditions on the existing laws and taking up a new theoretical perspective on this altered framework. What exempted the Church of Urban VIII from the alternative of change or eventual extinction? Nothing. And for centuries more, that is what it did to redress its action against Galileo and its opposition to his science — nothing. Who knows what moribund cravings lurked behind those official facades of the Vatican? Just as the Platonists clung to perfect circles, the Church's self-image as a static mechanism caused it to cling to a circular epistemology: authority and revelation (e.g., Scriptures) locked in reciprocal validation. It tried to use that self-referential approach to impose catechism answers to questions about the eternal — with dogma, with Inquisitions, and even up to the present day, with indoctrination of youth. But the impossibility of empirically validating those answers left them unsatisfying, especially by comparison to the visibly successful methods of science; hence, the progressive weakening of Church influence after Galileo. It is my personal observation that some persons have looked to the Church for answers to questions of a sort that relate to the origin and remote future of man and for which science may be useful while doctrine is not. Here is a domain of demand for which the Church is a receptive harbor for inquiries and is lacking as a supplier. And so, to address his supply-side problem, John Paul II is currently redefining the role of his Church — consistently with Galilean philosophy. He wants an alliance with science.

Just as the birth of modern science is visible in the difference between Galileo's *De Motu* and his *Discourses*, so is this revolution in the Church's view of its role visible in the difference between the Pope's 1983 and 1992 Addresses to scientists concerning the Galileo case. His older view is highlighted by the address to a gathering of scientists on the occasion of the 350th anniversary of Galileo's trial.

---

44. This section concerns the modern Roman Catholic Church and it is my expectation that almost any reader who is not a Catholic would be bored out of his skull by it.

45. No supernatural implication is intended; "ghost" in this context means a haunting memory, nothing more.

Address by John Paul II to Scientists (1983)

In this address,[46] the Pope spoke therapeutically toward the historical rift between the Church and science. The superficial effect was to convey a tone of apology to the memory of Galileo. Why would a Pope of the 20th century, who is in no way responsible for the mistake of a 17th century counterpart, undertake to redress that mistake? Has not the Church moved in a positive direction since the 17th century in that its Inquisition, its theocracy and its coercive power over literature have been eliminated or at least weakened? Perhaps his apologetic tone stems from the Church's self-image of a collective entity with an identity that persists over time. If instead its members viewed themselves as a group of autonomous individuals united only by a common goal, it might be said that the kind of positive action described at the end of this Chapter would be more appropriate than an apology. But if the apology is made, then it is important for the Pope and his advisors to get it right. Still, the Pope John Paul II of 1983 May 9 (or perhaps more accurately, his speechwriters) reinforced the view of the Church's role that Galileo had worked to overturn. He quoted his earlier remarks defending the role of Bellarmino (who forcibly protected the exegesis of Church fathers, unless science could conclusively demonstrate otherwise):

> " ... we can note that ... there were brilliant forerunners and freer minds, like St. Robert Bellarmine[47] in the case of Galileo Galilei, who wished that useless tensions and harmful rigidities between faith and science could be avoided."

Can this be the Cardinal Roberto Bellarmino, of "freer" mind, who failed to take adequate steps to protect the records of the Galileo case and to eliminate the unsigned memorandum from the case file?[48] Who muzzled Galileo and sent Bruno to the stake?[49] Apparently, his manner of avoiding "useless tensions and harmful rigidities" between faith and science was to prosecute or even execute dissidents.[50] If he had been a true "freer mind" who wished to avoid tensions and rigidities, Bellarmino would have protected Galileo's ability to defend a physical hypothesis which had not been demonstrated false. Instead, he treated it as false

---

46. I consult the official Vatican translation. See *Origins*, Catholic News Service, June 2, 1983, vol. 13 #3.

47. The spelling of Bellarmino's name was anglicized in the Vatican translation.

48. See Ch. 9 footnote 77 and Ch. 10. Bellarmino did not oppose the Decree of 1616, and worse still, did not explicitly supersede the Special Injunction, so that the Tribunal used both the Special Injunction and the Bellarmino affidavit against Galileo, saying that the latter was evidence he had been warned "that the said opinion is contrary to Holy Scripture". See also de Santillana, *The Crime of Galileo*, for evidence that the corruption of the Galileo case file occurred during the stewardship of Bellarmino.

49. Regarding Bellarmino's role in the execution of Giordano Bruno, see de Santillana, *Crime of Galileo*, p. 28 and Singer, *Bruno, His Life and Thought*, p. 177.

50. This is not to deny that there were role models in the early Church of prelates who did avoid such tensions and rigidities, but only to assert that Bellarmino was not one of them. Examples include (1) Gerbert, who became Pope in the year 999 under the name Sylvester II, wrote textbooks on arithmetic and geometry, and improved the abacus; (2) Pope Clement IV in his relationship with Roger Bacon.

and dangerous merely because it had not been proved true and, to him who had not bothered to understand Galileo's letters to Castelli and Cristina, seemed contradictory to Scripture. Is Bellarmino only defended because he has already been canonized?

On epistemology, the Pope said:

> "Your specialization of course imposes on you certain rules and indispensable limitations in investigation; but beyond these epistemological limits, let the inclination of your spirit carry you toward the universal and the absolute ... and you yourselves, be ready to seek all that is true, convinced that the realities of the spirit form part of what is real and part of the whole truth."

Does this sound familiar? Remember the Earth/heaven duality of Aristotle that Galileo worked to overturn? Scientists are "specializing" in a reality limited by scientific epistemology, according to this Pope in 1983. A different aspect, one that evidently concerns theologians, lies beyond these limits and is accessible not to science, he tells us, but to the "inclination of your spirit". The nature of this "inclination" as a gateway to another reality is unclear in the context of his address (presumably it does not refer to inclined plane experiments). What is clear in his remarks is that there are multiple realities, each with its own epistemology.

But there was a bright ray of hope in this Address, and a hint of better things to come. First, the bright ray:

> "Let us think of how the results of scientific research help us to know the universe better, to understand better the mystery of man; think of the advantages which the new means of communications and contact among people offer to society and to the church; let us think of the ability to produce incalculable economic and cultural wealth, and especially to promote the education of the masses, and to cure diseases previously thought incurable. What admirable achievements."

And the hint:

> "Yes, the church appeals to your capacities for research in order that no limit may be placed upon our common quest for knowledge."

Papal Commission Report and Papal Address of 1992

Fortunately, John Paul II was not finished. In July 1981 he had appointed a Papal Commission, not to retry Galileo, but to "undertake a calm and objective reflection, taking into account the historical and cultural context" in the words of the Commission's spokesman, Cardinal Poupard.[51] Whether or not objectively, the reflection seems to have been calmly undertaken indeed, for it took eleven years to reach a conclusion. The Commission was composed of four groups covering the exegetical, cultural, scientific/ epistemological, and historical/juridical aspects of the Galileo case. On 31 October 1992, Cardinal Poupard delivered the Commission's report to the Pope, after which the Pope on the same day addressed the Pontifical Academy of Sciences. What conclusion had the Papal Commission reached after eleven years of calm reflection, and what did the Pope have to say after hearing its report?

Cardinal Bellarmino, the Commission says, had allowed in his *Letter to Foscarini* for a future proof of Copernican theory, in which case theologians would have to review their biblical exegeses to avoid contradicting what had been proved. Galileo, it is alleged, had failed to prove the motion of Earth: the optical and mechanical proofs did not come for more than 150 years.[52] Benedict XIV in 1741 had the Holy Office grant an *imprimatur* to the first edition of the *Complete Works of Galileo*; this *imprimatur* constituted an implicit reform of the 1633 Sentence. The reform "became explicit in the decree of the Sacred Congregation of the Index that removed from the 1757 edition of the Catalogue of Forbidden Books works favoring the heliocentric theory". Pius VII confirmed this reform in 1822 when he overturned a refusal to grant the imprimatur to Canon Settele's *Elements of Optics and Astronomy*; the Pope's decision followed a report of Father Olivieri favorable to disbarring the *imprimatur* for works treating Copernican Astronomy as a thesis. After this spotty recounting of the Church's retreat from its condemnation of Galileo, the Commission concludes:

> "In conclusion, a rereading of the archival documents shows once more that all those involved in the trial, without exception, have a right to the benefit of good faith in the absence of extraprocedural documents showing the contrary. The philosophical and theological qualifications wrongly granted to the then new theories about the centrality of the sun and the movement of the earth were the result of a transitional situation in the field of astronomical knowledge and of an exegetical confusion regarding cosmology. Certain theologians, Galileo's contemporaries, being heirs of a unitarian concept of the world universally accepted until the dawn of the 17th century, failed to grasp the profound, non-literal meaning of the Scriptures when they describe the physical structure of the created universe. This led them unduly to transpose a question of factual observation into the realm of faith.

> It is in that historical and cultural framework, far removed from our own times, that Galileo's judges, incapable of dissociating faith from an age-old cosmology, believed quite wrongly that the adoption of the Copernican revolution, in fact not yet definitely proven,

---

51. For the official Vatican translation of the Commission Report and the Pope's Address on October 31, 1992, see *Origins*, pub. by Catholic News Service, Nov. 12, 1992, Vol. 22: No. 22.

52. This statement is an apparent reference to the Bessel (optical) and Foucault (mechanical) demonstrations; cf. Ch. 12. These demonstrations did not occur until about 200 years after the trial of Galileo, which is apparently what is meant by "more than 150 years".

was such as to undermine Catholic tradition, and that it was their duty to forbid its being taught. This subjective error of judgment, so clear to us today, led them to a disciplinary measure from which Galileo 'had much to suffer'. These mistakes must be frankly recognized, as you, Holy Father, have requested."

Pope John Paul II has demonstrated the leadership and courage to face a thorny issue that the Church has ducked during its 2 ½ century retreat from its mistake in the Galileo case, until his decision in 1981 to face the music, so to speak. He has certainly demonstrated a positive intent, and has done more than all his predecessors combined to redress that mistake. In his 1992 Address to the Pontifical Academy of Sciences, he explicitly acknowledges the theological utility of Galileo's distinction between Scripture and its interpretation, expressed in the *Letter to Cristina*. Galileo, says John Paul II, "showed himself to be more perceptive in this regard than the theologians who opposed him" (Bellarmino included). The intelligibility of the universe leads us, the Pope says, to "that transcendent and primordial Thought imprinted on all things". Galileo must have been a splendid theologian, it seems, since he certainly made the universe more intelligible. John Paul II acknowledges Galileo as the inspired founder of experimental method, and refers to Galileo's impact on Church theology:

> "The upset caused by the Copernican system thus demanded epistemological reflection on the biblical sciences, an effort which later would later [*sic*] produce abundant fruit in modern exegetical works and which has found sanction and a new stimulus in the dogmatic constitution *Dei Verbum* of the Second Vatican Council."

He goes on to say that Galileo's Sentence was reformed and the ensuing debate closed by the *imprimatur* granted to Canon Settele in 1820 (according to the Commission Report, the correct date is 1822).[53] He points out that after Einstein neither the sun nor Earth nor any particular place have the same significance as a reference point that they once had. He says that in Galileo's time the cosmos as then known was contained within the solar system alone. Really? Galileo in his reply to Ingoli made it clear that the fixed stars are far beyond the sun's planets, and this view had been held before Galileo. Nevertheless, he shows that he understands the effect of relativity theory on the Copernican debate better than those academics who believe the debate irrelevant in any context, when he says:

> "Thanks to his intuition as a brilliant physicist and by relying on different arguments, Galileo, who practically invented the experimental method, understood why only the sun could function as the center of the world as it was then known, that is to say, as a planetary system".[54]

---

53. And yet in our own century, a Catholic textbook defending the conviction of Galileo has been published with *imprimatur* and *nihil obstat*.

Inadequacy of the Reform of Galileo's Sentence

As encouraging as is this frank admission of error, I now ask the reader: What is wrong with this picture? If a kidnapper binds a man in chains, leaving him immobile for ten years, and then, having repented, releases the man, would it suffice to say that the act of release has "reformed" the action that bound him in chains? Galileo was physically imprisoned for ten years, and his book "imprisoned" on the Catalogue of Forbidden Books for at least a century. The Sentence against Galileo was more than a Sentence of Punishment; it contains a Verdict — Guilty of Vehement Suspicion of Heresy — based on false and misleading allegations written into the Sentence.[55]

The Papal Commission Report defaulted comment on these misstatements and added an evident falsehood to the record of Galileo's case — that the Church "reformed" Galileo's Sentence when the Congregation of the Index "removed from the 1757 edition of the Catalogue of Forbidden Books works favoring the heliocentric theory."[56] What wishy-washy language! They might better have said: *The Church forever and irrevocably abjures, curses and detests the Sentence that it pronounced upon Galileo.* That would be the appropriate language.

Instead, they point to mistakes made by "Galileo's judges". Was Galileo's punishment for revealing his discoveries to the world a personal vendetta of one crazed Pope acting through his hand-picked tribunal? Consider the continuity of Church behavior before and after the trial of 1633. Recall the host of those whose behavior led to orders for the violation of which Galileo was convicted: the reptilian Caccini, who denounced him; the zealous Lorini, who misrepresented him to the Inquisition; the mad-dog Bishop of Fiesole, who publicly insulted him; the two-faced Archbishop of Pisa, who schemed to obtain his original papers while feigning interest in his theories; the back-stabbing Pope Paul V, who (while

---

54. He adds unnecessarily "as it was then known". Galileo's argument did not assume and did not depend on the whole world's being a planetary system. Galileo argued for a correct relative reference point, not an absolute reference point, and his argument is not invalidated by relativity theory.

55. See Chapter 10 above.

56. This statement from Cardinal Poupard's Report on the Papal Commission Findings directly conflicts with the statement published by the Vatican Observatory Foundation: "... precisely because of this concern to avoid 'scandal to pious souls' and, even more so, out of fear at having finally to take a clear position with respect to the behavior of the Church, the books mentioned above [those of Copernicus, Diego Zuñiga, Foscarini, Kepler, and Galileo] remained on the Index of forbidden books right up to and including the edition of 1819"; Annibale Fantoli, *Galileo: For Copernicanism and for the Church*, trans. by Fr. George Coyne (Director, Vatican Observatory Research Group), 1994, p. 473. Has the Church officially decided that Galileo's book was not a "work favoring the heliocentric theory"? If not, the Church is saying that the *Dialogo* remained on the Index while having been removed from it!!! Is this another violation, for which the Church has given us ample precedent in the Galileo case, of Aristotle's Law of the Excluded Middle? Or, is this just another mystery, transcending human understanding, that must be accepted on faith?

inviting Galileo to visit Rome where he promised his support) prodded the Archbishop to obtain this evidence, and directed Bellarmino against Galileo; Bellarmino, who silenced Galileo with reassurances conveniently omitted from the files of the Inquisition; the Congregation of the Index, which condemned belief in the motion of Earth, the support of which was one of Galileo's highest goals. After the trial, the Church confirmed its action against Galileo by a most solemn and formal act — the Bull of 1664 — which is not erased by a 20th century Papal Commission treating it in stony silence as if it never happened. Compounding the error of attributing the condemnation of Galileo to his "judges", the Commission attributes the decision to "reform" Galileo's Sentence to an earlier Papacy which did not, in fact, reform it.

It is a most significant implication of this 1992 Papal address that the Church has finally come to agree with Galileo regarding theology; yet, as if theology and science have no connection, John Paul II repeats the denial that at the time of his condemnation Galileo had proved his case scientifically. We heard this denial in *Answer Wisely*, in the Report of the Papal Commission, and now a third time here. Really, this is an ill-advised position for the Church to take. First, there is no doubt that Galileo did indeed prove his case against the Church-sponsored system of Ptolemy. Secondly, the denial that Galileo proved his case for Copernicus displays a profound misunderstanding of the process of science. Science does not progress in vast and sudden quantum leaps from one apodeictic certainty to another. When one stifles or suppresses a hypothesis, one breaks the flow of progress to the next thesis. The Church clearly did break the flow of progress in the Galileo affair, yet John Paul II calls it a "myth" that "the Galileo case was the symbol of the Church's supposed rejection of scientific progress or of 'dogmatic' obscurantism opposed to the free search for truth". Here the Pope takes a position whose credibility is reminiscent of Galileo's denial that he defended the opinion of Copernicus.

Galileo's well supported disproof of the Church-sponsored hypothesis presented in the *Dialogo* was the object of forcible suppression by the Church. Ironically, he also had a proof of the Copernican hypothesis: his design of the experiment of Bessel. That the requisite instrumentation, and freedom of inquiry, for carrying it out did not exist in Galileo's lifetime does not alter his having merited credit for the proof. Of course, we see this in retrospect and it was not known to the Inquisitors who condemned him. But in arguing that he had not proved his case, the Inquisitors of 1633 failed to heed Galileo's words:

> *"before a physical proposition is condemned, it must be shown to be not rigorously demonstrated."*

Galileo's advice "to Cristina", accepted by John Paul II, inverts the position of Saint Bellarmino (who would have been willing to re-interpret Scripture if only he thought Galileo had proved his case) and puts the burden of scientific proof on those who would condemn a proposition. When this principle is ignored, science is impeded in favor of what the Pope calls "useless tensions and harmful rigidities".

Following the 1992 Papal address, an Associated Press journalist ventured a reckless prediction in these words:

> "Pope John Paul II on Saturday formally proclaimed that the Roman Catholic Church erred in condemning the astronomer Galileo for holding that the Earth was not the center of the universe. ... The speech was the Vatican's *final word* on the matter nearly four centuries after the astronomer was found guilty of violating church doctrine in contending the Earth revolved around the sun, and not vice versa".[57]

Eighty-five years earlier and with far-longer-range vision, Antonio Favaro had concluded the Preface to his volume *Galileo e L'inquisizione* in these words:

> "It [the Church] has atoned for that mistake [the condemnation of Galileo], *though perhaps not yet completely*, on the day it had to remove from the Index of Forbidden Books the condemned *Dialogue* of Galileo and to write in the same volumes of the 'Decreta' the permission to teach, uphold and defend the doctrine once declared 'absurd and false in philosophy and formally heretical'." [58]

Favaro's prophetic words "*... though perhaps not yet completely, ...*" are still true.

Reform of Church Epistemology

Nevertheless, the Church has offered to favor Galileo by adopting reforms that he outlined in his letters, by drawing conclusions from its catastrophic blunder and by acting upon those conclusions in such a way as to prevent another case like Galileo's from occurring. That is an enormous advance beyond precedent. Here, in these words from the 1992 Papal Address, is the beginning of John Paul II's revolution in ecclesiastical epistemology:

> "It is a duty for theologians to keep themselves regularly informed of scientific advances in order to examine, if such be necessary, whether or not there are reasons for taking them into account in their reflection or for introducing changes in their teaching.
> ...

---

57. *Kansas City Star*, Sunday, November 1, 1992, p. A-9. Emphasis mine.
58. Emphasis mine. See Appendix A for my translation of the entire Preface.

The seriousness of scientific knowledge will thus be the best contribution that the academy can make to the exact formulation and solution of the serious problems to which the church, by virtue of her specific mission, is obliged to pay close attention — problems no longer related merely to astronomy, physics, and mathematics, but also to relatively new disciplines such as biology and biogenetics. Many recent scientific discoveries and their possible applications affect man more directly than ever before, his thought and action, to the point of seeming to threaten the very basis of what is human."

As the scope of science advances into biological, psychological, behavioral, and other disciplines, the scope of what a theologian is prepared to teach without consulting scientific data approaches the vanishing point. While these disciplines become products of scientific method, the boundary between physics and Church philosophy keeps expanding across new ground. Therein lies the evanescence of the Church authority posture. Integral to scientific method, the feedback concept holds error not as something to be avoided but as a prerequisite of acquiring knowledge; a Galilean method of deriving knowledge from error is to control the error and reduce it convergently by repeatedly altering hypotheses according to the difference between what they predict and what

Ascanio Piccolomini, Archbishop of Siena
(Sketch by Thomas H. Norton)

experiments show. A declaration of divinely protected belief cuts off knowledge by blinding those under its "protection" to that error through the control of which knowledge can be attained. John Paul II is working to alter that posture — but will his vision be sustained? His call for change is no divine *fiat*; it can happen only through sustained action of those individuals who hear and heed the call. A new type of ecclesiastical prelate is required for the transition period to John Paul's vision — one who, like Piccolomini, understands and respects science and can think and operate as an individual.

Pope John Paul II will not easily be forgotten by Catholics and by all who are affected by his Church; the logical train of his initiatives cannot easily be derailed. Of all Popes in history, he is first to issue a mandate for Church teachings to conform to scientific findings, one that is motivated by Galileo's philosophy of science. This is nothing less than a clarion call for the eventual abandonment of indoctrination. Such a development would open the door to evolutes of ecclesiastical organizations that focus on the preservation of various studies: philosophy, history, arts and sciences, languages, and the artifacts of Church history, a role which the Medieval Church performed admirably in preserving elements of Greek civilization.

Included among these studies is theology: the study of God. To this study, too, Galileo has contributed, beyond the letter to Cristina, by personal example. Durability and divinity were linked conceptually in that ancient Greek culture from which we have benefitted by Galileo picking up loose threads of its science. That which endures is a god.[59] Galileo often referred to Archimedes, who served as a model for human progress, as "divine". Rather than making a god in his own image, a man can strive for a durable presence through his work. Also traditionally considered theological, questions about the origin and remote future of the human species relate to large-scale processes in the universe. These are addressed in modern times by the science of cosmology. Given the rejection of theocracy by theologians, theology and science are compatible because theology belongs to science. Therefore, theological knowledge is attainable through Galilean methods of discovery and validation.

Having freed the long-suppressed works of their prisoner, the gatekeepers of *Il Gioiello* are now unlocking the jewel of rational epistemology. They call upon the now freed prisoner to unbind them from a static faith inaccessible to science. Galileo has become an inspiration and guiding force behind improvements in the system of his persecutors. If properly positioned individuals sustain this trend, it leads to the abandonment of indoctrination and the application of Galilean scientific reasoning to diverse areas of human knowledge and endeavor. This is the justice for which the unrequited abuse of Galileo cries out, over four centuries of ingratitude, to all of us, his beneficiaries.

---

59. Cf. G. M. A. Grube, *Plato's Thought* (Methuen, 1935), p. 150, as quoted by W. K. C. Guthrie, *The Greek Philosophers from Thales to Aristotle*, Harper & Row, NY, 1975: "Any power, any force, we see at work in the world, which is not born with us and will continue after we are gone could thus be called a god, and most of them were."

APPENDIX A: Author's Translation[1] of Favaro's Preface to *Galileo e l'Inquisizione*

This National Edition of the works of Galileo addresses a need of students that has waited too long for history to do justice to these events. As a necessary consequence of producing it, I embarked upon a search into the painful events of his life. My research has been conducted in such depth as no one has previously even contemplated and has been extended to details hardly touched upon up until now. All of this could not have been done without my obtaining from the supreme Church authority the concession of pushing the research up to where the eye of the non-clergyman has never penetrated, except in cases of cheating.

Woefully revealed in the Inquisition documents are those prickly piles of thorns strewn along the brutal path by which Galileo led himself to the heights of glory. His startling astronomical discoveries that so brilliantly confirmed the Copernican doctrine of the system of the universe, though they earned unconditional and enthusiastic recognition of the true students of nature, nevertheless invited the attention of theologians, who then started to watch the consequences with a suspicious eye. These theologians had well understood where the discoverer was aiming his presentations, and while all of Rome, challenged by Galileo to verify the results he had announced, gave free reign to proper admiration in greeting this new "Columbus of the heavens", Cardinal Bellarmino, one of the Inquisitors General, secretly addressed himself to the mathematicians of the *Collegio Romano* to obtain their support, while the Inquisition still more secretly wrote the name of the bold new author into the dreaded book of suspected heretics.

When the Church began to announce some serious actions against the book of Copernicus, and in Florence (both from the pulpit and in the Court of the Grand Duke) Church apologists pointed to contradictions between the sacred scriptures and the accused doctrine, Galileo decided to intervene against a powerful threat: that prohibition of the doctrine and the victory of the conventional (Ptolemaic) ideas with the Grand Duke would block him from fighting for that truth in whose triumph he was from now on placing the goal of his entire life.

---

1. A literal translation of this 1907 document necessarily preserves much of Favaro's rather Ciceronian style, more attuned to the ear of his time than to the modern ear. Nevertheless, in order to increase the reliability of this translation as a source document, I have aimed to make it as literal as possible without impeding comprehension.

The memorable letter to his faithful confidant Castelli, enlarged upon in another famous letter, to the Grand Duchess Cristina, in which he clearly and masterfully marks the limits between science and the faith, exasperated the theological community that already had declared itself against Galileo through the well-known denunciation by the Dominican Tommaso Caccini in the Church of Santa Maria Novella. A religious brother, Lorini, who had already revealed himself as an anti-Copernican at San Marco, denounced to the Inquisition the letter to Castelli as that which contained suspicious propositions and defended opinions contrary to the interpretation that the Church Fathers had given of sacred scriptures. Having sniffed this out, Galileo (forgetting about himself and the danger which he was confronting) rushed to Rome to foil the plots that were being hatched against the system of which, with the *Letters on Sunspots*, he had made himself by now an open proponent.

In Rome, he busied himself making new converts. To stem the dangerous current, the Inquisition secretly hastened its procedure. Expecting to be called to defend others, Galileo deluded himself in the belief that the treacherous Tribunal wanted to be enlightened by him. But while he prepared himself to bring forth the weightiest arguments, they were actually proceeding against him as the principal accused, one so dangerous as to have to deny him even the right of self-defense.

In the short span of a week, the trial is concluded; the doctrine is declared stupid, absurd, and formally heretical, and by order of the Pope, Galileo is called by Cardinal Bellarmino, and before the Commissary of the Holy Office and before witnesses, of whom the official presence had been for the most part hidden from him, he is ordered altogether to abandon the condemned opinion, and in no way still to hold it, teach it, and[2] defend it; otherwise, procedures would be initiated against him in the Holy Office. Galileo promised to obey and, on the very same day on which Bellarmino announced to the Congregation of the Holy Office that the admonition had been imposed, read the decree of prohibition of the works of Copernicus and of the other proponents of the same doctrine.

Having returned to Florence, and sequestered himself a little later on the hills of Bellosguardo, Galileo seemed preoccupied by his studies (in which he applied the eclipses of the Medicean satellites of Jupiter to the determination of longitude at

---

2. Favaro's use of the conjunction "**and**" rather than "**or**" deviates from the language of the 26 Feb 1616 Special Injunction: "*... quovis modo, teneat, doceat aut defendat, ...*" (" ... not to hold, teach, or defend it in any way whatever, ..."). It would obviously be more difficult for the Inquisitors to use an "and" clause for any subsequent conviction; therefore, in inferring Galileo's innocence from all the data of the case, including the language of this document, I make the assumption that is less favorable to my case.

sea) and by the controversy with Grassi about comets: a question which gave rise to that insuperable jewel of polemic writing, *The Assayer*, dedicated by the Lincei to the new Pope Urban VIII, who as a Cardinal had been both in prose and verse an extremely great admirer and supporter of our philosopher.

The most fervent desire that Galileo experienced was that of reuniting with the old patron who had now risen to the Papal throne. Then too, the new Urban VIII had maintained his beneficent posture toward Galileo, who was urged by the solicitations of his friends, and especially the pleas of his daughter Sister Maria Celeste, to reinforce that relationship. These factors bolstered Galileo's own voluntary decision not to let slip away a fine opportunity without taking some step in favor of the Copernican doctrine.

Happily welcomed in the course of nearly six weeks at Rome, the eternal city, during which he had at least six audiences with the Pope, Galileo received from him gifts of indulgences, medallions, a Lamb of God, and an honorable title and promise. But as for the opinion of Copernicus, in reaction to the Catholic fear about the dangers that the faith would have encountered in case the condemned doctrine might have turned out to be the very truth, Galileo did not obtain anything but the sole expressed declaration that it wasn't to be feared that someone should ever be about to demonstrate it necessarily true.[3]

If the principal purpose of this venture had been unsuccessful, it is nevertheless appropriate to believe that Galileo, who often completely deceived himself about that which he took greatly to heart, had drawn his conviction that the decree of prohibition would not be upheld in strict severity. For that reason, a short time after having returned from Rome, he took courage to respond to Ingoli, who had addressed to him a refutation of the Copernican system. The knowledge that his response to Ingoli, circulated in manuscript form, had been read and appreciated by the Pope himself greatly confirmed him in his delusion.[4]

---

3. In this purposely rather literal translation it is difficult for me to avoid the ambiguity of this clause, as expressed by Favaro: *"che non era da temere che alcuno fosse mai per dimostrarla necessariamente vera."* It was originally my opinion that the passage makes much more sense taken to mean that no one need fear such a demonstration rather than to mean that such a demonstration is impossible, since the evidence indicates that Urban wanted to encourage Galileo to write a dialogue (in which the Ptolemaic system would be given at least equal force of argument). However, rdc has offered an alternative explanation with which I have come to agree. Under this interpretation, the necessity to which Urban refers is not Galileo's necessity of conforming to all the theoretical and physical considerations, but God's necessity of being free to do otherwise than human reason can explain.

These same, not unwarranted hopes induced him to resume that formidable work undertaken in the happy years at Padova, already announced to Kepler, also promised in the *Sidereus Nuncius*, and many times suspended but never abandoned, in which, with the help of the new astronomy and the integrated physics, the incontestability of the doctrine of Earth's motion was to be demonstrated with all the evidence.[5] And these hopes had reasonably become a near certitude, after he had heard how (to Campanella who reported to the new Pope that the prohibition of the book had obstructed the conversion of certain German Protestant noblemen) Urban VIII had responded in these exact words:

**"It was never our intention, and if it had been up
to us, this decree would not have been made."**

With more ardor than ever, Galileo dedicated himself to the completion of his work, from which, beginning at the instant that he heard this, nothing could distract him. But, however much he believed that he could place his trust both in the good intention of the new Pope and in his personal favor, and despite these new positive developments, that fatal decree had been issued, and that awful and exact injunction had been given to him. As a consequence, his writings could never obtain the necessary approval if he should openly support the condemned doctrine. So, he had to stretch his inventive power and to submit himself to the torment of putting forth as a hypothesis that which he felt to be absolute truth.

Having obtained (or more accurately, seized with powerful mediation) the approval of the censors, after he accepted all the imposed variations to his text, including that most fortunate variation of the title,[6] and having published his *Dialogue*, the true intentions of the author appeared to everyone only too clear.

---

4. In evaluating Favaro's opinion that Galileo deluded himself, *a posteriori* knowledge of the Galileo case is a disadvantage. Note that Favaro somewhat paradoxically agrees that Galileo's hopes are "not unwarranted" (in the following sentence).

5. The phrase "with all the evidence" appears to refer to Galileo's application of empirical data from his astronomical and physical discoveries to the conclusion already partially demonstrated by Copernicus on purely mathematical grounds.

6. Galileo had originally titled the work *De fluxu et refluxu maris* ("On the ebb and flow of the sea"), but that was apparently too physical sounding for Urban, who wanted to hear only mathematical (suppositional) language; the Master of the Sacred Palace therefore insisted that Galileo change it to delete any reference to the tides; cf. Drake's *Galileo at Work*, pp. 319-320. No doubt Favaro's "most fortunate" refers to the fact that the argument based on the tides occupied only one day of the four-day dialogue and turned out to be incorrect.

And since the unfortunate conclusion of the work placed in the mouth of the speaker (who always put forth oppositions that were mostly inconclusive and empty scholastic subtleties) an argument that the same Pope had suggested to Galileo, it was an easy thing to persuade the vain and proud Urban VIII that in this ridiculous character the rash author had wanted to represent him; and this sufficed to change him suddenly from the friend and protector into an implacable enemy, so that he induced himself to believe that this book was more loathsome and pernicious to the Holy Church than the writings of Calvin and of Luther. The recommendations of the Grand Duke and the offices of the Tuscan Ambassador were just sufficient to obtain the concession that the case would be deferred to the examination of one particular Congregation; but hardly had this Congregation given this matter its opinion, of which there was already no doubt, than it was ordered that the Inquisitor of Florence request Galileo to appear before the Commissar of the Holy Office in Rome.

The cruel anxieties and the fear of the worst, harming the already shaken health of the unhappy philosopher to whom any postponement was refused, made him sick. Three doctors called to his bedside declared his life to be jeopardized by any small cause of worry from the world outside, but Urban VIII saw this as a pretext for evading his order and had a letter written to the Inquisitor demanding that the Congregation of the Holy Office send, **at the expense of Galileo**, an official accompanied by Church physicians, and that if they will find him in condition to travel, they will bind him in chains and transport him bound to Rome.

Terrorized by the ferocity of the Pope, who according to Ambassador Niccolini was threatening some extraordinary measures, this same Grand Duke did not know how to resist any further and caused Galileo to understand that he should by all means obey and leave for Rome.

And in the harshest of winter weather, amid the dangers of the plague that was spreading all over Italy, of that same Bubonic plague of which the fearful memory is immortalized in the pages of the great historical novel *I Promessi Sposi* by Manzoni, Galileo left for Rome. Urban VIII finally brings him under his control.

It would turn out to be completely superfluous, step by step to follow here the hapless philosopher along the tortured course of this second court case. In every detail, and with a cruelty that no commentary could exaggerate, the documents cover the whole sordid drama, which narrowly escaped killing its victim. The National Edition brings these documents to light wholly as the archives preserve them for us, reproducing the signatures of Galileo to the fearful cross-examinations. Even by the way the letters of the signature are formed, the growing anxiety of the spirit of the dignified old man breaks through.

These documents, like all those related to the trial that were used in the tribunal of the Inquisition, were originally maintained in the archives of the Holy Office in Rome and issued in two parallel series: one of which, having the title of "Decreta", contained reports or abstracts of reports and the decisions of the Congregation; in the other of which, there were, or at least should have been, all the reports of the procedures against the accused, the cross-examinations of the accused and of the witnesses, the related correspondence, and the sentences and renunciations. Of these two series, that of the '"Decreta" can still be found in the archives of the Holy Office, open in exceptional circumstances at the illustrious discretion of Pope Leo XIII. From the other, which was also archived there, the trials of Galileo, repeatedly studied by the same Inquisition, were removed — we don't know either when or how — and combined into one volume, which after various diversions passed into the Secret Vatican Archive. There, before having been seen by myself and by Isodoro Del Lungo, who also in this part of our common labors was to me a most precious collaborator, it was seen by others and made known repeatedly to scholars.

Questions that, for about 30 years, were raised concerning the authenticity and integrity of the documents contained in the Vatican volume made it reasonable to decide that some rules ought to be established for the publication of those documents in order that one might follow the discussions, which continued to debate this most serious and delicate matter with extraordinary zeal. But today, we don't believe there is any longer a single person who honestly suspects that some artful alterations have been introduced into these documents. And to the reaching of such a conclusion, the entire publication of the "Decreta" was useful, as it shows the continuous and perfect correspondence of the two series of reports concerning the same matter.

We must grant that the volume transferred to the Secret Vatican Archive does not contain the really complete collection of all the documents pertinent to the trial of Galileo. Other documents (which may have been originally contained in this volume, but now may be dispersed among another series, because one finds other matters treated in them) could be bound up with the papers related to different matters, or they could have been partially destroyed because, following the old custom of the Holy Office, not everything was saved. Or again, they may have been lost in transportation, to damage or to theft, to which the archives of Rome were most subject (especially those of the Inquisition) in the time of Napoleon and of the Second Roman Republic.

Despite these lacunae, there were not, nor could there have been among the original reports, additional documents of the highest importance that illustrate the day-by-day events that transpired. The National Edition includes, for example, the Correspondences which, relative to this dismal year 1633, occupy one entire volume. By revealing the secret manipulations behind the scenes, they permit us to reconstruct the great drama in the most minute detail.

Thus, the story of the condemnation of Galileo can be written now in all its completeness. Apart from those esoteric discussions on the authority that delivered it, the story does not need rhetorical speeches or invective in order to bring to light how it represents, if not the greatest, one of the very greater mistakes of the Roman Curia. It has atoned for that mistake, though perhaps not yet completely, on the day it had to remove from the Index of Forbidden Books the condemned *Dialogue* of Galileo and to write in the same volumes of the "Decreta" the permission to teach, uphold and defend the doctrine once declared "absurd and false in philosophy and formally heretical".

In Grateful Memory

In the foregoing translation, the author acknowledges with gratitude the expert assistance of Professor Anna Maria Jardini whose comments on the somewhat arcane turn-of-the-century literary style of Antonio Favaro proved invaluable. Sadly, Professor Jardini died in 1994, less than a month after completing her review of this translation. A native of Milan and for many years an instructor in languages and in Old English literature at various California colleges, she exemplified the highest standards of what an educator should be. She is and will continue to be keenly missed by the author, one of her last students.

APPENDIX B: On the Unpublished Introduction to
Drake's Translation of Galileo's *Dialogo* [1]

For these sixteen years during which Professor Drake's remarks have remained unpublished, readers of the *Dialogo* have lost the benefit of his contribution to understanding Galileo's physics, his theory of the tides, and his motivation in creating them. That loss applies to my own understanding. Before gaining access to his proposed introduction, I had written the entire text of this book, excepting this appendix and the portion of my epilogue that concerns its suppression. Rather than expand what is written here, I have chosen to leave this book intact so as to convey to the reader my sense of surprise at certain of his unpublished comments. Professor Drake alerts the reader to his intent at the outset:

> *"Though Galileo's science was elementary in the sense of its being but a beginning, it was highly advanced in other ways, along lines abandoned in favor of dynamics by Newton and his successors. To be fully understood, it requires explanation to Newtonians of the present as it did to Aristotelians of Galileo's time, but of a kind not previously offered."*

Galileo's real motivation in creating the *Dialogo*, he tells us, was to present his own physics rather than to sell Copernican ideas. *Galileo re-structured the presentation of his new physics and altered the title of the* Dialogo *in order to render his Church a service.*[2] He asserts as a matter of probability that Galileo sold Urban VIII on the importance of writing the *Dialogo* from the perspective of bestowing intellectual legitimacy on the edict of 1616, by showing that the Church understood the arguments in favor of Copernicus when it issued the edict:

> *"When Cardinal Zollern's report of dissatisfaction with the 1616 edict on the part of foreign Catholic sympathizers was heard by the pope in 1624, and Urban expressed his opinion that the edict should not have been issued, Galileo probably proposed a way in which his personal interests could be made to serve those of the Church and of Italian intellectual leadership. Publication of his tide theory could be used as the pretext for discussing the earth's motions as pure hypothesis, showing to the world in that way the true intention and meaning of the 1616 edict. ...*

---

1. Stillman Drake (1980). "Translator's Preface to Galileo's *Dialogue* (third edition)." Typescript. University of Toronto Library collection. I am grateful to Mrs. Florence Drake for permission to use these excerpts from Professor Drake's manuscript.
2. Drake asserts that in writing the *Dialogo,* Galileo was motivated more by zeal for Catholicism than for Copernicanism. I consider it equally likely that Galileo found the appearance of Catholic zeal convenient to his goal of showing the relevance of his new physics to cosmological questions like the Copernican one. (Here and throughout this appendix, paraphrases of Drake are included in the text in italics. All direct quotations are enclosed in quotation marks and indented from both margins.)

> *The proposal which I suppose Galileo to have offered could hardly fail to appeal to Urban VIII, himself an intellectual who desired the support of thinkers and who shared Galileo's pride in Italy and her reputation of scientific leadership. While specific documentary evidence is lacking that Galileo left Rome with the pope's permission to proceed with his proposed book, all scholars have agreed that he probably did. Galileo certainly believed himself to have received such permission, to judge from his subsequent actions and letters. ..."*

Did Galileo write the *Dialogo* in defiance or in defense of the edict of 1616? Drake continues:

> *"By putting his Catholic motivation in first place, when altering his preface [to remove the reference to the "ebb and flow of the seas"], Galileo laid grounds for skepticism on the part of later historians though of course he could not foresee the later trial from which they were to deduce his motives by agreeing with the Inquisition that Galileo had acted deceitfully. Having the endorsement of so high an authority, historians viewed the preface as hypocritical, or at best as written tongue-in-cheek. ...*
>
> *The common idea that the* Dialogue *was meant to circumvent the 1616 edict took root because no constructive purpose remained evident for the book after its title was changed, other than Copernican propaganda. Mention of the edict in the opening sentence of a book assumed to circumvent it appeared cynical, its author's defense of the edict hypocritical."*

Drake attacks the myth that Galileo failed his ostensible purpose of demonstrating Copernican motion. The Church has held (right up to the present day) that Galileo did not prove his case for Copernicus, but Drake counters:

> *"Historians and philosophers of science now like to say that no direct proof of the earth's annual motion existed until the 1830's, when the parallax of some fixed stars finally became measurable, and that direct proof of the daily rotation was lacking until 1850, when the Foucault pendulum was devised. That is a kindly thing to say on behalf of the Church, which had vacated the 1616 edict shortly before those events, but it has no scientific significance. Both motions of the earth were established from a scientific standpoint in 1687 by the consequences Newton drew from his law of universal gravitation, already confirmed by innumerable observations and measurements. No scientist was surprised, let alone converted to belief in motions of the earth, by stellar parallax or the Foucault pendulum.*
>
> *It is my own view that the Copernican motions were in effect still earlier demonstrated from a scientific standpoint — by Galileo in his*

> Dialogue, *though he could not there claim to have demonstrated them because of the 1616 edict. I allude to his explanation in the Fourth Day of the necessary existence of tides, which was Galileo's personal reason for wanting to publish this book."*

This remarkable suggestion that through an erroneous theory of tides Galileo had made his case for Copernicus confronts the current position of the Church head on. The Papal Commission convened by Pope John Paul II to re-examine the Galileo case said in 1992: "Galileo had not succeeded in proving irrefutably the double motion of the Earth — its annual orbit round the sun and its daily rotation on the polar axis — when he was convinced that he had found proof of it in the ocean tides, the true origin of which only Newton would later demonstrate." Does Drake's statement last quoted above answer this contention? Progress in this debate depends partly on understanding the relevance (to the Copernican question) of Galileo's tidal theory, as far as it goes, and of the underlying physics. After considering the fundamentals that distinguish Galilean from Newtonian physics, Drake will return to the main argument, and so shall we here.

What is the perspective of this new physics, the source of Galileo's real and covert motivation for writing the *Dialogo* around a tidal theory for which Drake has advanced his bold claim? A new science may grow from and beyond an earlier development, eventually diverging from it in some way. Drake here describes a divergence of Newtonian from Galilean mechanics, about which Galilean science is misunderstood as was Aristotle's:

> *"... long-accepted opinions may be 90% satisfactory and 100% misleading. That was the case with Aristotle's science, opposed by Galileo, and it is again the case with the prevailing view of Galileo's science, as I see it."*

How did Galileo advance beyond that point of divergence? *To approach an answer, it needs to be realized that Galileo's* Dialogo, *despite the long form of its title, is a work on physics contrasting two cosmologies, not two astronomical systems:*

> *"What interested Galileo about systems of the world was the general question of the 'constitution' or arrangement of the 'integral world bodies,' a cosmological rather than an astronomical interest."*

This is no reflection on Galileo's capability to calculate astronomical orbits. He did so repeatedly, not for planets, but for satellites:

> *"... there are literally thousands of original and accurate astronomical*

> *calculations to be found in Galileo's working papers ... Galileo neglected planetary astronomy in Kepler's sense because he was a physicist rather than an astronomer, and he trusted astronomers to do their work as well as he could, or better — much as physicists trust engineers to tabulate data and build bridges."*

Whereas Galileo adopted certain kinematic principles that obviated the concept of force in his physics, Newton established a point of divergence with Galilean mechanics by adding the concept of a universal force field. Professor Drake describes the overlapping domains of the Galilean and Newtonian developments:

> *"Newtonian preconceptions entail a physics of forces, applicable to the entire universe, whereas Galileo's physics of natural motion (to be defined below) was applicable only to heavy bodies situated on or near the earth's surface and moving at measurable speeds through measurable distances. That very restricted domain is of course amenable to understanding in Newton's universal physics by considering gravity as a force. The same domain lends itself also to understanding in a physics that does not consider gravity as a force, and such a physics was created by Galileo. It was consistent with Newton's <u>only</u> in the restricted domain mentioned, and Galileo took care to remain within that in applying his physics. ...*
>
> *Newton's creation of a universal physics required the introduction of the concept of force, and especially what we call 'the force of gravity.' The concept of force was excluded from Galileo's physics as from Aristotle's almost by definition, since physics started as the science of nature (physis) and force was then conceived as the very antithesis of nature. ... Our [post-Newtonian] physics associates (and even identifies) accelerated motions with forces; Galileo's did not. ...*
>
> *Like Aristotle's, Galileo's physics was what we call kinematics; Buridan's physics, like Newton's, added dynamics. Galileo is often misleadingly called the father of modern dynamics, an idea that could only arise after Newton associated acceleration with force and only because Galileo had fathered the modern physical treatment of acceleration."*

How was Galileo able to create a science of motion lacking the concept of a force field? Professor Drake approaches this question by enumerating the fundamental principles of that science:

> *"We should first find out whether the dynamic concept of inertia as*

> *framed by Newton had a counterpart in Galileo's kinematics, and if it did, assign a suitable name to it. ... a limitation of scope was imposed by Galileo on his physics, depriving it of the universal character of Aristotelian, Cartesian, and Newtonian natural philosophy."*

The *Dialogo* discusses three principles of Galilean physics mentioned by Drake:

> *(1) "... Galileo took geocentric <u>speed</u> to be conserved near the earth's surface in circles that were definitely not geocentric, and even passed through the earth's center. "[3]*

> *(2) "Galileo took the approach* [of heavy bodies toward Earth's center] *to be straight toward the center and to be accelerated in such a way that distances traversed from rest are proportional to the square of the times elapsed."*

> *(3) "A third principle was adopted for motions of projectiles; that is, for heavy bodies to which motion had been imparted by force before they began to move by themselves with respect to the earth. I call this third principle "conservation of rectilineal motion"; it is indistinguishable from inertial motion, since both speed and direction were specified. But because Galileo did not give it any dynamic quality, it might be misleading to call this "the principle of inertia" in the Newtonian sense.*

So let us call it the principle of inertia in the Galilean sense, or simply, Galilean inertia. A physics from the standpoint of which the Earth itself is considered a moving body must be able to predict the future positions of objects for which the initial conditions of motion are known. From a strictly kinematic inertial principle, how could Galileo predict the paths of falling objects and projectiles? Professor Drake approaches this question by contrasting Galilean inertia with conservation of geocentric speed and distinguishing the situations in which each applies:

> *"A principle of conservation of geocentric speed would be superfluous in Newton's dynamics, but it was not so in Galileo's kinematics. Newton's physics is correct and induces in us a preconception that no other description of the phenomena of natural motion can also be correct and internally consistent, making us feel that Galileo's first principle must have been unwarranted and inconsistent with his third. But what this really means is only that we have not yet succeeded in grasping Galileo's standpoint in physics. ... If Euclid had omitted his 'parallel postulate,' he*

---

3. Drake may have meant curvilinear speed along the Earth's surface. (jm)

could have derived many of his propositions without it, as was done later. Without the concept of force, Galileo derived many physical propositions that agreed with the observations and measurements it was possible for him to make, and by considering that physics alone we may arrive at his standpoint."

"... let us consider the quite elementary case of a body dropped from the mast of a moving ship. This illustrates Galileo's first principle, for he said that during that brief fall the body preserves the speed it had along the circle traced by the mast around the earth's center. ... In cases where a body falls through any considerable distance, it became possible by the nineteenth century to detect an advance in the direction of the earth's rotation; but since Galileo had already noted that implication of his principle on p. 233, the later experimental result does not vindicate our choice of the tangent against Galileo's use of the arc. Within the range of phenomena to which Galileo restricted physics, what he said did not contradict experience. Outside that range his physics was not 'mistaken,' but inapplicable, viewed from his standpoint."

Given that it had an initial speed with respect to Earth's surface, would the stone falling from the mast continue in its forward motion? Evidently so:

"Continued projectile motion would be straight, Galileo said, if it were not deflected downward by weight."

And what of an object resting on Earth's surface? Geocentric speed tends to be conserved as its mass tends toward Earth's center in accordance with the law of falling bodies. But wouldn't the motion along the tangent line more than offset its downward tendency if the object were able to participate in a sufficiently high speed of rotation? *On a perfectly smooth frictionless Earth, a body resting on it would tend to slip backward against increasing rotational speed, but a rooted object such as a tree could be projected if the object is brought to a higher speed than is suitable for its orbiting at the radius of Earth.* Of course, the real Earth is not smooth and spherical like Aristotle's heavenly bodies, and so even unattached bodies resting on its surface would be susceptible of projection from its surface if its rotation were indefinitely accelerated, as they receive higher increments of speed from contact with surface irregularities.[4]

Professor Drake notes that modern critics have gone astray by failing to note the observational basis of Galileo's "theory of semi-circular fall" on pp. 164-166 of his translation of the *Dialogo*:

---

4. See Ch. 9 "Galileo's Close Brush with New Discovery"

> *"The descent of a body dropped from a tower appears to us to be straight along the side of the tower. If the earth rotates, the path of such a body will not be straight absolutely, but only relatively to an observer on earth. The question then arises what the path would be to an observer watching the body fall as the earth rotates, but not sharing in that motion. Assuming that its geocentric speed is conserved and also that the body remains between the same two radii bounding the arc which would represent that speed if the body remained atop the tower, Galileo opined that its absolute path from top to base of tower would be another circular arc which, if prolonged, would pass through the center of the earth, or very nearly to it. ...*
>
> *The diagram on p. 165 has distracted attention from Galileo's text since his own day. ... He did not and could not employ a tangent line, because he already knew that projection could not occur. A heavy body at rest with respect to the earth conserves its geocentric speed. Unsupported, however, it cannot conserve geocentric <u>motion</u>, as it would by 'circular inertia', since it would not remain in the arc ... but must fall.*
>
> *Now, it is a very curious fact that during the first four seconds of fall, which is as long as free fall could actually be observed from towers available to Galileo, there is no measurable difference between ratios of distance under the times-squared law and ratios of the versine relationship[5] in his diagram as far as the earth's surface. If fall could continue into the earth, great differences would soon be evident. The initial similarity of ratios was acknowledged by Mersenne, and again by a group of mathematicians who debated Galileo's proposition during the 1660's. By neglecting it, modern critics have failed to note that Galileo's position was in accord with practicable tests.[6]*

In these words, Stillman Drake wraps up his discussion of Galileo's physics (apart from the tides), ending with a subtle reminder that physics can be advanced by historians who merely elucidate the problems discussed by physicists of the past:[7]

---

5. $\text{versin}(x) = 1 - \cos(x)$
6. In 1636, Fermat used Galileo's own law of falling bodies to identify the spiral $\rho^2 = a\theta$ as Galileo's "semicircle". (This equation implies simply a curve starting at the origin and spiralling outward so that its distance from the origin at any elapsed angle is proportional to the square root of that angle.) Fermat penned his result in an interesting memoir: "When Galileo (a man of great genius) dubiously asserted in his *Dialogues* that, under the supposition of a diurnal motion of the earth, a naturally falling body would describe a semicircle, we had occasion to inquire more closely into the truth. Now, having already demonstrated the error of this opinion, we give the true curve, which, unless I am mistaken, is the same as that which, according to Pappus, Menelaus called 'marvelous' ..." See Michael Sean Mahoney, *The Mathematical Career of Pierre de Fermat*, Princeton U. Press, 1994 (2nd ed.), p. 227.

> *"He was more interested in presenting it* [his physics], *as something all his own, than he was in gaining support for the Copernican astronomy as such by writing and publishing the* Dialogue *after many years of silence on that subject. I am content, however, to have tried to shed some new light for others on problems that have been debated with regard to Galileo's physics."*

Turning to the tides, Professor Drake characterizes the prevailing misconception of Galileo's theory by the mainstream of historians thus:

> *"It remains for me to explain why Galileo was deeply attached to his theory of the tides. That is far from evident at present, because historians usually treat that theory as a pathetic thing — or rather, they present as 'Galileo's tide theory' a truly pathetic thing in which no scientist, and certainly not Galileo, could reasonably have taken pride. The prevailing caricature of Galileo's thought on this matter may be briefly sketched as follows: Seas are disturbed by the earth's combined daily rotation and annual revolution, which act on them in opposite directions by day and in the same direction at night. This (it is said) would produce a single daily low tide everywhere at midnight and high tide at noon. Confronted by the fact that observed tide periods do not conform, Galileo is said to have brushed that aside as a result of factors which deserve no serious consideration by historians."*

Professor Drake acknowledges that the explanations provided in the Fourth Day of the *Dialogo* are incorrect. And yet, he points out, Galileo's pride in it was justifiable. Why? He continues ...

> *"The* Dialogue *presented the first explanation of tidal phenomena that may properly be called 'mechanical,' in which motions of the seas are traced to other motions and to observed hydraulic phenomena — as well as the first integrated theory explaining all three main tide cycles. For this, Galileo required not one but two principal causes, of which the one he described as the more powerful is simply omitted from the usual caricature sketched above. The cause there brushed aside as unworthy of consideration still plays a role in modern theories with regard to tides in gulfs, channels, and seas closed at one or both ends. Indeed, Galileo's*

---

7. A passage like this could possibly bring upon the Professor criticism for overly identifying with his hero. Whoever would disconnect the mission of a scientific historian from that of science itself is mistaken about the source of scientific progress. Might such criticism might originate from those who, in Drake's own words, "are more interested in discovering sources, plagiarisms, errors, and inconsistencies in Galileo's thought than in understanding it"?

> *entire approach bears interesting resemblances to aspects of the post-Laplacian tide theories which long ago displaced Newton's explanation."*

The Keplerian/Newtonian dynamic view of the tides posited bulges in the water caused by the force of the moon's gravitational attraction:

> *"Kepler, adopting the attractive power of the moon, envisioned a lunar bulge of water accumulating as the moon rose well above the horizon; this bulge then turned to follow the moon westward until either the moon set and released it, or the watery bulge encountered a shore and began to subside. ... He opined that if the seas were not held back by the earth, they would flow to the moon, and that any two bodies isolated in space would move towards one another through distances inversely proportional to their sizes. That afterthought, removing from Kepler's theory the objectionable concept of a specific and exclusive attraction of the moon for water alone, foreshadowed the concept of universal gravitation that Newton put to use when he had correctly determined the proportionality at which Kepler had incorrectly guessed. Newton then explained the existence of two tides per day by noting that the moon ... pulls the water nearest to it harder than it pulls the whole earth, acting on its center, and pulls the whole earth harder than it pulls the water opposite to it, causing two bulges of water outward from the earth. By following the attracting bodies around the earth, those bulges would cause rises and falls of water along our coasts."*

Nevertheless, Professor Drake argues, the bulge theories are not a part of modern tide theory:

> *"The problems raised by bulge theories are both theoretical and observational. From the theoretical side, the main trouble is that seawater has hardly any viscosity. ... the actual bulge is seldom more than one foot high and never more than three feet, even in the largest ocean, at the equator, with sun and moon pulling in the same direction. By an error in calculation, Newton supposed watery bulges to rise as high as twelve feet, enough to account for the magnitude of tides known to him. ... Modern tide theory assumes tide-producing forces of gravitational origin (among others), acting laterally to produce flows of water rather than vertically to produce bulges. Great rises and falls are implied only along extended coasts obstructing flows, not in mid-sea. ... Galileo's explanation of tides depended on flows occasioned by different (and continually changing) speeds of the waters in different parts of large seas. In that regard Galileo's theory resembled ours, though the forces required in modern theory (or in Newton's) were absent from it, ...*
>
>   *Yet there was also a dynamic aspect to Galileo's explanation, in*

that the weight of water was essential to it. ... In Galileo's theory, motions of sea-basins were needed to account for the <u>existence</u> of flows, after which sheer weight of water overpowered them and governed the actually observed tidal <u>periods</u>."

The first of Galileo's two principal causes of the tides, the Copernican motions of Earth, though insufficient to explain the observed tides, was correct in principle:

"Each point [on the earth] *must then* [as it rotates] *pass through a daily cycle of differing accelerations toward the sun, except at the two poles, whose distances from the sun remain constant. Points on the earth's surface which can move freely, including all parts of the seas, will respond to these changing accelerations by moving. ... Galileo was correct in saying that given the two Copernican motions of the earth, continual disturbances in its seas would follow. ...*

It is of interest that Kepler, who seems to have adumbrated Newton's explanation, was so far from associating tides with the earth's motions that he denied, in a letter to a friend, any possibility that tides resulted from terrestrial motions. Even Newton's tide theory did not logically require the daily rotation, since tidal bulges could be towed around a resting earth by the attracting bodies; it was left to Laplace, ushering in modern tide theory, to establish the essential role of the diurnal rotation. Yet Galileo had noted, on p. 427, that neither diurnal rotation nor annual revolution alone could disturb the seas, while both combined must do so."

What was Galileo's principle of the tides, generally ignored by historians? Drake introduces it by noting that Galileo's tidal theory concerned motions of parts of a large sea and not of seas as a whole. He continues:

"Inertial deformation of water with change of speed was no less a principal part of Galileo's tidal explanation than the long, oscillatory behavior of piled-up water in the process of regaining level equilibrium. Flows near coasts have their counterparts throughout a large sea, by reason of the differences between absolute speeds at any two widely separated places. The ever-changing absolute speeds of <u>parts</u> of large seas, invoked by Galileo, is ignored in modern attacks against him adducing proofs that seas <u>as wholes</u> do not move with respect to the earth — as if Galileo or anyone else had ever said they do. ...

Galileo described the inertial deformation of water by change of speed as a simple fact of observation; water rises against one end of a container when it is moved from rest. Once out of level, water rocks back and forth in a period depending on its length and depth."

Were Copernican motions scientifically demonstrated by the *Dialogo*, as Professor Drake has claimed? In the *Dialogo* Galileo rendered the motion of Earth a nonzero probability by dispensing with the arguments against it, and a high probability by pointing out the synchronism of annual variations in sunspot motion and of terrestrial tides with the revolution and rotation of Earth, respectively. He recognized that when sea water is accelerated, the conservation of a positive geocentric speed would cause a fluid ocean to encroach upon the shores of a solid earth, and he noted the dependency of ocean tides on the size and shape of the basin. For instance, the Red Sea, which extends "lengthwise toward the poles" and is "narrow in the other direction", is lacking in tides.[8] He assigned the tides two principal causes. The primary cause is the cycle of acceleration and retardation of the parts of Earth that "consists in the additions and subtractions which the diurnal whirling makes with respect to the annual motion"; i.e., the interaction of the Earth's rotation with its motion around the sun.[9] The second cause is the cycle of consequent oscillations set up by the mass of the water attempting to restore itself to equilibrium.

Galileo's explanation of the tides is perhaps less purely kinematic than Drake has suggested. Besides invoking the weight of water in his second cause, a dynamic factor that Drake admits, Galileo speaks also of a *force* that moves Earth and its moon around the sun.[10] Comparing the Earth-moon system to the mechanism of a wheel clock, in which timing is regulated by the variable length of a counterpoise, he refers to the weight of the moon as having an effect on the tides.[11] Alterations in the motion of Earth, he says, depend on the moon, and hence, the oscillations of the tides have a monthly component.

From the point of view of the modern theory of tides, Galileo's mistake is not, as is commonly believed, ignoring the moon's mass, but rather assigning it a secondary

---

8. *Dialogo*, Drake's translation, op. cit., p. 433.

9. *Dialogo*, Drake's translation, op. cit., p. 446. Were Ptolemy's immobile Earth set in west to east rotation, the fluid parts which lie largely along latitudinal lines would tend to slip backward against the solid earth, resulting in an east to west motion of the water relative to the land. (He notes that principal air currents are east to west for a similar reason.)

10. "Now if it is true that the force which moves the earth and the moon around the sun always retains the same strength, and if it is true that the same moving body moved by the same force but in unequal circles passes over similar arcs of smaller circles in shorter times, then it must necessarily be said that the moon when at its least distance from the sun (that is, at conjunction) passes through greater arcs of the earth's orbit than when it is at its greatest distance (that is, at opposition and full moon)." *Dialogo*, Drake's translation, p. 453.

11. "... what happens in this matter is just what happened to the rate of the clock, the moon representing uto us that weight which is attached now farther from the center, in order to make the vibrations of the stick less frequent, and now closer, in order to speed them up." ibid., p. 453.

role of setting up an oscillation in the primary effect. (In saying this, I have the benefit of hindsight.) But the components of his theory are also the significant components of the modern theory. Is the effect of Galileo's durable *impeto* on the tides significant? Modern tide theory describes the equilibrium tide in equations for tidal force and elevation of the water surface that first assume an "inertia-less" system, and are then corrected for inertial effects. If one compares the equilibrium tide with the real ocean tide, "the tide in the ocean *deviates markedly* from the equilibrium tide, which is not surprising if one recalls that the equilibrium tide is based on neglect of the inertial forces."[12] Modern tide theory would be as incomplete without the Galilean parameters as the Newtonian/Laplacian ones; Galilean inertia is an essential concept in understanding the tides as is Kepler's notion of interplanetary gravitational force.

The contribution of Galileo to modern tide theory does not end there. Laplace's theory describing the motion of a molecule of seawater analyzes the ebb and flow of the tides as harmonic[13] functions of annual, diurnal, and semidiurnal cycles of oscillation.[14] For the obvious reason that water is a fluid, this effect is far more significant for ocean tides than for land tides. In modern (post-Laplacian) tide theory, the centrifugal force of Earth in its orbital motion offsets the tide-generating attractive forces of sun and moon. Hence, a theory that considers only gravitational effects on tides would tend to overestimate their magnitude. The Laplacian theory led to the modern theory of tides by combining Newtonian gravitational effects with every component of the Galilean theory of tides: effects due to the dual motion of the earth, to inertial deformation of water, to complex harmonic motion, and to the physical topography of the ocean basin. Just as in the case of the parallactic proof of the earth's motion (Bessel), Galileo's pioneering tidal theory plainly readable in the *Dialogo* waited for others to give it the force of mature theory and observation bound in harmony. Clearly, Galileo's tidal theory was incorrect in the way that it treated the effect of the moon's mass and ignored

---

12. *McGraw-Hill Encyclopedia of Science and Technology*, 1992, vol. 7, p. 372. Emphasis mine.

13. Cf. Gillespie, op. cit., vol. xv pp. 297-298. "In effect, he [Laplace] did treat the determination of the location of a molecule of seawater relative to its equilibrium position like that of the displacement of a point on a string vibrating so as to produce a beat of tones and overtones."

14. Cf. Galileo's *Dialogo* (Drake translation, op. cit.) p. 434: " ... it remains now for us to make another important reflection upon the two principal causes of the tides, thereafter compounding them and mixing them together. The first and simplest of these, as I have often said, is the definite acceleration and retardation of the parts of the earth from which the waters receive a determinate period, running toward the east and returning to the west within a space of twenty-four hours. The other depends upon the water's own weight, which, once moved by the primary cause, tries then to restore itself to equilibrium by repeated oscillations which are not determinate as to one preëstablished time alone, but which have differences of duration according to the different lengths and depths of the containers and basins of the oceans. In so far as they depend upon this second principle, some would flow and return in one hour, some in two, in four, in six, in eight, in ten, etc."

that of the sun. But seen in the full context of Newton's and Laplace's complementary developments to it, Galileo's theory of tides is compatible with modern theory and observations of the tides, whereas the Ptolemaic view of an immobile earth is not.

Galileo's treatment of the tides does not need to be defended in order to realize that he demonstrated the case for Copernicus. In that sense, Professor Drake's emphasis on the tide theory may be misplaced, though it is insightful and a welcome antidote to the "prevailing caricature" of Galileo's thought regarding the tides. The 17th century theologians do not seem to have noticed that in the very *Dialogo* that they condemned, Galileo not only dispensed with the reasons to doubt the motion of the earth, but gave a very powerful argument for it that had nothing to do with the tides: the annual variations in sunspot motion. He also suggested that astronomers apply themselves to noting "what variation ought to be produced in the fixed stars by the annual movement of the earth"[15], a proof of Copernican motion finally carried out by Bessel in his observation of 61 Cygni. I conclude that Galileo did scientifically demonstrate Copernican motions in the *Dialogo*. From the tidal theory, as from Galileo's design of Bessel's experiment, this may perhaps be seen only in retrospect, but from his remaining arguments, it was visible in his own time.

Today, empirical data on the behavior of various parts of Earth's waters, more than the theory of tides, is the basis of tide prediction. Galileo made the first serious attempt at analyzing the behavior of fluids in various basins on a rotating sphere, one that supplies identifiable components of modern tidal theory. Further analysis of the correspondence between tidal data and theory would not only improve that theory but spill over into the science of weather prediction for which better models are always being sought.

In our discussion of Zeno in Chapter One, we noted that to have a positive effect on the progress of knowledge, it suffices to pose well chosen questions that are difficult to answer given the state of knowledge of one's own time. Professor Drake has posed questions that require scientific historians to develop better tools and freer channels of communication with which to answer them. He wrote the rejected document in order that the questions he raised and his views on them would have the accessibility provided by an introduction to a widely used translation. That goal was thwarted by the refusal to publish it there. Whether or not one agrees with his conclusions, Drake's analysis is important (not just to academic scholars, but to readers of his *Dialogo* translation) because consideration of that analysis promotes the understanding of Galileo's science.

---

15. Galileo's *Dialogo* (Drake's translation, op. cit.) p. 372.

Is the quest for this understanding merely a matter of justice to Galileo who "moved" the Earth with a lever supplied by him and extended by his successors? No. Galileo symbolizes for us the importance of freedom of inquiry. To understand his struggle to inquire and to publish freely as it actually happened can improve our understanding of the struggle that persists today. Just as it was retarded by censorship in the time of Galileo and by indoctrination after him, the propagation of ideas remains throttled by a tradition that would have them filtered through the professional biases of an academic oligarchy. Tradition operates to distort decisions. One who attends a lecture reveals his decision to those who sponsor it. The purchaser of a new book communicates a decision to its publisher that affects what future books are published. Publishers deliver what they can sell; hence, the importance of readers making their interests visible. The history of Drake's attempts to deliver the fruits of his Galileo studies to a general audience illustrates the opportunity that individuals have to nurture intellectual expression. You and I can participate in the development and positive application of science, which withers without free expression concerning its meaning, its methods and its history. Each of us can function as an active and visible source of demand for the work of thinkers who are operating on the horizons of knowledge.

APPENDIX C:
Galileo, Huygens, and Kepler Laws
Related to Newton's Inverse-Square Law

Though he seems to have overlooked it, Galileo had access to Kepler's Third Law of Planetary Motion, according to which the square of the period P of a planet is proportional to the cube of its mean separation s from the sun; i.e., $P^2 = cs^3$, where c is the constant of proportionality. Using that law, Galileo's principle of accelerated motion reformulated by Newton as Lex II, and Huygens' law of centripetal acceleration, it is possible from a modern viewpoint to represent the gravitational attraction F_g between the sun and Earth (or any planet) as varying inversely with the square of their separation distance s. To simplify the mathematics involved in illustrating this connection, we will adopt the Copernican/Galilean assumption of a circular orbit, as illustrated below.

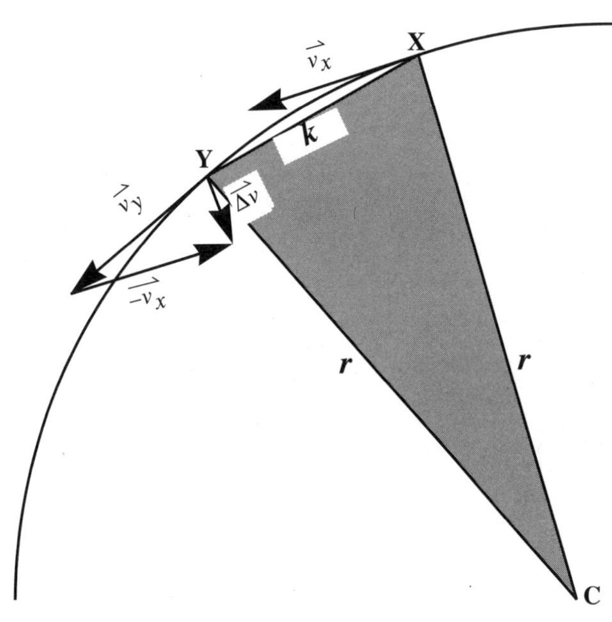

Fig. C-1. A Galilean Planetary Orbit
(Viewed from the North Celestial Pole)

Let's define the acceleration of an object moving circularly at constant speed as a function of its velocity and the radius of motion. Refer to Figure C-1. At two successive positions X and Y of an orbiting body separated by a time Δt, the change-in-velocity vector of the object is $\Delta v = v_y - v_x$, where both vectors are of magnitude v. The isosceles triangle formed by these two vectors and their difference is similar to the triangle connecting the center C with X and Y, their respective equal sides being mutually perpendicular. The similarity yields the proportion:

$$\frac{\Delta v}{v} = \frac{k}{r},$$

where k is the length of the chord connecting X and Y, and r is Earth's orbital radius, from which we have

$$\Delta v = v\frac{k}{r}.$$

Just as Galileo will argue[1] that as smaller and smaller increments in the region of time t are considered, the area between the circumference, the tangent line, and a radius vector extended to the tangent line is bounded by any arbitrarily small function of time, the difference between the chord length k and the arc length $v\Delta t$ along which the planet travels becomes arbitrarily small. Therefore, k tends to $v\Delta t$ from which in the limit we have $\Delta v = (v^2\Delta t)/r$, and since the object's acceleration is defined by $\overrightarrow{\Delta v}/\Delta t$, its magnitude is:

$$a = v^2/r$$

along the tangent line.[2]

Recalling that c is the proportionality constant in Kepler's Third Law:[3]

$$F_g = ma = m \cdot \frac{v^2}{r} = \frac{m}{r} \cdot \left(\frac{2\pi r}{P}\right)^2 = \frac{m \cdot 4\pi^2 r^2}{r \cdot cr^3} = \frac{4\pi^2 m}{cr^2} = c'\frac{m}{r^2}$$

This exercise illustrates the importance of Vieta's development of algebra to Galileo's successors Huygens and Newton.

---

1. Cf. Drake's translation of Galileo's *Dialogo*, pp. 199-200. Galileo is there referring to an object at Earth's surface that moves in an arc due to Earth's rotation; similar logic applies here.

2. This result was discovered in a different but equivalent form by Huygens circa 1659.

3. In the first step, we invoke Galileo's principle of accelerated motion, reformulated later by Newton as the Second Law of Motion and, where mass is invariant with time, usually written in the shorthand f = ma. In the second step, we invoke Huygens' relationship (just demonstrated) between centripetal acceleration and the circular velocity of an object moving at radius r from the center. In the third step, we express circular velocity as the rate of completing each orbit. In the fourth step, we invoke Kepler's Third Law, $P^2 = c \cdot r^3$. The remaining two steps are just algebraic reduction.

APPENDIX D:
Extension of Galileo's Law of Falling Bodies for a Viscous Medium

Assume a sphere of weight W falling through a viscous medium. During fall, the sphere incurs the resultant of the downward force, its weight W, and the two upwardly directed forces: R, and the buoyancy B. It makes sense that when a terminal velocity is reached (since the motion is thereafter unaccelerated), the two upward forces would balance the downward one, and that is what George Gabriel Stokes found: W = R+B. Since W and B are calculated from the volume of the sphere and the densities of the sphere and of the fluid, ρ and ρ', respectively, we have $W = (4/3)\pi r^3 \rho g$ and $B = (4/3)\pi r^3 \rho' g$. From the relation W-R-B = 0, we can use (12-1) to solve for the terminal velocity $v = v_t$:

(D-2)
$$v_t = (2/9)\frac{r^2 g(\rho - \rho')}{\eta}$$

Applying Law II of the mechanics of Galileo-Newton to express the net downward force up to and including the time at which terminal velocity is attained:

(D-3)
$$m\frac{dv}{dt} = W - (B + R) = 6\pi\eta r(v_t - v)$$

Rearranging terms to isolate functions of velocity and time, since v = 0 when t = 0,

(D-4)
$$\int_0^v \frac{dv}{v_t - v} = \frac{6\pi\eta r}{m}\int_0^t dt \, ,$$

from which we see that:

(D-5)
$$v = v_t\left[1 - e^{\frac{-6\pi\eta r}{m} \cdot t}\right] .$$

Combining (D-2) and (D-5) gives us the sought after extension, under the assumed invariants, of Galileo's law of falling bodies (v = gt) for viscous fluid media:

(D-6)
$$v = \frac{2}{9} \cdot \frac{r^2 g(\rho - \rho')}{\eta} \cdot \left[1 - e^{\frac{-6\pi\eta r}{m} \cdot t}\right]$$

BIBLIOGRAPHY

The author regrets, and apologizes to the reader, that information about historical persons and events came to him from so many and diverse sources over so long a period of time, predating his intention to write this book, that his attempt to list them is bound to be incomplete. Had he known, when this youthful compulsion to collect this information took hold of him, that he would one day write this book, he would have kept much better records. Listed here are some of the more important references.

Ball, W. W. Rouse   *A Primer of the History of Mathematics*, MacMillan, London, 1927

Berry, Arthur   *A Short History of Astronomy*, John Murray, 1898 and Dover, 1961

Boyer, Carl   *A History of Mathematics*, John Wiley & Sons, NY, 1968 and 1991

Dantzig, Tobias   *The Bequest of the Greeks*, Charles Scribner's Sons, NY, 1955

de Santillana, Giorgio   *The Crime of Galileo*, U. of Chicago Press, 1955

Dijksterhuis, E. J.   *Archimedes*, trans. by C. Dikshoorn, Princeton U. Press, 1987

Drake, Stillman   *Galileo at Work: His Scientific Biography*, U. of Chicago Press, 1978

Galileo, Oxford University Press, 1980

Discoveries and Opinions of Galileo, Doubleday & Co., NY, 1957

Galileo: Pioneer Scientist, U. of Toronto Press, 1990

History of Free Fall, Wall and Thompson, Toronto, 1989

Durant, Will and Ariel   *The Story of Civilization*, VII: *The Age of Reason Begins*, Simon and Schuster, New York, 1961

Fahie, J.J.   *Galileo, His Life and Works*, James Potts Co. N.Y. 1903

Favaro, Antonio   *Galileo e L'Inquisizione*, Barbera, Firenze, 1907

Fermi, L.
and Bernardini, G.   *Galileo and the Scientific Revolution*, Basic Books, NY1961

Finocchiaro, M.   *The Galileo Affair*, U. of Ca. Press, 1989

Galilei, Galileo   *DIALOGO SOPRA I DUE MASSIMI SISTEMI DEL MONDO*, Ed. by Liberio Sosio, orig. pub. 1632, pub. by Giulio Einaudi, Torino, 1970; and in English translation as:

Dialogue Concerning the Two Chief World Systems, Stillman Drake, U. of Ca. Press, orig. pub. 1953, 2nd rev. ed. 1967, and as *Dialogue On the Great World Systems*: In the Salusbury Translation, U. of Chicago Press, 1953

Operations of the Geometric and Military Compass, English trans. by Stillman Drake, Dibner Library, 1978

DISCORSI E DIMOSTRAZIONI MATEMATICHE intorno a due nuove Scienze Attenati alla Mecanica & I Movimenti Locali, in Leida, Appresso gli Elsevirii, 1638; republished in 1966 by Impression Anastaltique, Culture et Civilisation, 115 Avenue Gabriel Lebon, Bruxelles; and in English translation as:
Two New Sciences, trans. by Stillman Drake, Wall & Thompson, Six O'Connor Drive, Toronto, Ontario, Canada M4K2K1, 1989

Various papers contained in English translation in Drake's *Discoveries and Opinions of Galileo,* op. cit.

Grasshoff, Gerd — *The History of Ptolemy's Star Catalogue*, Springer-Verlag, NY, 1990

Heath, Sir Thomas — *Aristarchus of Samos: The Ancient Copernicus*, Dover, NY, 1981

Loria, Gino — *Galileo Galilei,* Seconda Edizione Aumentata, Hoepli, Milano, 1938

MacLachlan, James — *Children of Prometheus*, Wall & Emerson, Toronto, 1988

McMullin, et al. — *Galileo: Man of Science*, Basic Books, Inc., NY, London, 1967

Poupard, et al. — *Galileo Galilei: 350 Anni di Storia*, Piemme, Roma, 1984

Shepherd, Walter [1] — *Outline History of Science*, Philosophical Library, New York, 1965

Thomas, Ivor — *Greek Mathematical Works*, William Heinemann, Ltd., London, 1980

White, A. D. — *A History of the Warfare of Science with Theology in Christendom*, Prometheus Books, Buffalo, NY, reprinted 1993

Young, G. F. — *The Medici*, Random House, The Modern Library, NY, 1930 (Preface dated 1910)

---

1. In 1981, the publisher advised me that this excellent little book was out of print; I then contacted University Microfilms and, through the courtesy of Philosophical Library, University Microfilms, and the West Coast University library (where I found a copy), it was microfilmed on acid-free paper and subsequently became available in hardcover edition from University Microfilms, Ann Arbor, Michigan.

OTHER TITLES WHICH CAN BE
REQUESTED FROM

ADASI Publishing Company
1465 Woodbury Ave., Suite 261
Portsmouth, N. H. 03801

After Math by Ed Barbeau. A book of puzzles and brainteasers by a mathematics professor who received the David Hilbert Award for contributions to the enrichment of mathematics learning through the stimulation of mathematics challenges. 198 pages, soft cover.

Two New Sciences: *Including Centers of Gravity and Force of Percussion*, by Galileo Galilei. Professor Stillman Drake's classic translation of Galileo's final and authoritative masterpiece that set forth the new sciences of motion and of material strengths and established physics as a modern science. Originally published in 1638 as *Discourses and Mathematical Demonstrations Concerning Two New Sciences Pertaining to Mechanics and Local Motions*. 327 pages, soft cover.

Children of Prometheus: *A History of Science and Technology*, Collegiate Edition by James MacLachlan. A highly readable account, developed from scripts for a course presented on radio. 320 pages, hardbound.

History of Free Fall: *Aristotle to Galileo*, with an Epilogue on π *in the sky*, by Stillman Drake. Succinct and fascinating account of developments leading to the discovery of the laws of free fall, their reception, and application to Kepler's problem -- to understand the scheme of the placement of the planetary orbits. 99 pages, soft cover.

Symbols

M